T0269382

Bayesian Data Analysis in Ecology Using Linear Models with R, BUGS, and Stan

Bayesian Data Analysis in Ecology Using Linear Models with R, BUGS, and Stan

Fränzi Korner-Nievergelt
Tobias Roth
Stefanie von Felten
Jérôme Guélat
Bettina Almasi
Pius Korner-Nievergelt

AMSTERDAM • BOSTON • HEIDELBERG • LONDON
NEW YORK • OXFORD • PARIS • SAN DIEGO
SAN FRANCISCO • SINGAPORE • SYDNEY • TOKYO

Academic Press is an imprint of Elsevier

Academic Press is an imprint of Elsevier
32 Jamestown Road, London NW1 7BY, UK
525 B Street, Suite 1800, San Diego, CA 92101-4495, USA
225 Wyman Street, Waltham, MA 02451, USA
The Boulevard, Langford Lane, Kidlington, Oxford OX5 1GB, UK

ISBN: 978-0-12-801370-0

British Library Cataloguing in Publication Data
A catalogue record for this book is available from the British Library

Library of Congress Cataloging-in-Publication Data
A catalog record for this book is available from the Library of Congress

For information on all Academic Press Publications
visit our website at **http://store.elsevier.com/**

Printed and bound in the USA

Contents

Digital Assets

Thank you for selecting Academic Press' *Bayesian Data Analysis in Ecology Using Linear Models with R, BUGS, and Stan*. To complement the learning experience, we have provided a number of online tools to accompany this edition.

To view the R-package "blmeco" that contains all example data and some specific functions presented in the book, visit www.r-project.org.

The full R-Code and exercises for each chapter are provided at www.oikostat.ch/blmeco.htm.

Acknowledgments

The basis of this book is a course script written for statistics classes at the International Max Planck Research School for Organismal Biology (IMPRS)—see www.orn.mpg.de/2453/Short_portrait. We, therefore, sincerely thank all the IMPRS students who have used the script and worked with us. The text grew as a result of questions and problems that appeared during the application of linear models to the various Ph.D. projects of the IMPRS students. Their enthusiasm in analyzing data and discussion of their problems motivated us to write this book, with the hope that it will be of help to future students. We especially thank Daniel Piechowski and Margrit Hieber-Ruiz for hiring us to give the courses at the IMPRS.

The main part of the book was written by FK and PK during time spent at the Colorado Cooperative Fish and Wildlife Research Unit and the Department of Fish, Wildlife, and Conservation Biology at Colorado State University in the spring of 2014. Here, we were kindly hosted and experienced a motivating time. William Kendall made this possible, for which we are very grateful. Gabriele Engeler and Joyce Pratt managed the administrational challenges of tenure there and made us feel at home. Allison Level kindly introduced us to the CSU library system, which we used extensively while writing this book. We enjoyed a very inspiring environment and cordially thank all the Fish, Wildlife, and Conservation Biology staff and students who we met during our stay.

The companies and institutions at which the authors were employed during the work on the book always positively supported the project, even when it produced delays in other projects. We are grateful to all our colleagues at the Swiss Ornithological Institute (www.vogelwarte.ch), oikostat GmbH (www.oikostat.ch), Hintermann & Weber AG (www.hintermannweber.ch), the University of Basel, and the Clinical Trial Unit at the University of Basel (www.scto.ch/en/CTU-Network/CTU-Basel.html).

We are very grateful to the R Development Core Team (http://www.r-project.org/contributors.html) for providing and maintaining this wonderful software and network tool. We appreciate the flexibility and understandability of the language R and the possibilitiy to easily exchange code. Similarly, we would like to thank the developers of BUGS (http://www.openbugs.net/w/BugsDev) and Stan (http://mc-stan.org/team.html) for making all their extremely useful software freely available. Coding BUGS or Stan has helped us in many cases to think more clearly about the biological processes that have generated our data.

Example data were kindly provided by the Ulmet-Kommission (www.bnv.ch), the Landschaft und Gewässer of Kanton Aargau, the Swiss Ornithological Institute (www.vogelwarte.ch), Valentin Amrhein, Anja Bock, Christoph Bühler, Karl-Heinz Clever, Thomas Gottschalk, Martin Grüebler, Günther Herbert, Thomas Hoffmeister, Rainer Holler, Beat Naef-Daenzer, Werner Peter, Luc Schifferli, Udo Seum, Maris Strazds, and Jean-Luc Zollinger.

For comments on the manuscript we thank Martin Bulla, Kim Meichtry-Stier and Marco Perrig. We also thank Roey Angel, Karin Boos, Paul Conn, and two anonymous reviewers for many valuable suggestions regarding the book's structure and details in the text. Holger Schielzeth gave valuable comments and input for Chapter 10, and David Anderson and Michael Schaub commented on Chapter 11. Bob Carpenter figured out essential parts of the Stan code for the Cormack–Jolly–Seber model. Michael Betancourt and Bob Carpenter commented on the introduction to MCMC and the Stan examples. Valentin Amrhein and Barbara Helm provided input for Chapter 17. All these people greatly improved the quality of the book, made the text more accessible, and helped reduce the error rate.

Finally, we are extremely thankful for the tremendous work that Kate Huyvaert did proofreading our English.

Chapter 1

Why do we Need Statistical Models and What is this Book About?

Chapter Outline

1.1 WHY WE NEED STATISTICAL MODELS

There are at least four main reasons why statistical models are used: (1) models help to describe how we think a system works, (2) data can be summarized using models, (3) comparison of model predictions with data helps with understanding the system, and (4) models allow for predictions, including the quantification of their uncertainty, and, therefore, they help with making decisions.

A statistical model is a mathematical construct based on probability theory that aims to reconstruct the system or the process under study; the data are observations of this system or process. When we speak of "models" in this book, we always mean statistical models. Models express what we know (or, better, what we think we know) about a natural system. The difference between the model and the observations shows that what we think about the system may still not be realistic and, therefore, points out what we may want to think about more intensively. In this way, statistical models help with understanding natural systems.

Analyzing data using statistical models is rarely just applying one model to the data and extracting the results. Rather, it is an iterative process of fitting a model, comparing the model with the data, gaining insight into the system from specific discrepancies between the model and the data, and then finding a more realistic model. Analyzing data using statistical models is a learning process. Reality is usually too complex to be perfectly represented by a model. Thus, no model is perfect, but a good model is useful (e.g., Box, 1979). Often, several models may be plausible and fit the data reasonably well. In such cases, the inference can be based on the set of all models, or a model that performs best for

Bayesian Data Analysis in Ecology Using Linear Models with R, BUGS, and Stan
http://dx.doi.org/10.1016/B978-0-12-801370-0.00001-0.

a specific purpose is selected. In Chapter 11 we have compiled a number of approaches we found useful for model comparisons and multimodel inference.

Once we have one or several models, we want to draw inferences from the model(s). Estimates of the effects of the predictor variables on the outcome variables, fitted values, or derived quantities that are of biological interest are extracted, together with an uncertainty estimate. In this book we use, except in one example, Bayesian methods to assess uncertainty of the estimates.

Models summarize data. When we have measured the height of 100 trees in a forest and we would like to report these heights to colleagues, we report the mean and the standard deviation instead of reporting all 100 values. The mean and the standard deviation, together with a distributional assumption (e.g., the normal distribution) represent a statistical model that describes the data. We do not need to report all 100 values because the 2 values (mean and standard deviation) describe the distribution of the 100 values sufficiently well so that people have a picture of the heights of the 100 trees. With increasing complexity of the data, we need more complex models that summarize the data in a sensible way.

Statistical models are widely applied because they allow for quantifying uncertainty and making predictions. A well-known application of statistical models is the weather forecast. Additional examples include the prediction of bird or bat collision risks at wind energy turbines based on some covariates, the avalanche bulletins, or all the models used to predict changes of an ecosystem when temperatures rise. Political decisions are often based on models or model predictions. Models are pervasive; they even govern our daily life. For example, we first expected our children to get home before 3:30 p.m. because we knew that the school bus drops them off at 3:24, and a child can walk 200 m in around 4 min. What we had in mind was a model child. After some weeks observing the time our children came home after school, we could compare the model prediction with real data. Based on this comparison and short interviews with the children, we included "playing with the neighbor's dog" in our model and updated the expected arrival time to 3:45 p.m.

1.2 WHAT THIS BOOK IS ABOUT

This book is about a broad class of statistical models called linear models. Such models have a systematic part and a stochastic part. The systematic part describes how the outcome variable (y, variable of interest) is related to the predictor variables (x, explanatory variables). This part produces the fitted values that are completely defined by the values of the predictor variables. The stochastic part of the model describes the scatter of the observations around the fitted values using a probability distribution. For example, a regression line is the systematic part of the model, and the scatter of the data around the regression line (more precisely: the distribution of the residuals) is the stochastic part.

Linear models are probably the most commonly used models in biology and in many other research areas. Linear models form the basis for many statistical methods such as survival analysis, structural equation analysis, variance components analysis, time-series analysis, and most multivariate techniques. It is of crucial importance to understand linear models when doing quantitative research in biology, agronomy, social sciences, and so on. This book introduces linear models and describes how to fit linear models in R, BUGS, and Stan. The book is written for scientists (particularly organismal biologists and ecologists; many of our examples come from ecology). The number of mathematical formulae is reduced to what we think is essential to correctly interpret model structure and results.

Chapter 2 provides some basic information regarding software used in this book, important statistical terms, and how to work with them using the statistical software package R, which is used in most chapters of the book.

The linear relationship between the outcome y and the predictor x can be straightforward, as in linear models with normal error distribution (normal linear model, LM, Chapter 4). But the linear relationship can also be indirect via a link function. In this case, the direct linear relationship is between a transformed outcome variable and the predictor variables, and, usually, the model has a nonnormal error distribution such as Poisson or binomial (generalized linear model, GLM, Chapter 8). Generalized linear models can handle outcome variables that are not on a continuous and infinite scale, such as counts and proportions.

For some linear models (LM, GLM) the observations are required to be independent of each other. However, this is often not the case, for example, when more than one measurement is taken on the same individual (i.e., repeated measurements) or when several individuals belong to the same nest, farm, or another grouping factor. Such data should be analyzed using mixed models (LMM, GLMM, Chapters 7 and 9); they account for the nonindependence of the observations. Nonindependence of data may also be introduced when observations are made close to each other (in space or time). In Chapter 6 we show how temporal or spatial autocorrelation is detected and we give a few hints about how temporal correlation can be addressed. In Chapter 13, we analyze spatial data using a species distribution example.

Chapter 14 contains examples of more complex analyses of ecological data sets. These models should be understandable with the theory learned in the first part of the book. The chapter presents ideas on how the linear model can be expanded to more complex models. The software BUGS and Stan, introduced in Chapter 12, are used to fit these complex models. BUGS and Stan are relatively easy to use and flexible enough to build many specific models. We hope that this chapter motivates biologists and others to build their own models for the particular process they are investigating.

Throughout the book, we treat model checking using graphical methods with high importance. Residual analysis is discussed in Chapter 6. Chapter 10 introduces posterior predictive model checking. Posterior predictive model checking is used in Chapter 14 to explore the performance of more complex models such as a zero-inflated and a territory occupancy model. Finally, in Chapter 15, we present possible ways to assess prior sensitivity.

The aim of the checklist in Chapter 16 is to guide scientists through a data analysis. It may be used as a look-up table for choosing a type of model depending on the data type, deciding whether to transform variables or not, deciding which test statistic to use in posterior predictive model checking, or understanding what may help when the residual plots do not look as they should. Such look-up tables cannot be general and complete, but the suggestions they make can help when starting an analysis.

For the reasons explained in Chapter 3, we use Bayesian methods to draw inference from the models throughout the book. However, the book is not a thorough introduction to Bayesian data analysis. We introduce the principles of Bayesian data analysis that we think are important for the application of Bayesian methods. We start simply by producing posterior distributions for the model parameters of linear models fitted in the widely used open source software R (R Core Team, 2014). In the second part of the book, we introduce Markov chain Monte Carlo simulations for non-mathematicians and use the software OpenBUGS (Lunn et al., 2013) and Stan (mc-stan.org). The third part of the book includes, in addition to the data analysis checklist, example text for the presentation of results from a Bayesian data analysis in a paper. We also explain how the methods presented in the book can be described in the methods section of a paper. Hopefully, the book provides a gentle introduction to applied Bayesian data analysis and motivates the reader to deepen and expand knowledge about these techniques, and to apply Bayesian methods in their data analyses.

FURTHER READING

Gelman and Hill (2007) teach Bayesian data analysis using linear models in a very creative way, with examples from the social and political sciences. Kruschke (2011) gives a thorough and very understandable introduction to Bayesian data analysis. McCarthy (2007) concisely introduces Bayesian methods using WinBUGS. Kéry (2010) gives an introduction to linear models using Bayesian methods with WinBUGS. Stauffer (2008) works practically through common research problems in the life sciences using Bayesian methods.

Faraway (2005, 2006) and Fox and Weisberg (2011) provide applied introductions to linear models using frequentist methods in R. Note that there is an extensive erratum to Faraway (2006) on the web. Zuur et al. (2009, 2012) are practical and understandable introductions to linear models in R with a particular focus on complex real ecological data problems such as nonindependent data. Zuur et al. (2012) also introduce Bayesian methods. A more theoretical approach, including R code, is Aitkin et al. (2009). We can also recommend the chapters introducing generalized linear models in Wood (2006).

Chapter 2

Prerequisites and Vocabulary

Chapter Outline

2.1 SOFTWARE

In most chapters of this book we work with the statistical software R (R Core Team, 2014). R is a very powerful tool for statistics and graphics in general. However, it is limited with regard to Bayesian methods applied to more complex models. In Part II of the book (Chapters 12–15), we therefore use Open BUGS (www.openbugs.net; Spiegelhalter et al., 2007) and Stan (Stan Development Team, 2014), using specific interfaces to operate them from within R. OpenBUGS and Stan are introduced in Chapter 12. Here, we briefly introduce R.

2.1.1 What Is R?

R is a software environment for statistics and graphics that is free in two ways: free to download and free source code (www.r-project.org). The first version of R was written by Robert Gentleman and Ross Ihaka of the University of Auckland (note that both names begin with "R"). Since 1997, R has been governed by a core group of R contributors (www.r-project.org/contributors.html). R is a descendant of the commercial S language and environment that was developed at Bell Laboratories by John Chambers and colleagues. Most code written for S runs in R, too. It is an asset of R that, along with statistical analyses, well-designed publication-quality graphics can be produced. R runs on all operating systems (UNIX, Linux, Mac, Windows).

Bayesian Data Analysis in Ecology Using Linear Models with R, BUGS, and Stan
http://dx.doi.org/10.1016/B978-0-12-801370-0.00002-2.

R is different from many statistical software packages that work with menus. R is a programming language or, in fact, a programming environment. This means that we need to write down our commands in the form of R code. While this may need a bit of effort in the beginning, we will soon be able to reap the first fruits. Writing code enforces us to know what we are doing and why we are doing it, and enables us to learn about statistics and the R language rapidly. And because we save the R code of our analyses, they are easily reproduced, comprehensible for colleagues (especially if the code is furnished with comments), or easily adapted and extended to a similar new analysis. Due to its flexibility, R also allows us to write our own functions and to make them available for other users by sharing R code or, even better, by compiling them in an R package. R packages are extensions of the slim basic R distribution, which is supplied with only about eight packages, and typically contain R functions and sometimes also data sets. A steadily increasing number of packages are available from the network of CRAN mirror sites (currently over 5000), accessible at www.r-project.org.

Compared to other dynamic, high-level programming languages such as Python (www.python.org) or Julia (Bezanson et al., 2012; www.julialang.org), R will need more time for complex computations on large data sets. However, the aim of R is to provide an intuitive, "easy to use" programming language for data analyses for those who are not computer specialists (Chambers, 2008), thereby trading off computing power and sometimes also precision of the code. For example, R is quite flexible regarding the use of spaces in the code, which is convenient for the user. In contrast, Python and Julia require a stricter coding, which makes the code more precise but also more difficult to learn. Thus, we consider R as the ideal language for many statistical problems faced by ecologists and many other scientists.

2.1.2 Working with R

If you are completely new to R, we recommend that you take an introductory course or work through an introductory book or document (see recommendations in the Further Reading section at the end of this chapter). R is organized around functions, that is, defined commands that typically require inputs (arguments) and return an output. In what follows, we will explain some important R functions used in this book, without providing a full introduction to R. Moreover, the list of functions explained in this chapter is only a selection and we will come across many other functions in this book. That said, what follows should suffice to give you a jumpstart.

We can easily install additional packages by using the function `install.packages` and load packages by using the function `library`.

Each R function has documentation describing what the function does and how it is used. If the package containing a function is loaded in the current R session, we can open the documentation using `?`. Typing `?mean` into the R console will open the documentation for the function `mean` (arithmetic mean). If we are looking for a specific function, we can use the function `help.search` to search for functions within all installed packages. Typing `help.search("linear model")`, will open a list of functions dealing with linear models (together with the package containing them). For example, `stats::lm` suggests the function `lm` from the package `stats`. Shorter, but equivalent to `help.search("linear model")` is `??"linear model"`. Alternatively, R's online documentation can also be accessed with `help.start()`. Functions/packages that are not installed yet can be found using the specific search menu on www.r-project.org. Once familiar with using the R help and searching the internet efficiently for R-related topics, we can independently broaden our knowledge about R.

Note that whenever we show R code in this book, the code is printed in orange font. Comments, which are preceded by a hash sign, #, and are therefore not executed by R, are printed in green. R output is printed in blue font.

2.2 IMPORTANT STATISTICAL TERMS AND HOW TO HANDLE THEM IN R

2.2.1 Data Sets, Variables, and Observations

Data are always collected on a sample of objects (e.g., animals, plants, or plots). An observation refers to the smallest observational or experimental unit. In fact, this can also be a smaller unit, such as the wing of a bird, a leaf of a plant, or a subplot. Data are collected with regard to certain characteristics (e.g., age, sex, size, weight, level of blood parameters), all of which are called variables. A collection of data, a so-called "data set," can consist of one or many variables. The term *variable* illustrates that these characteristics vary between the observations.

Variables can be classified in several ways, for instance, by the scale of measurement. We distinguish between nominal, ordinal, and numeric variables (see Table 2-1). Nominal and ordinal variables can be summarized as categorical variables. Numeric variables can be further classified as discrete or continuous. Moreover, note that categorical variables are often called factors and numeric variables are often called covariates.

Now let us look at ways to store and handle data in R. A simple, but probably the most important, data structure is a vector. It is a collection of ordered elements of the same type. We can use the function `c` to combine these elements, which are automatically coerced to a common type. The type of elements determines the type of the vector. Vectors can (among other things) be used to represent variables. Here are some examples:

```
v1 <- c(1,4,2,8)
v2 <- c("bird","bat","frog","bear")
v3 <- c(1,4,"bird","bat")
```

TABLE 2-1 Scales of Measurement

Scale	Examples	Properties	Typical coding in R
Nominal	Sex, genotype, habitat	Identity (values have a unique meaning)	Factor
Ordinal	Altitudinal zones (e.g., foothill, montane, subalpine, alpine zone)	Identity and magnitude (values have an ordered relationship, some values are larger and some are smaller)	Ordered factor
Numeric	Discrete: counts Continuous: body weight, wing length, speed	Identity, magnitude, and equal intervals (units along the scale are equal to each other) and possibly a minimum value of zero (ratios are interpretable)	Integer Numeric

R is an object-oriented language and vectors are specific types of objects. The class of objects can be obtained by the function `class`. A vector of numbers (e.g., v1) is a numeric vector (corresponding to a numeric variable); a vector of words (v2) is a character vector (corresponding to a categorical variable). If we mix numbers and words (v3), we will get a character vector.

```
class(v1)
[1] "numeric"
class(v2)
[1] "character"
class(v3)
[1] "character"
```

The function `rev` can be used to reverse the order of elements.

```
rev(v1)
[1] 8 2 4 1
```

Numeric vectors can be used in arithmetic expressions, using the usual arithmetic operators `+`, `-`, `*`, and `/`, including `^` for raising to a power. The operations are performed element by element. In addition, all of the common arithmetic functions are available (e.g., `log` and `sqrt` for the logarithm and the square root). To generate a sequence of numbers, R offers several possibilities. A simple one is the colon operator: `1:30` will produce the sequence 1, 2, 3, …, 30. The function `seq` is more general: `seq(5,100,by = 5)` will produce the sequence 5, 10, 15, …, 100.

R also knows logical vectors, which can have the values TRUE or FALSE. We can generate them using conditions defined by the logical operators `<`, `<=`, `>`, `>=` (less than, less than or equal to, greater than, greater than or equal to), `==` (exact equality), and `!=` (inequality). The vector will contain TRUE where the condition is met and FALSE if not. We can further use `&` (intersection, logical "and"), `|` (union, logical "or"), and `!` (negation, logical "not") to combine logical expressions. When logical vectors are used in arithmetic expressions, they are coerced to numeric with FALSE becoming 0 and TRUE becoming 1.

Categorical variables should be coded as factors, using the function `factor` or `as.factor`. Thereby, the levels of the factor can be coded with characters or with numbers (but the former is often more informative). Ordered categorical variables can be coded as ordered factors by using `factor(..., ordered = TRUE)` or the function `ordered`. Other types of vectors include "Date" for date and time variables and "complex" for complex numbers (not used in this book).

Instead of storing variables as individual vectors, we can combine them into a data frame, using the function `data.frame`. The function produces an object of the class "data.frame," which is the most fundamental data structure used for statistical modeling in R. Different types of variables are allowed within a single data frame. Note that most data sets provided in the package blmeco, which accompanies this book, are data frames.

Data are often entered and stored in spreadsheet files, such as those produced by Excel or LibreOffice. To work with such data in R, we need to read them into R. This can be done by the function `read.table` (and its descendants), which reads in data having various file formats (e.g., comma- or tab-delimited text) and generates a data frame object. It is very important to consider the specific structure of a data frame and to use the same layout in the original spreadsheet: a data frame is a data table with observations in rows and variables in columns. The first row contains the header, which contains the names of the variables. This format is standard practice and should be compatible with all other statistical software packages, too.

Now we combine the vectors v1, v2, and v3 created earlier to a data frame called "dat" and print the result by typing the name of the data frame:

```
dat <- data.frame(v1, v2, v3)
dat
  v1   v2   v3
1  1 bird    1
2  4  bat    4
3  2 frog bird
4  8 bear  bat
dat <- data.frame(number = v1, animal = v2, mix = v3)
dat
```

```
  number animal  mix
1      1   bird    1
2      4    bat    4
3      2   frog bird
4      8   bear  bat
```

By default, the names of the vectors are taken as variable names in dat, but we can also give them new names. A useful function to quickly generate a data frame in some situations (e.g., if we have several categorical variables that we want to combine in a full factorial manner) is `expand.grid`. We supply a number of vectors (variables) and `expand.grid` creates a data frame with a row for every combination of elements of the supplied vectors, the first variables varying fastest. For example:

```
dat2 <- expand.grid(number = v1, animal = v2)
dat2
   number  animal
1       1    bird
2       4    bird
3       2    bird
4       8    bird
5       1     bat
6       4     bat
7       2     bat
8       8     bat
9       1    frog
10      4    frog
11      2    frog
12      8    frog
13      1    bear
14      4    bear
15      2    bear
16      8    bear
```

Using square brackets allows for selecting parts of a vector or data frame, for example,

```
v1[v1>3]
[1] 4 8
```

```
dat2[dat2$animal=="bat",]
    number  animal
5        1     bat
6        4     bat
7        2     bat
8        8     bat
```

Because `dat2` has two dimensions (rows and columns), we need to provide a selection for each dimension, separated by a comma. Because we want all values along the second dimension (all columns), we do not provide anything after the comma (thereby selecting "all there is").

Now let us have a closer look at the data set "cortbowl" from the package blmeco to better understand the structure of data frame objects and to understand the connection between scale of measurement and the coding of variables in R. We first need to load the package blmeco and then the data set. The function `head` is convenient to look at the first six observations of the data frame.

```
library(blmeco)        # load the package
data(cortbowl)         # load the data set
head(cortbowl)         # show first six observations
     Brood    Ring  Implant  Age     days   totCort
1      301  898331        P   49       20     5.761
2      301  898332        P   29        2     8.418
3      301  898332        P   47       20     8.047
4      301  898333        C   25        2    25.744
5      302  898185        P   57       20     8.041
6      302  898188        C   28   before     6.338
```

The data frame cortbowl contains data on 151 nestlings of barn owls *Tyto alba* (identifiable by the variable Ring) of varying age from 54 broods. Each nestling either received a corticosterone implant or a placebo implant (variable Implant with levels C and P). Corticosterone levels (variable totCort) were determined from blood samples taken just before implantation, or 2 or 20 days after implantation (variable days). Each observation (row) refers to one nestling measured on a particular day. Because multiple measurements were taken per nestling and multiple nestlings may belong to the same brood, cortbowl is an example of a hierarchical data set (see Chapter 7). The function `str` shows the structure of the data frame (of objects in general).

```
str(cortbowl) # show the structure of the data.frame
'data.frame':    287 obs. of 6 variables:
$ Brood   : Factor w/ 54 levels "231","232","233",..: 7 7 7 7 8 8 ...
$ Ring    : Factor w/ 151 levels "898054","898055",..: 44 45 45 46 ...
$ Implant : Factor w/ 2 levels "C","P": 2 2 2 1 2 1 1 1 2 1 ...
$ Age     : int  49 29 47 25 57 28 35 53 35 31 ...
$ days    : Factor w/ 3 levels "2","20","before": 2 1 2 1 2 3 1 2 1 1 ...
$ totCort : num  5.76 8.42 8.05 25.74 8.04 ...
```

`str` returns the number of observations (287 in our example) and variables (6), the names and the coding of variables. Note that not all nestlings could be measured on each day, so the data set only contains 287 rows (instead of 151 nestlings $\times$ 3 days $=$ 453). Brood, Ring, and Implant are nominal categorical variables, although numbers are used to name the levels of Brood and Ring. While character vectors such as Implant are by default transformed to factors by the functions `data.frame` and `read.table`, numeric vectors are kept numeric. Thus, if a categorical variable is coded with numbers (as are Brood and Ring), it must be explicitly transformed to a factor using the functions `factor` or `as.factor`. Coding as factor ensures that, when used for modeling, these variables are recognized as nominal. However, using words rather than numbers to code factors is good practice to avoid erroneously treating a factor as a numeric variable. The variable days is also coded as factor (with levels "before", "2", and "20"). Age is coded as an integer with only whole years recorded, although age is clearly continuous rather than discrete in nature. Counts would be a more typical case of a discrete variable (see Chapter 8). The variable totCort is a continuous numeric variable. Special types of categorical variables are binary variables, with only two categories (e.g., implant and sex, with variables coded as no/yes or 0/1).

We often have to choose whether we treat a variable as a factor or as numeric: for example, we may want to use the variable days as a nominal variable if we are mainly interested in differences (e.g., in totCort) between the day before the implantation, day 2, and day 20 after implantation. If we had measured totCort on more than three days, it may be more interesting to use the variable days as numeric (replacing "before" by day 1), to be able to look at the temporal course of totCort.

2.2.2 Distributions and Summary Statistics

The values of a variable typically vary. This means they exhibit a certain distribution. Histograms provide a graphical tool to display the shape of distributions. Summary statistics inform us about the distribution of observed values in a sample and allow communication of a lot of information in a few numbers. A statistic is a sample property, that is, it can be calculated from the observations in the sample. In contrast, a parameter is a property of the population from which the sample was taken. As parameters are usually unknown (unless we simulate data from a known distribution), we use statistics to estimate them. Table 2-2 gives an overview of some statistics, given a sample x of size n, $\{x_1, x_2, x_3, \ldots, x_n\}$, ordered with x_1 being the smallest and x_n being the largest value, including the corresponding R function (note that the ordering of x is only important for the median).

There are different measures of location of a distribution, that is, the value around which most values scatter, and measures of dispersion that describe the spread of the distribution. The most important measure of location is the arithmetic mean (or average). It describes the "center" of symmetric distributions (such as the normal distribution). However, it has the disadvantage of being sensitive to extreme values. The median is an alternative measure of location that is generally more appropriate in the case of asymmetric distributions; it is not sensitive to extreme values. The median is the central value of the ordered sample (the formula is given in Table 2-2). If n is even, it is the arithmetic mean of the two most central values.

TABLE 2-2 Most Common Summary Statistics with Corresponding R Functions and Formulae

Statistic	Parameter	R function	Formula
Arithmetic mean	Population mean (μ)	`mean`	$\hat{\mu} = \bar{x} = \frac{1}{n}\sum_{i=1}^{n} x_i$
Median	Population median	`median`	$median = x_{(n+1)/2}(for\ uneven\ n)$ $median = \frac{1}{2}(x_{n/2} + x_{(n/2+1)})(for\ even\ n)$
Sample variance	Population variance (σ^2)	`var`	$\widehat{\sigma^2}\,(\text{usually}) = s^2 = \frac{1}{n-1}\sum_{i=1}^{n}(\bar{x} - x_i)^2$
Sample standard deviation	Population standard deviation (σ)	`sd`	$\hat{\sigma}(\text{usually}) = s = \sqrt{s^2}$

Note: The second column contains the names and notation of the corresponding (unknown) parameter for which the statistic is usually used as an estimate. The hat sign above a parameter name indicates that we refer to its estimate. The maximum likelihood estimate (Chapter 5) of the variance corresponds to the variance formula using n instead of n-1 in the denominator, see, e.g., Royle & Dorazio (2008).

The spread of a distribution can be measured by the variance or the standard deviation. The variance of a sample is the sum of the squared deviations from the sample mean over all observations, divided by $(n - 1)$. The variance is hard to interpret, as it is usually quite a large number (due to squaring). The standard deviation (SD), which is the square root of the variance, is easier. It is approximately the average deviation of an observation from the sample mean. In the case of a normal distribution, about two thirds of the data are expected within one standard deviation around the mean.

Quantiles inform us about both location and spread of a distribution. The p-quantile is the value x with the property that a proportion p of all values are less than or equal to x. The median is the 50% quantile. The 25% quantile and the 75% quantile are also called the lower and upper quartiles, respectively. The range between the 25% and the 75% quartile is called the interquartile range. This range includes 50% of the distribution and is also used as a measure of dispersion. The R function `quantile` extracts sample quantiles. The median, the quartiles, and the interquartile range can be graphically displayed using box-and-whisker plots (boxplots for short).

When we use statistical models, we need to make reasonable assumptions about the distribution of the variable we aim to explain (outcome or response variable). Statistical models, of which a variety is dealt with in this book, are based on certain parametric distributions. "Parametric" means that these distributions are fully described by a few parameters. The most important parametric distribution used for statistical modeling is the normal distribution, also known as the Gaussian distribution. The Gaussian distribution is introduced more technically in Chapter 5. Qualitatively, it describes, at least approximately, the distribution of observations of any (continuous) variable that tends to cluster around the mean. The importance of the normal distribution is a consequence of the central limit theorem. Without going into detail about this, the practical implications are as follows:

- The sample mean of any sample of random variables (also if these are themselves not normally distributed), tends to have a normal distribution. The larger the sample size, the better the approximation.
- The binomial distribution and the Poisson (Chapter 8) distribution can be approximated by the normal distribution under some circumstances.
- Any variable that is the result of a combination of a large number of small effects (such as phenotypic characteristics that are determined by many genes) tends to show a bell-shaped distribution. This justifies the common use of the normal distribution to describe such data.
- For the same reason, the normal distribution can be used to describe error variation (residual variance) in linear models. In practice, the error is often the sum of many unobserved processes.

If we have a sample of n observations that are normally distributed with mean μ and standard deviation σ, then it is known that the arithmetic mean $\bar{x}$ of the sample is normally distributed around μ with standard deviation $SD_{\bar{x}} = \sigma/\sqrt{n}$. In practice, however, we do not know σ, but estimate it by the sample standard deviation s. Thus, the standard deviation of the sample mean is estimated by the "standard error of the mean" (SEM), which is calculated as $SEM = s/\sqrt{n}$. While the standard deviation s $(= \hat{\sigma})$ describes the variability of individual observations (Table 2-2), SEM describes the uncertainty about the sample mean as an estimate for μ. Due to the division by $\sqrt{n}$, SEM is smaller for large samples and larger for small samples.

We may wonder about the division by $(n - 1)$ in the formula for the sample variance in Table 2-2. This is due to the infamous "degrees of freedom" issue. A quick explanation why we divide by $(n - 1)$ instead of n is that we need $\bar{x}$, an estimate of the sample mean, to calculate the variance. Using this estimate costs us one degree of freedom, so we divide by $n - 1$. To see why, let us assume we know that the sum of three numbers is 42. Can we tell what the three numbers are? The answer is no because we can freely choose the first and the second number. But the third number is fixed as soon as we know the first and the second: it is 42 − (first number + second number). So the degrees of freedom are $3 - 1 = 2$ in this case, and this also applies if we know the average of the three numbers instead of their sum. Another explanation is that, because $\bar{x}$ is estimated from the sample, it is exactly in the middle of the data whereas the true population mean would be a bit off. Thus, the sum of the squared differences, $\sum_{i=1}^{n}(\bar{x}-x_i)^2$ is a little bit smaller than what it should be (the sum of squared differences is smallest when taken with regard to the sample mean), and dividing by $(n-1)$ instead of n corrects for this. In general, whenever k parameters that are estimated from the data are used in a formula to estimate a new parameter, the degrees of freedom for this estimation are $n-k$ (n being the sample size).

2.2.3 More on R Objects

Most R functions are applied to one or several objects and produce one or several new objects. For example, the functions `data.frame` and `read.table` produce a data frame. Other data structures are offered by the object classes "array" and "list."

An array is an n-dimensional vector. Unlike the different columns of a data frame, all elements of an array need to be of the same type. The object class "matrix" is a special case of an array, one with two dimensions. The function `sim`, which we introduce in Chapter 3, returns parts of its results in the form of an array. A very useful function to do calculations based on an array (or a matrix) is `apply`. We simulate a data set to illustrate this: two sites were visited over five years by each of three observers who counted the number of breeding pairs of storks. We simulate the number of pairs by using the function `rpois` to get random numbers from a Poisson distribution (Chapter 8). We can create a three-dimensional array containing the numbers of stork pairs using site, year, and observer as dimensions.

```
Sites <- c("Site1", "Site2")
Years <- 2010:2014
Observers <- c("Ben","Sue","Emma")
pairs <- rpois(n = 2*5*3, lambda = 10)
birds <- array(data = pairs, dim = c(2,5,3),
               dimnames = list(site = Sites, year = Years,
               observer = Observers))
birds
, , observer = Ben


       year
site    2010  2011  2012  2013  2014
  Site1   13    13     5    10     5
  Site2    6    14    11    10     9

, , observer = Sue


       year
site    2010  2011  2012  2013  2014
  Site1   14     6     5    12     7
  Site2    9     9     8     9    12
```

```
, , observer = Emma

        year
site     2010  2011  2012  2013  2014
  Site1    10    11    13     8    13
  Site2    13     6     9     8     8
```

Using apply, we can easily calculate the sum of pairs per observer (across all sites and years) by choosing MARGIN = 3 (for observer) or the mean number of pairs per site and year (averaged over all observers) by choosing MARGIN = c(1,2) for site and year:

```
apply(birds, MARGIN = 3, FUN = sum)
Ben  Sue  Emma
123  100    92
apply(birds, MARGIN = c(1,2), FUN = mean)
     year
site           2010       2011  2012      2013      2014
  Site1     6.00000   9.333333     9  13.66667 14.000000
  Site2    10.33333  11.666667    12  10.33333  8.666667
```

Yet another and rather flexible class of object are lists. A list is a more general form of vector that can contain various elements of different types; often these are themselves lists or vectors. Lists are often used to return the results of a computation. For example, the summary of a linear model produced by lm is contained in a list.

2.2.4 R Functions for Graphics

R offers a variety of possibilities to produce publication-quality graphics (see recommendations in the Further Reading section at the end of this chapter). In this book we stick to the most basic graphical function plot to create graphics, to which more elements can easily be added. The plot function is a generic function. This means that the action performed depends on the class of arguments given to the function. We can add lines, points, segments, or text to an existing plot by using the functions lines or abline, points, segments, and text, respectively.

Let's look at some simple examples using the data set cortbowl:

```
plot(totCort ~ Implant, data = cortbowl)
plot(totCort ~ Age, data = cortbowl[cortbowl$Implant == "P",])
points(totCort ~ Age, data = cortbowl[cortbowl$Implant == "C",],
pch = 20)
plot(cortbowl)
```

Using plot on a numeric dependent variable (totCort) and a factor as predictor (Implant) produces a box-and-whisker plot (Figure 2-1, left). If the

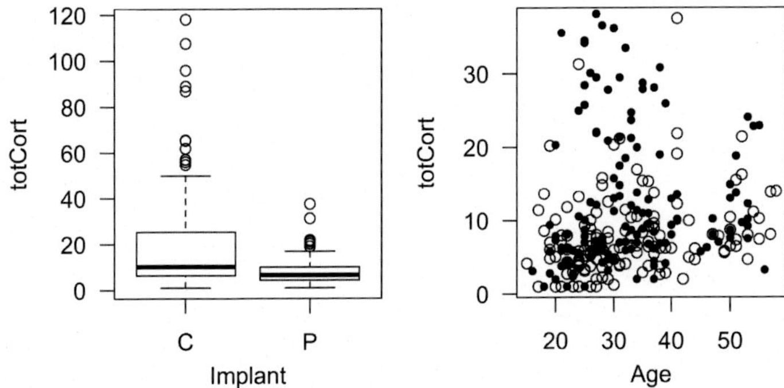

FIGURE 2-1 *Left:* Boxplot of blood corticosterone measurements (totCort) for corticosterone (C) and placebo (P) treated barn owl nestlings. Bold horizontal bar = median; box = interquartile range. The whiskers are drawn from the first or third quartile to the lowest or to the largest value within 1.5 times the interquartile range, respectively. Circles are observations beyond the whiskers. *Right:* Blood corticosterone measurements (totCort) in relation to age. Open symbols = placebo-treated nestlings, closed symbols = corticosterone-treated nestlings.

predictor is numeric, too (Age), the result is a scatterplot. Note that the scatterplot first only shows the points for nestlings that received the placebo because we select only the rows for which Implant=="P", using the squared brackets. A different symbol (defined by pch = 20) is used for the nestlings that received corticosterone (Implant == "C"). These symbols are added using the function `points` (Figure 2-1, right). It is also possible to plot the whole data set. The result is a scatterplot of every variable against every other variable.

Further, we sometimes use the functions `boxplot` and `hist` to explicitly draw box-and-whisker plots and histograms.

2.2.5 Writing Our Own R Functions

R allows us to write our own functions. Here we write a function that calculates the standard error of the mean (SEM). We define the following function:

```
sem <- function(x) sd(x)/sqrt(length(x))
```

`function(x)` means that sem is a function of x, x being the only argument required by the function. The function `sd` calculates the standard deviation of the sample, `length` extracts the number of elements of the sample (sample size), and `sqrt` calculates its square root. We can define a vector x and apply the function sem as follows:

```
x <- c(10,7,5,9,13,2,20,5)
sem(x)
[1] 1.994971
```

This is a good time to mention the topic of missing values. Biological data sets often have missing values (e.g., due to organisms that died, or failure to take measurements). In R, missing values are coded as NA and need to be treated explicitly. If we add a missing value to x and apply sem again, sem produces NA, because the function sd produces NA:

```
x <- c(x, NA)
x
[1] 10 7 5 9 13 2 20 5 NA
sem(x)
[1] NA
sd(x)
[1] NA
```

Unless we specify how to handle missing values, any calculation in R that involves missing values will produce a missing value again. An adapted version of sem could look as follows:

```
sem <- function(x) sd(x, na.rm = TRUE)/sqrt(sum(!is.na(x)))
sem(x)
[1] 1.994971
```

Despite the missing value we have added to x we get the same result as before. The argument na.rm = TRUE within the function sd causes R to ignore missing values.

FURTHER READING

If you are interested in an introductory statistics book (on frequentist statistics) that works with R, we recommend Crawley (2005), Crawley (2007), Dalgaard (2008), or Logan (2010). The books by Crawley and Logan use biological examples whereas most of the examples in Daalgard (2008) come from the medical sciences.

Basic statistical knowledge can be acquired from any classical textbook in statistics. Quinn and Keough (2009) provide a nice introduction to experimental design and data analysis, focusing on the close link between design and analysis. They work through a large number of ecological examples, unfortunately omitting modern mixed-effects models. If you really have a hard time in finding the motivation to read about statistics, you may want to look at Larry Gonick & Woollcott Smith (2005).

If you need a more thorough introduction to R, we recommend Venables et al. (2014; downloadable from *http://cran.r-project.org/doc/manuals/R-intro.pdf*), Zuur et al. (2009), or Chambers (2008).

To learn more about how to generate specialized, publication-quality graphics, you might want to read a book focusing on R graphics such as Murrell (2006) or Chang (2012), or read the reference books on the R graphics packages *lattice* (Sarkar, 2008) or *ggplot* (Wickham, 2009).

Chapter 3

The Bayesian and the Frequentist Ways of Analyzing Data

Chapter Outline

3.1 SHORT HISTORICAL OVERVIEW

Reverend Thomas Bayes (1701 or 1702–1761) developed the Bayes theorem. Based on this theorem, he described how to obtain the probability of a hypothesis given an observation, that is, data. However, he was so worried whether it would be acceptable to apply his theory to real-world examples that he did not dare to publish it. His methods were only published posthumously (Bayes, 1763). Without the help of computers, Bayes' methods were applicable to just simple problems.

Much later, the concept of null hypothesis testing was introduced by Ronald A. Fisher (1890–1962) in his book *Statistical Methods for Research Workers* (Fisher, 1925) and many other publications. Egon Pearson (1895–1980) and others developed the frequentist statistical methods, which are based on the probability of the data given a null hypothesis. These methods are solvable for many simple and some moderately complex examples.

The rapidly improving capacity of computers in recent years now enables us to use Bayesian methods also for more (and even very) complex problems using simulation techniques (Smith et al., 1985; Gelfand & Smith, 1990; Gilks et al., 1996).

Bayesian Data Analysis in Ecology Using Linear Models with R, BUGS, and Stan
http://dx.doi.org/10.1016/B978-0-12-801370-0.00003-4.

3.2 THE BAYESIAN WAY

Bayesian methods use Bayes' theorem to update prior knowledge about a parameter with information coming from the data to obtain posterior knowledge. The prior and posterior knowledge are mathematically described by a probability distribution (prior and posterior distributions of the parameter).

Bayes' theorem for discrete events says that the probability of event A given event B has occurred, $P(A|B)$, equals the probability of event A, $P(A)$, times the probability of event B conditional on A, $P(B|A)$, divided by the probability of event B, $P(B)$:

$$P(A|B) = \frac{P(A)P(B|A)}{P(B)} \tag{3-1}$$

When using Bayes' theorem for drawing inference from data, we are interested in the probability distribution of one or several parameters, θ (called "theta"), after having looked at the data y, that is, the posterior distribution, $p(\theta|y)$. To this end, Bayes' theorem (3-1) is reformulated for continuous parameters using probability distributions rather than probabilities for discrete events:

$$p(\theta|y) = \frac{p(\theta)p(y|\theta)}{p(y)} \tag{3-2}$$

The *posterior distribution*, $p(\theta|y)$, describes what we know about the parameter (or about the set of parameters), θ, after having looked at the data and given the prior knowledge and the model. The *prior distribution* of θ, $p(\theta)$, describes what we know about θ before having looked at the data. This is often very little but it can include information from earlier studies. The probability of the data conditional on θ, $p(y|\theta)$, is called *likelihood.*

The word *likelihood* is used in Bayesian statistics with a slightly different meaning than it is used in frequentist statistics. The frequentist likelihood, $L(\theta|y)$, is a relative measure for the probability of the observed data given specific parameter values. The likelihood is a number often close to zero (see also Chapter 5). In contrast, Bayesians use likelihood for the density distribution of the data conditional on the parameters of a model. Thus, in Bayesian statistics the likelihood is a distribution (i.e., the area under the curve is 1) whereas in frequentist statistics it is a scalar. The *prior probability of the data,* $p(y)$, equals the integral of $p(y|\theta)p(\theta)$ over all possible values of θ; thus $p(y)$ is a constant.

The integral can be solved numerically only for a few simple cases. For this reason, Bayesian statistics were not widely applied before the computer age. Nowadays, a variety of different simulation algorithms exist that allow sampling from distributions that are only known to proportionality (Gilks et al.,

1996; Brémaud, 1999). Dropping the term $p(y)$ in Equation 3-2 leads to a term that is proportional to the posterior distribution: $p(\theta|y) \propto p(\theta)p(y|\theta)$. Simulation algorithms such as Markov chain Monte Carlo simulation (MCMC) can, therefore, sample from the posterior distribution without having to know $p(y)$. A large enough sample of the posterior distribution can then be used to draw inference about the parameters of interest.

3.2.1 Estimating the Mean of a Normal Distribution with a Known Variance

The purpose of this section is to illustrate the Bayesian method using a theoretical example. It has only limited practical value. Therefore, feel free to skip this chapter if you are afraid of mathematical meditations. One of the simplest examples is to estimate the mean of a normal distribution with known variance based on a sample of n measurements.

The model of the data is $y \sim Norm(\theta, \sigma^2)$, which means "$y$ is normally distributed with mean θ and variance σ^2." θ is the true (population) mean and σ^2 the variance of the population from which the data are a random sample. Given the data contain three measurements, $y_1 = 27.1$, $y_2 = 14.6$, $y_3 = 14.6$, and we know that the variance, σ^2, is 20, what do we know about the mean θ of the population from which the data are a random sample? Based on Bayes' theorem and the normal prior distribution it is possible to see that the posterior distribution of the mean θ is itself a normal distribution with mean μ_n and variance σ_n, $p(\theta|y) = Norm(\mu_n, \sigma_n)$, where

$$\mu_n = \frac{\frac{\mu_0}{\sigma_0^2} + \frac{n\bar{y}}{\sigma^2}}{\frac{1}{\sigma_0^2} + \frac{n}{\sigma^2}}, \text{ and } \frac{1}{\sigma_n^2} = \frac{1}{\sigma_0^2} + \frac{n}{\sigma^2}.$$

μ_0 and σ_0 are the mean and the variance of the prior distribution for θ. Gelman et al. (2014) provide the derivation of these formulas. When we know in advance that μ cannot be very far from 0, but we have very little knowledge about θ, we might assume a flat normal distribution around 0—for example, $Norm(0, \sigma_0^2 = 200)$—as the prior distribution for θ (the dotted line in Figure 3-1; only part of the right tail is visible). This results in a posterior distribution $p(\theta|y) = Norm(18.2, \sigma_0^2 = 6.5)$ (the solid line in Figure 3-1). This posterior distribution expresses what we know about θ based on our prior knowledge, the data, and assuming our model adequately describes the process that generated the data (the data and the model are formalized in the likelihood, which, in this example, is a normal distribution with mean equal to the arithmetic mean of the data and variance equal to 20, the dashed line in Figure 3-1).

The posterior distribution is a combination of the information of the data and the prior; the data and the prior are weighted according to the information they contain. This information content is measured by the precision (which is

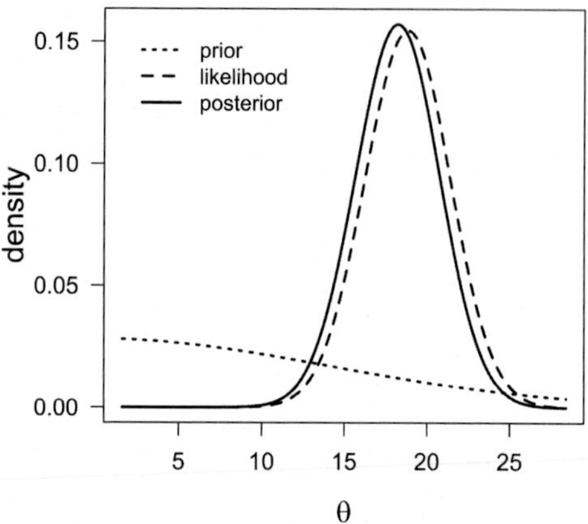

FIGURE 3-1 Prior distribution, likelihood, and posterior distribution of the mean θ. The plot has been drawn using the R function `triplot.normal.knownvariance` provided in the R package blmeco.

equal to the inverse of the variance). If we use a completely flat (i.e., non-informative) prior distribution, the posterior distribution equals the likelihood (upper left panel in Figure 3-2). In this case the inference drawn does not differ from the inference drawn with frequentist methods. The more we know *a priori* about θ, the stronger the influence the prior has on the posterior (Figure 3-2). When informative prior distributions are used, the inference drawn with Bayesian methods differ from the ones drawn with frequentist methods (instead of saying "prior distribution" and "posterior distribution" one often only says "prior" and "posterior").

Note that the likelihood is the same in all panels of Figure 3-2 because the same data set and the same model is used.

3.2.2 Estimating Mean and Variance of a Normal Distribution Using Simulation

The estimation of a mean θ and a variance σ^2 of a normal distribution, where both parameters are unknown, is more complicated than estimating the mean only, because the posterior distribution is two-dimensional. Such a posterior distribution is called a *joint posterior distribution*. Nevertheless, it is still possible to obtain the joint posterior distribution analytically (Albert, 2007; Gelman et al., 2014). Rather than following the analytical route, we now demonstrate how we can simulate a posterior distribution using the `sim` function in the package arm.

First, we obtain parameter estimates using classical methods such as least-squares or maximum likelihood. Then, `sim` uses the results from the model fit to calculate the posterior distribution assuming flat prior distributions (Gelman

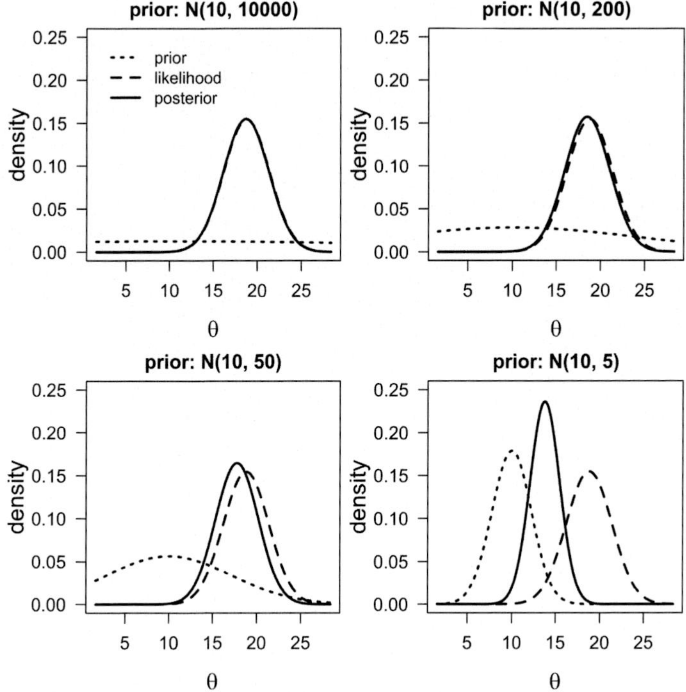

FIGURE 3-2 Prior, likelihood, and posterior distribution of the mean θ using different prior distributions for θ.

and Hill, 2007). Because the rather complicated formula of the joint posterior distribution of the model parameters is not of much help for many users (such as us biologists), `sim` samples pairs of random values of θ and σ from this distribution.

Given enough random samples, uncertainty measurements for each model parameter can be obtained. Often we calculate an interval within which we expect the true parameter value to be with a probability of 0.95, the so-called 95% *credible interval* (CrI). Note that the frequentist confidence interval does not allow such a straightforward interpretation (see Section 3.3). Credible intervals can be defined in different ways: two commonly used CrIs are the symmetric CrI and the highest posterior density interval. The symmetric 95% CrI is the interval between the 2.5% and 97.5% quantiles of the posterior distribution. For skewed posterior distributions, the density values at the 2.5% and 97.5% quantiles can be very different. The highest posterior density 95% interval is a 95% interval that is chosen so that, for any value within the interval, the density function is equal to or higher than for any value outside the interval.

Let's use a simple example. What is the mean height of humans? A random sample of 10 people was drawn and their height measured. A normal model with unknown mean and unknown variance, $y \sim Norm(\theta, \sigma^2)$, can be assumed for human height, and the model is fitted to the data by the least-squares method in R using the `lm` function to get estimates for the mean and standard deviation:

```
# simulate hypothetical body height measurements
true.mean <- 165        # population mean
true.sd <- 10           # standard deviation
y <- round(rnorm(10, mean=true.mean, sd=true.sd))
mod <- lm(y~1)          # least-squares fit
mod                     # least-squares estimate of θ
Call:
lm(formula = y ~ 1)

Coefficients:
(Intercept)
   165.7

summary(mod)$sigma  # least-squares estimate of σ
[1] 13.30038
```

From the R output we obtain least-squares estimates for θ and σ. We can describe our data distribution as $y \sim Norm(\widehat{\theta} = 165.7, \widehat{\sigma^2} = 13.3^2 = 176.9)$. The hats above the model parameters indicate that these parameters were estimated from data, that is, their true values are unknown. Estimates should always be given together with a measurement of their uncertainty. In Bayesian statistics, the uncertainty is measured by the variance of the posterior distribution $p(\theta,\sigma|y)$. It describes which pairs of values of θ and σ are plausible given the data, the prior, and the model. The function `sim` draws pairs of values from the joint posterior distribution of the two parameters.

```
library(arm)
nsim <- 5000
bsim <- sim(mod, n.sim=nsim)
str(bsim)
  Formal class 'sim' [package "arm"] with 2 slots
    ..@ coef : num [1:5000, 1] 169 166 169 159 153 ...
    .. ..- attr(*, "dimnames")=List of 2
    .. .. ..$ : NULL
    .. .. ..$ : NULL
    ..@ sigma: num [1:5000] 12.3  11.7  15.1  15.9  22.1 ...
```

The function `sim` produces an object of class "sim" that contains two slots, "coef" (containing 5000 simulated values for θ) and "sigma" (containing the 5000 corresponding values for σ). Note that the order matters: the coef and sigma values form pairs of reasonable combinations of θ and σ values; we are talking about a joint posterior distribution of the two parameters. A scatterplot

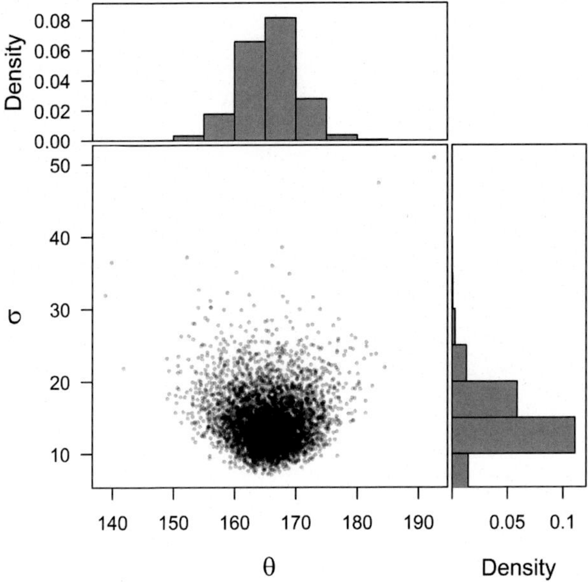

FIGURE 3-3 There are 5000 draws from the joint posterior distribution of θ and σ. *Lower left panel*: every dot is one draw from the joint posterior distribution of θ and σ. The histograms in the upper and right panels give the marginal posterior distributions of θ and σ, respectively.

allows us to visualize the joint posterior distribution, $p(\theta,\sigma|y)$ (lower left panel in Figure 3-3). We can use the simulated values to draw our conclusions (see the following).

If we look at the simulated values of each parameter separately (the histograms in Figure 3-3), we get an approximation of the *marginal posterior distribution* for each parameter. In the language of mathematicians: the marginal posterior distribution of θ is the joint posterior distribution, $p(\theta,\sigma|y)$, integrated over all possible values of σ. Using simulations, we can do this by simply plotting a histogram of all the simulated θ values (or, vice versa, of σ to show the marginal posterior of σ, i.e., $p(\sigma|y)$). Analytical descriptions of these distributions would be a scaled-inverse chi-square distribution for σ and a t distribution for θ, respectively. This might be interesting to know but it is not needed to draw conclusions.

The *conditional posterior distribution* is the posterior distribution of one parameter given a specific value for the other parameter. The conditional distribution of θ, $p(\theta|\sigma,y)$, is not the same for all values of σ.

The @ Sign

To extract the elements (called slots) "coef" and "sigma" of the `sim` object we have to use the @ sign (not the $ sign). This is because `sim` is written in the S4 style rather than in the S3 style. The S4 style is a slightly more flexible programming

Continued

The @ Sign—cont'd

style than the S3 style (Chambers, 1998; Chambers, 2008). An alternative method to extract the slot "coef" from the `sim` object is to use the function `coef`: `coef(sim object)`. There is currently no function to directly extract the slot "sigma" other than by using bsim@sigma (or, `slot(bsim, "sigma")`).

To look at large S4 objects, use the `max.level` argument of `str`. This suppresses all lower-level elements.

```
str(object, max.level=2)
```

Summary statistics such as the median and other quantiles describe the posterior knowledge about the parameters θ and σ.

```
quantile(bsim@coef, prob=c(0.025, 0.5, 0.975))
    2.5%        50%      97.5%
155.8396  165.6774  174.5724
quantile(bsim@sigma, prob=c(0.025, 0.5, 0.975))
    2.5%         50%       97.5%
9.106862  13.858589  23.918661
```

We can report our estimates for the population mean body size = 165.7 cm with a 95% CrI of 155.8 to 174.6 cm, and the standard deviation = 13.3 cm (95% CrI: 9.1–23.9 cm). Alternatively, the highest posterior density interval can be computed using `HPDinterval` from the coda package. This function can only handle objects that are designed to contain the results of a Markov chain Monte Carlo simulation (see Chapter 12). Therefore, the `sim` object is first converted to an mcmc-object before the calculation of the interval.

```
library(coda)
HPDinterval(as.mcmc(bsim@coef))
        lower       upper
var1 155.8757   174.5816
attr(,"Probability")
[1] 0.95
```

Because the posterior distribution is symmetric in this case, the two credible intervals are similar. Based on the simulated values from the posterior distribution of the model parameters, meaningful quantities can be derived. For example, how sure are we that the population mean of human heights is larger than 160? We can estimate this probability by the proportion of values that are larger than 160 among the simulated values from the posterior distribution.

```
sum(bsim@coef>160)/nsim
[1] 0.8942
```

The ">" sign produces a vector of TRUE and FALSE values. `sum` counts the number of TRUEs, that is, the number of simulated values that are larger than

160 cm. The division by nsim gives the proportion of the posterior distribution that is larger than 160. Alternatively, `mean(bsim@coef>160)` may be used.

We, therefore, infer from our data that there is a probability of 0.89 that the true mean of the population is larger than 160 cm. Note that this is not the proportion of individuals among the population that are over 160 cm high. It merely expresses what we can infer about the mean of the population from the actual sample of 10 persons.

One important advantage of a simulated joint posterior distribution is that the propagation of uncertainty from the estimated model parameters to derived parameters is very easy. For example, if we would like to compare the variance in body size between humans and elephants, it does not make sense to compare the standard deviations because the mean body size is so much different. Rather, we would like to compare the coefficient of variation, which is defined by the ratio of the standard deviation to the mean: $CV = \sigma/\theta$. A sample of the posterior distribution of CV is obtained simply by calculating this ratio for every pair of σ and θ values in their joint posterior distribution.

```
cvsim <- bsim@sigma/bsim@coef
quantile(cvsim, prob=c(0.025, 0.5, 0.975))
       2.5%        50%      97.5%
0.05504881 0.08374738 0.14492354
```

The estimated coefficient of variation for body size of humans is 0.08 (0.06–0.14) and this may be compared to a similar estimate from a random sample of individual elephants.

In the example just described, we have not defined any prior distribution, because `sim` uses flat prior distributions when applied to an `lm` object. It is possible to specify priors when using the function `bayesglm` from the package arm instead of using `lm` to fit the model. However, when using informative priors in this book, we fit the models using BUGS or Stan, because we like the intuitive way a model is specified in these programming languages and their flexibility.

BUGS and Stan will only be introduced in the second part of the book. Before that we will use `sim` with flat priors because it is simple, fast, and safe, and it provides all advantages of simulated joint posterior distributions. There may be no need to assess prior influence when using `sim` on `lm` objects in most cases. But we would like to stress that it is important to assess prior influence when we use informative priors and in more complex models where it is often tricky to define noninformative priors.

3.3 THE FREQUENTIST WAY

Frequentists interpret probability as a relative frequency of an event rather than a description of knowledge (or "degree of belief") as in the Bayesian framework. The frequentist way of looking at data is always to ask: how would the parameters change if the experiment were repeated many times (therefore "frequentist")?

Back to the human height example: A random sample of 10 people yielded a sample mean $\bar{y} = 165.7$ cm, and the measurements scattered around the mean with a standard deviation of $sd(y) = 13.3$ cm. If we want to use $\bar{y}$ as an estimate for the true mean of human height, θ, how close can we expect $\bar{y}$ to be to this true mean? Or, in other words, if we take many additional random samples of 10 persons, what is the variance of these sample means (Figure 3-4)? The square root of this variance of many sample means is the *standard error* of the mean. It is computed by dividing the sample standard deviation by the square root of the sample size: $\text{se}(\bar{y}) = \frac{sd(y)}{\sqrt{n}}$. It decreases when sample size n increases. The standard error of the example data here is $\frac{13.3}{\sqrt{10}} = 4.2$. From the standard error, a 95% *confidence interval* (CI) can be obtained: 95% CI $= \bar{y} \pm 1.96^{*}\text{se}(\bar{y})$, which yields a CI of 157.5 to 173.9 cm for human height in our example. The interpretation of the CI is: If we repeat the study under the same conditions using the same sample size many times, 95% of these CI will contain the true population mean. Recall the interpretation of a Bayesian credible interval (CrI): we are 95% sure that the true mean is within the CrI.

To assess the significance of an effect, for example, to assess whether the estimated population mean body size is larger than 160 cm, frequentists apply null hypothesis tests. The result of such a hypothesis test is a p-value, which is the probability of observing the observed test statistic, or a more extreme one, given the null hypothesis (a "more extreme observation" is one that is even farther away from the null hypothesis than the actual observation). The null hypothesis is the situation when no effect is present, for example, the mean of the population is exactly 160 cm.

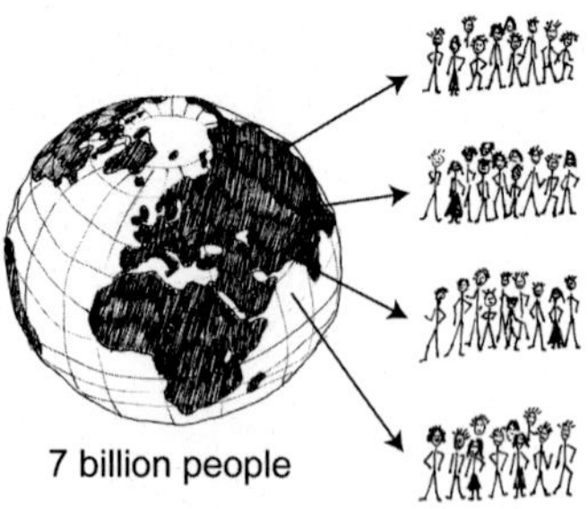

population	random sample of 10 persons	data	sample mean	sample standard deviation
		165, 170, 152, 169, 143, 171, 170, 152, 188, 177	165.7	13.30
		171, 173, 173, 165, 162, 162, 170, 171, 152, 151	165.0	8.22
		171, 164, 157, 147, 160, 160, 169, 166, 181, 180	165.6	10.38
7 billion people		152, 178, 169, 163, 166, 176, 170, 169, 155, 156	165.4	8.82

FIGURE 3-4 Frequentist way of describing uncertainty in a parameter estimate (here the average human height): Consider the variance in the parameter estimates (here the sample mean and sample standard deviation) that is due to random sampling.

```
t.test(y, mu=160)
       One Sample t-test

data: y
t = 1.3552, df = 9, p-value = 0.2084
alternative hypothesis: true mean is not equal to 160
95 percent confidence interval:
  156.1855 175.2145
sample estimates:
mean of x
    165.7
```

Because the *p*-value is larger than 0.05, the result of the one-sample *t*-test is: "The mean of the population is not significantly different from 160 cm."

Null hypothesis testing has been widely criticized for many years (e.g., Berkson, 1938; Cohen, 1994; Anderson et al., 2000; Matthews, 2000; Ridley et al., 2007; Burnham & Anderson, 2014). Null hypotheses are (almost) never meaningful, and it is not a desirable property of a technique to have an absurd starting point. Further, a very precise result (small CI) with effect size close to zero greatly contributes to the update of our knowledge and, therefore, should be published even if the *p*-value is larger than 0.05. In the literature just cited, we find many more arguments why one should avoid drawing conclusions based only on a *p*-value. Though this principle is very well known, it is still disregarded at times.

Note that similar criticism applies to analyses that use Bayesian credible intervals solely to find "significant" effects based on whether 0 is in the 95% interval or not. Thus, both frequentists and Bayesians should not use an arbitrary cutoff point to make decisions. Such decisions can only be done with regard to the relevance of the result in the field of the study. In this book, we cannot present how to make these decisions because the decision depends on what consequences such a decision has. Therefore, other information than the data is needed to decide what effects are important. It is also a topic of decision theory (e.g., Wolfson et al., 1996; Williams et al., 2002; Marescot et al., 2013; Yokomizo et al., 2014). A prerequisite for making such decisions is that the information in a data set is appropriately described by estimates and CIs or CrIs of the parameters of interests. The latter is what we do in this book.

Finite Populations

Here (and throughout the whole book) we assume that the population from which the data were drawn has infinite size, or at least is much larger than the sample size. If the sample includes a large proportion of the population, then the standard error for a parameter is smaller than under the assumption of infinite population size. If, for example, we have sampled all seven billion humans, the variance of the sample mean will become zero, because we have measured the true population mean. In some applied examples we might be interested in a finite population parameter, for example, when we have measured the oak tree density on one-third of the area of the canton of Lucerne (a county of Switzerland) and we want to know the average oak tree density specifically for the whole canton of Lucerne. In such cases, the error measurements for the parameters can be reduced by a finite population correction; see, for example, Thompson (2002) or Valliant et al. (2000).

3.4 COMPARISON OF THE BAYESIAN AND THE FREQUENTIST WAYS

An important difference between the Bayesian and frequentist ways of analyzing data is how conclusions are drawn (Table 3-1). Bayesians are interested in the probability of a (meaningful) hypothesis (e.g., that an effect is larger than a relevant size, say, $H: \theta > e$) given the data, $p(H|y)$, or they describe what we know about a parameter after having looked at the data by the posterior distribution $p(\theta|y)$. Frequentist methods assess the probability of the data given a null hypothesis, $p(y|H_0)$, or they give estimates with uncertainty measures for specific model parameters.

The second important difference is that Bayesians combine prior information with information that is obtained from data, whereas frequentists solely infer from the data at hand. Because prior information may influence inference, some people have been resistant to using Bayesian statistics. However, if prior influence is reported transparently, the degree of information in the data becomes even clearer than without any prior influence. Bayesian statistics provides the theory to combine information from different sources and, therefore, meta-analyses become relatively simple using Bayesian analysis.

Doing science consists of updating prior knowledge with information from experiments or observations to obtain new (posterior) knowledge about a parameter. The Bayesian way of analyzing data formalizes the learning process

TABLE 3-1 Characteristics of Bayesian and Frequentist Statistics

	Bayesian statistics	Frequentist statistics
Interpretation of *probability*	Description of what we know about a parameter	Relative frequency of an event
Information used	Prior distribution $p(\theta)$ and likelihood $p(y|\theta)$—that is, data	Data only
Uncertainty measurement	Credible interval (CrI)	Confidence interval (CI)
Assessing significance	Probability of meaningful hypotheses $p(H|y)$	Null hypothesis tests $p(y|H_0)$

of science. One of the reasons for the development of frequentist statistics was that it was not feasible to do the simulation part of Bayesian statistics before the computer age (because of the earlier application of frequentist statistical methods, they are sometimes called the "classical" statistical methods, even though the idea of Bayes is older). Nowadays everybody can use his or her own computer to simulate from posterior distributions very easily. The technical difficulties that impeded scientists to use Bayesian methods have largely disappeared. However, frequentist statistics is still commonly used and taught. Some scientists even reject Bayesian statistics mainly because they argue that using prior information makes a study subjective (see e.g., discussion in the Forum of Ecological Applications, 19, 2009).

Frequentist analysis is subjective too! Subjectivity enters at many levels, from deciding what questions to investigate, which data to collect, which results to publish, and so on. Using prior information is not the only subjective step during a data analysis—the key thing is to report what has been done. A famous statistician once asked us whether we had ever laughed out loud at a statistical result because it absolutely did not make any sense? If yes, this is clear evidence that we have prior knowledge; clearly it is reasonable to report this prior knowledge and use it in the analysis.

In this book, we primarily use Bayesian methods because they allow more meaningful inferences (i.e., probabilities of specific hypotheses, posterior distributions for derived quantities). In fact, in the case of generalized linear mixed models, they are the only exact way to draw inference (Bolker et al., 2008). Bayesian methods make the fit of more complex models (such as in the second part of this book) feasible. Further, they make meta-analysis relatively simple, and we believe that they are the methods of the future.

While our book clearly focuses on Bayesian analyses, we also briefly introduce the most important classical frequentist tests associated with linear models because they are still widely used, and because almost 100 years of scientific literature has used these techniques that we, as readers, would like to understand.

FURTHER READING

In a very informative discussion, McCarthy (2007) compares the frequentist (null hypothesis testing), the information-theoretic, and the Bayesian methods. Kruschke (2011) is an easy-to-read introduction to Bayesian data analysis. Gelman et al. (2014) provides the mathematical background of applied Bayesian data analysis; the first edition of this book (1995) may have been seminal in making the Bayesian methods available for applied statisticians. Those who are not mathematicians may prefer starting with the more applied introduction to Bayesian modeling by Gelman and Hill (2007). A mixture of mathematical theory, R code, and WinBUGS code for doing Bayesian analyses is provided in the book by Christensen et al. (2011). Several recent applied statistics text books use Bayesian methods—for example, Stauffer (2008), Zuur et al. (2009), and Kéry (2010)—illustrating how these methods allow fitting of more complex models and drawing more meaningful inferences. We also like the introduction of Royle and Dorazio (2008).

Chapter 4

Normal Linear Models

Chapter Outline

4.1 LINEAR REGRESSION

4.1.1 Background

Linear regression is the basis of a large part of applied statistical analysis. Analysis of variance (ANOVA) and analysis of covariance (ANCOVA) can be considered special cases of linear regression, and generalized linear models are extensions of linear regression. We, therefore, start with a rather detailed introduction to linear regression.

Typical questions that can be answered using linear regression are: How does y change with changes in x? How is y predicted from x? The first step in an ordinary linear regression (i.e., one x and one y variable) is a scatterplot of y against x. Then, we search for the line that fits the best and describe how the observations scatter around this regression line, that is, we describe the distribution of the residuals $\varepsilon_i = y_i - \widehat{y}_i$ (Figure 4-1). The model formula of a simple linear regression is:

$$\begin{aligned} y_i &\sim Norm(\widehat{y}_i, \widehat{\sigma}^2) \\ \widehat{y}_i &= \widehat{\beta}_0 + \widehat{\beta}_1 x_i \end{aligned} \tag{4-1}$$

Bayesian Data Analysis in Ecology Using Linear Models with R, BUGS, and Stan
http://dx.doi.org/10.1016/B978-0-12-801370-0.00004-6.

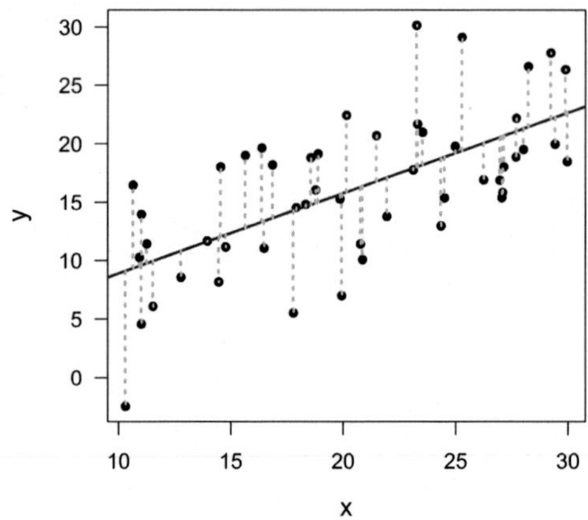

FIGURE 4-1 Linear regression. Black dots = observations, blue solid line = regression line, orange dotted lines = residuals. The fitted values lie where the orange dotted lines touch the blue regression line.

In other words: the observation y_i stems from a normal distribution with mean $\hat{y}_i$ and variance σ^2. The mean, $\hat{y}_i$, equals the sum of the intercept (β_0) and the product of the slope (β_1) and the predictor value, x_i. Equivalently, the regression could be written as:

$$\begin{aligned} y_i &= \hat{\beta}_0 + \hat{\beta}_1 x_i + \hat{\varepsilon}_i \\ \hat{\varepsilon}_i &\sim Norm\left(0, \sigma^2\right) \end{aligned} \tag{4-2}$$

where $\hat{\varepsilon}_i = y_i - \hat{y}_i$ are the residuals. We prefer the notation in Equation 4-1 because, in this formula, the stochastic part (first row) is nicely separated from the deterministic part (second row) of the model, whereas, in the second notation (Eq. 4-2) the first row contains both stochastic and deterministic parts. Equation 4-1 can also be written in one row:

$$\mathbf{y} \sim Norm\left(\mathbf{X}\boldsymbol{\beta}, \sigma^2 \mathrm{I}\right)$$

where **I** is the $n \times n$ identity matrix (it transforms the variance parameter to a $n \times n$ matrix with its diagonal elements equal σ^2; n is the sample size).

The multiplication by **I** is necessary here because we use vector notation, **y** instead of y_i. Here, **y** is the vector of all observations, whereas y_i is a single observation, i. When using vector notation, we can write the linear predictor of the model, $\beta_0 + \beta_1 x_i$, as a multiplication of the vector of the model coefficients

$$\boldsymbol{\beta} = \begin{pmatrix} \beta_0 \\ \beta_1 \end{pmatrix}$$

times the model matrix

$$\mathbf{X} = \begin{pmatrix} 1 & x_1 \\ \ldots & \ldots \\ 1 & x_n \end{pmatrix}$$

where $x_1, ..., x_n$ are the observed values for the covariate, x. The first column of **X** contains only ones because the values in this column are multiplied with the intercept, β_0. To the intercept, the product of the second element of $\boldsymbol{\beta}$, β_1, with each element in the second column of **X** is added to obtain the fitted value for each observation, $\widehat{\mathbf{y}}$ (Figure 4-1):

$$\boldsymbol{\beta}\mathbf{X} = \begin{pmatrix} \beta_0 \\ \beta_1 \end{pmatrix} \times \begin{pmatrix} 1 & x_1 \\ \ldots & \ldots \\ 1 & x_n \end{pmatrix} = \begin{pmatrix} \beta_0 + \beta_1 x_1 \\ \ldots \\ \beta_0 + \beta_1 x_n \end{pmatrix} = \begin{pmatrix} \widehat{y_1} \\ \ldots \\ \widehat{y_n} \end{pmatrix} = \widehat{\mathbf{y}} \tag{4-3}$$

The hat sign ($\widehat{\ }$) above a letter indicates that it is an estimate rather than the true value for the specific parameter. Thus, $\widehat{y}_1$ is the value on the regression line at the place where $x = x_1$, and $\widehat{\beta}_0$ is a specific value (the point estimate) of β_0 that has been estimated by fitting a regression line to a specific data set. We apologize for sometimes forgetting the hat-sign, for example, above every β-value in Equation 4-3. However, we try to use the hat sign where it is essential. For example, we always use the hat sign when we speak of fitted values (e.g., $\widehat{y}_1$), whereas we normally omit the hat sign for model coefficients β_k, because we do not know the true values of the model coefficients anyway, so there is no danger of confusing it with observed data. However, we do use hat signs for the model coefficients when we mean the specific value that has been estimated based on data rather than the theoretical coefficient.

In Equation 4-1, all parameters that are unknown have a hat sign. The fitted values $\widehat{y}_1$ are directly defined by the model coefficients, $\widehat{\beta}_0$ and $\widehat{\beta}_1$. Therefore, when we can estimate $\widehat{\beta}_0$, $\widehat{\beta}_1$, and $\widehat{\sigma}$, the model is fully defined (β_0, β_1, and σ are the model parameters). The last parameter $\widehat{\sigma}$ describes how the observations scatter around the regression line and relies on the assumption that the residuals are normally distributed. The estimates for the model parameters of a linear regression are obtained by searching for the best fitting regression line. To do so, we search for the regression line that minimizes the sum of the squared residuals. This model fitting method is called the least-squares method, abbreviated as LS. It has a very simple solution using matrix algebra (see e.g., Aitkin et al., 2009). Note that we can apply LS techniques independent of whether we use a Bayesian or frequentist framework to draw inference.

In Bayesian statistics, Equation 4-1 is called the data model, because it describes mathematically the process that has (or, better, that we think has) produced the data. This nomenclature also helps to distinguish data models from models for parameters such as prior distributions.

4.1.2 Fitting a Linear Regression in R

The least-squares estimates for the model parameters of a linear regression are obtained in R using the function `lm`. For illustration, we first simulate a data set and then fit a linear regression to these simulated data. The advantage of simulating data is that the following analyses can be reproduced without having to read data into R.

```
n <- 50          # sample size
sigma <- 5       # standard deviation of the residuals
b0 <- 2          # intercept
b1 <- 0.7        # slope
x <- runif(n, 10, 30) # sample values of the covariate
yhat <- b0 + b1*x
y <- rnorm(n, yhat, sd=sigma)
# plot the data (Fig. 4.1)
plot(x,y, pch=16, las=1, cex.lab=1.2)
abline(lm(y~x), lwd=2, col="blue") # insert regression line
# add the residuals
segments(x, fitted(lm(y~x)), x, y, lwd=2, col="orange", lty=3)
```

Then, we fit a linear regression to the data.

```
mod <- lm(y~x)
mod

Call:
lm(formula = y ~ x)

Coefficients:
(Intercept)        x
      2.005    0.688

summary(mod)$sigma
[1] 5.04918
```

The object "mod" produced by `lm` contains the estimates for the intercept, β_0, and the slope, β_1. The residual standard deviation σ is extracted using the function `summary`.

Conclusions drawn from a model depend on the model assumptions. When model assumptions are violated, estimates usually are biased and inappropriate conclusions can be drawn. We devote Chapter 6 to the assessment of model assumptions, given its importance.

4.1.3 Drawing Conclusions

To answer the question about how strongly y is related to x, or to predict y from x, and because we usually draw inference in a Bayesian framework, we are interested in the joint posterior distribution of $\boldsymbol{\beta}$ (vector that contains β_0 and β_1) and σ^2, the residual variance. The function `sim` does this for us. To somewhat demystify the `sim` function we briefly explain what `sim` does. The principle is to first draw a random value from the marginal posterior distribution of σ^2, and then to draw random values from the conditional posterior distribution for $\boldsymbol{\beta}$ (Gelman et al., 2014).

The conditional posterior distribution of the parameter vector $\boldsymbol{\beta}$, $p(\boldsymbol{\beta}|\sigma^2,\mathbf{y},\mathbf{X})$ is the posterior distribution of $\boldsymbol{\beta}$ given a specific value for σ^2. This conditional distribution can be analytically derived. With flat prior distributions, it is a uni- or multivariate normal distribution $p(\boldsymbol{\beta}|\sigma^2,\mathbf{y},\mathbf{X}) = Norm(\widehat{\boldsymbol{\beta}}, V_\beta\sigma^2)$ with

$$\widehat{\boldsymbol{\beta}} = \left(\mathbf{X}^T\mathbf{X}\right)^{-1}\mathbf{X}^T\mathbf{y} \tag{4-4}$$

and $V_\beta = (\mathrm{X}^T\mathrm{X})^{-1}$. For models with the normal error distribution, the LS estimates for $\boldsymbol{\beta}$ (given by Eq. 4-4) equal the maximum likelihood (ML) estimates (see Chapter 5).

The marginal posterior distribution of σ^2 is independent of specific values of $\boldsymbol{\beta}$. It is, for flat prior distributions, an inverse chi-square distribution $p(\sigma^2|\mathbf{y},\mathbf{X}) = \text{Inv-}\chi^2(n\text{-}k, s^2)$, where $s^2 = \frac{1}{n-k}(\mathbf{y} - \mathbf{X}\widehat{\boldsymbol{\beta}})^T(\mathbf{y} - \mathbf{X}\widehat{\boldsymbol{\beta}})$, and k is the number of parameters. The marginal posterior distribution of $\boldsymbol{\beta}$ can be obtained by integrating the conditional posterior distribution $p(\boldsymbol{\beta}|\sigma^2,\mathbf{y},\mathbf{X}) = Norm(\widehat{\boldsymbol{\beta}}, V_\beta\sigma^2)$ over the distribution of σ^2. This results in a uni- or multivariate t-distribution. However, it is not necessary to do this analytically. Using the function `sim`, we can draw samples from $p(\boldsymbol{\beta}|\sigma^2,\mathbf{y},\mathbf{X})$ and describe the marginal posterior distributions of $\boldsymbol{\beta}$ using the simulated values.

```
nsim <- 1000
bsim <- sim(mod, n.sim=nsim)
```

The function `sim` simulates (in our example) 1000 values from the joint posterior distribution of the three model parameters. These simulated values are shown in Figure 4-2.

The posterior distributions describe the range of plausible parameter values given the data and the model. They express our uncertainty about the model parameters; they show what we know about the model parameters after having looked at the data and given the model is realistic.

The 2.5% and 97.5% quantiles of the marginal posterior distributions can be used as 95% credible intervals of the model parameters. The function `coef` extracts the simulated values for the beta coefficients, returning a matrix with nsim rows and the number of columns corresponding to the number of parameters. In our example, the first column contains the simulated values from the posterior distribution of the intercept and the second column contains values from the posterior distribution of the slope. The "2" in the second argument of the `apply`-function (see Chapter 2) indicates that the `quantile` function is applied columnwise.

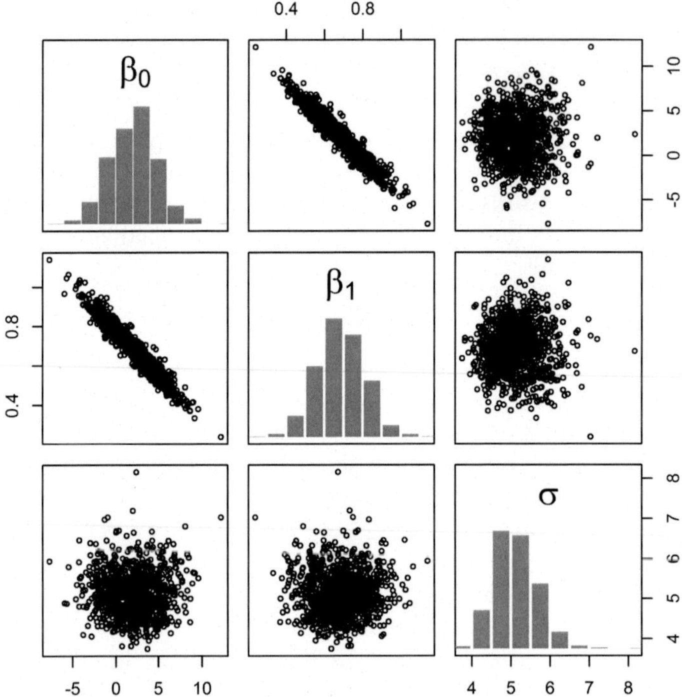

FIGURE 4-2 Joint (scatterplots) and marginal (histograms) posterior distribution of the model parameters. The six scatterplots show, using different axes, the three-dimensional cloud of 1000 simulations from the joint posterior distribution of β_0, β_1, and σ.

```
apply(coef(bsim), 2, quantile, prob=c(0.025, 0.975))
          [,1]     [,2]
2.5%   −3.043    0.439
97.5%   7.231    0.921
```

We also can calculate a credible interval of the estimated residual standard deviation, $\hat{\sigma}$.

```
quantile(bsim@sigma, prob=c(0.025, 0.975))
   2.5%    97.5%
  4.203    6.300
```

Using Bayesian methods allows us to get a posterior probability for specific hypotheses, such as "The slope parameter is larger than 1" or "The slope parameter is larger than 0.5". These probabilities are the proportion of simulated values from the posterior distribution that are larger than 1 and 0.5, respectively.

```
sum(coef(bsim)[,2]>1)/nsim
[1] 0.008
sum(coef(bsim)[,2]>0.5)/nsim
[1] 0.935
```

From this, there is very little evidence in the data that the slope is larger than 1, but we are quite confident that the slope is larger than 0.5 (given our model is realistic).

We often want to show the effect of x on y graphically, with information about the uncertainty of the parameter estimates included in the graph. To draw such effect plots, we use the simulated values from the posterior distribution of the model parameters. From the deterministic part of the model, we know the regression line $\widehat{y}_i = \widehat{\beta}_0 + \widehat{\beta}_1 x_i$. The simulation from the joint posterior distribution of β_0 and β_1 gives 1000 pairs of intercepts and slopes that describe 1000 different regression lines. We can draw these regression lines in an x-y plot to show the uncertainty in the regression line estimation (Figure 4-3, left) using the following code:

```
plot(x,y, pch=16, las=1, cex.lab=1.2)        # plot observations
for(i in 1:nsim) abline(coef(bsim)[i,1], coef(bsim)[i,2],
        col=rgb(0,0,0,0.05)) # add semitransparent regression lines
```

A more convenient way to show uncertainty is to draw the 95% credible interval, CrI, of the regression line. To this end, we first define new x-values for which we would like to have the fitted values (about 100 points across the range of x will produce smooth-looking lines when connected by line segments). We save these new x-values within the new data frame "newdat." Then, we create a new model matrix that contains these new x-values ("newmodmat") using the function `model.matrix`. We then calculate the 1000 fitted values for each element of the new x (one value for each of the 1000 simulated regressions, Figure 4-3), using matrix multiplication (`%*%`). We save these values in the matrix "fitmat". Finally, we extract the 2.5% and 97.5% quantiles for each x-value from fitmat, and draw the lines for the lower and upper limits of the credible interval (Figure 4-3, right).

```
newdat <- data.frame(x=seq(10, 30, by=0.1))
newmodmat <- model.matrix(~x, data=newdat)
fitmat <- matrix(ncol=nsim, nrow=nrow(newdat))
for(i in 1:nsim) fitmat[,i] <- newmodmat %*% coef(bsim)[i,]
plot(x,y, pch=16, las=1, cex.lab=1.2)
abline(mod, lwd=2)

lines(newdat$x, apply(fitmat, 1, quantile, prob=0.025), lty=3)
lines(newdat$x, apply(fitmat, 1, quantile, prob=0.975), lty=3)
```

The interpretation of the 95% credible interval is straightforward: We are 95% sure that the true regression line is within the credible interval. The larger the sample size, the narrower the interval, because each additional data point increases information about the true regression line.

The credible interval measures uncertainty of the regression line, but it does not describe how new observations would scatter around the regression line. If we want to describe where future observations will be, we have to

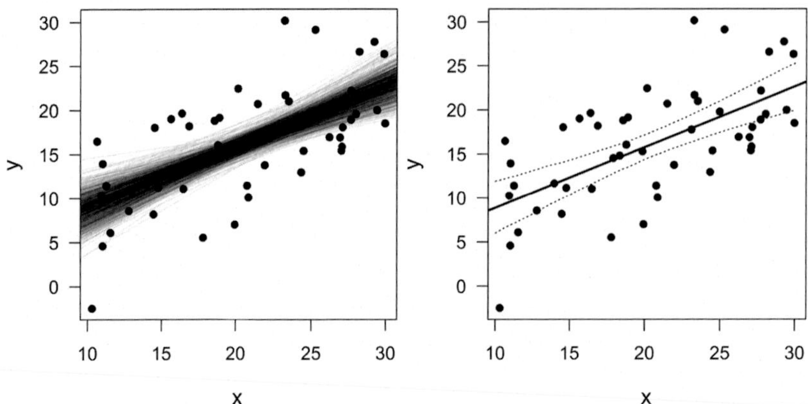

FIGURE 4-3 Regression with visualization of the uncertainty in the parameter estimates. *Left*: 1000 draws from the joint posterior distribution for the intercept and slope parameters. *Right*: 95% credible interval of the posterior distribution of the fitted values.

report the posterior predictive distribution. We can get a sample of random draws from the posterior predictive distribution $y^*|\beta, \sigma^2, X \sim Norm(X\beta, \sigma^2)$ using the simulated joint posterior distributions of the model parameters, thus taking the uncertainty of the parameter estimates into account. We draw a new y^*-value from $Norm(X\beta,\sigma^2)$ for each simulated set of model parameters. Then, we can visualize the 2.5% and 97.5% quantiles of this distribution for each new x-value.

```
# prepare matrix for simulated new data
newy <- matrix(ncol=nsim, nrow=nrow(newdat))
# for each simulated fitted value, simulate one new y-value
for(i in 1:nsim) newy[,i] <- rnorm(nrow(newdat), mean=fitmat[,i],
                                   sd=bsim@sigma[i])
lines(newdat$x, apply(newy, 1, quantile, prob=0.025), lty=2)
lines(newdat$x, apply(newy, 1, quantile, prob=0.975), lty=2)
```

Future observations are expected to be within the interval defined by the broken lines in Figure 4-4 with a probability of 0.95. Increasing sample size will not give a narrower, but a more precise predictive distribution.

The way we produced Figure 4-4 is somewhat tedious compared to how easy we could have obtained the same figure using frequentist methods: `predict(mod, newdata=newdat, interval=c("prediction"))` would have produced the y-values for the lower and upper lines in Figure 4-4 in one R-code line. However, once we have a simulated sample of the posterior predictive distribution, we have much more information than is contained in the frequentist prediction interval. For example, we could give an estimate for the proportion of observations greater than 20, given $x = 25$.

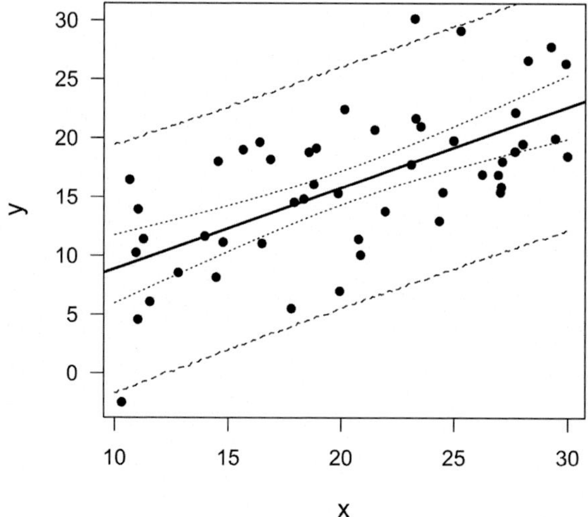

FIGURE 4-4 Regression line with 95% credible interval (dotted line) and the 95% interval of the simulated predictive distribution (broken line). Note that we increased the number of simulations to 50,000 to produce smooth lines.

```
sum(newy[newdat$x==25,]>20)/nsim
[1] 0.44504
```

Thus, we expect 44% of future observations with $x = 25$ to be higher than 20. We can extract similar information for any relevant threshold value.

Another reason to learn the more complicated R code we presented here, compared to the frequentist methods, is that, for more complicated models such as mixed models, the frequentist methods to obtain confidence intervals of fitted values are much more complicated than the Bayesian method just presented. The latter can be used with only slight adaptations for mixed models and also for generalized linear mixed models.

4.1.4 Frequentist Results

The solution for $\widehat{\boldsymbol{\beta}}$ is the Equation 4-3 defined earlier. Most statistical software, including R, return an estimated frequentist standard error for each $\widehat{\beta}_k$. We extract these standard errors together with the estimates for the model parameters using the `summary` function.

```
summary(mod)
Call:
lm(formula = y ~ x)

Residuals:
      Min       1Q   Median     3Q     Max
-11.5777 -3.6280 -0.0532 3.9873 12.1374
```

```
Coefficients:
             Estimate  Std. Error  t value    Pr(>|t|)
(Intercept)    2.0050      2.5349    0.791       0.433
x              0.6880      0.1186    5.800  5.07e-07 ***
---
Signif. codes: 0 '***' 0.001 '**' 0.01 '*' 0.05 '.' 0.1 ' ' 1
Residual standard error: 5.049 on 48 degrees of freedom
Multiple R-squared: 0.412, Adjusted R-squared: 0.3998
F-statistic: 33.63 on 1 and 48 DF, p-value: 5.067e-07
```

The summary output first gives a rough summary of the residual distribution. However, we will do more rigorous residual analyses in Chapter 6. The estimates of the model coefficients follow. The column "Estimate" contains the estimates for the intercept $\widehat{\beta}_0$ and the slope $\widehat{\beta}_1$. The column "Std. Error" contains the estimated (frequentist) standard errors of the estimates. The last two columns contain the t-value and the p-value of the classical t-test for the null hypothesis that the coefficient equals zero. The last part of the summary output gives the parameter σ of the model, named "residual standard error" and the residual degrees of freedom.

We try to avoid the name "residual standard error" and use "sigma" instead, because σ is not a measurement of uncertainty of a parameter estimate like the standard errors of the model coefficients are. σ is a model parameter that describes how the observations scatter around the fitted values, that is, it is a standard deviation. It is independent of sample size, whereas the standard errors of the estimates for the model parameters will decrease with increasing sample size. Such a standard error of the estimate of σ, however, is not given in the summary output. Note that, by using Bayesian methods, we could easily obtain the standard error of the estimated σ by calculating the standard deviation of the posterior distribution of σ. The R^2 and the adjusted R^2 are explained in Section 10.2.

4.2 REGRESSION VARIANTS: ANOVA, ANCOVA, AND MULTIPLE REGRESSION

4.2.1 One-Way ANOVA

The aim of analysis of variance (ANOVA) is to compare means of an outcome variable y between different groups. To do so in the frequentist's framework, variances between and within the groups are compared (hence the name "analysis of variance"). If the variance between the group means is larger than expected by chance given the variance not explained by the grouping structure (i.e., the variance within the groups), we reject the null hypothesis of no differences between the groups. When doing an ANOVA in a Bayesian way, inference is based on the posterior distributions of the group means and the differences between the group means.

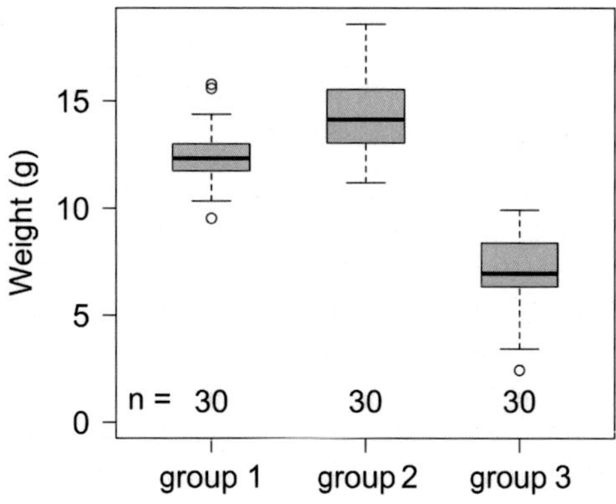

FIGURE 4-5 Weights (g) of the 30 individuals in each group. The dark horizontal line is the median, the box contains 50% of the observations (i.e., the interquartile range), the whiskers mark the range of all observations that are less than 1.5 times the interquartile range away from the edge of the box.

One-way ANOVA means that we only have one explanatory variable (a factor). We illustrate the one-way ANOVA based on an example of simulated data (Figure 4-5). We have measured weights of 30 virtual individuals for each of 3 groups. Possible research questions could be: How big are the differences between the group means? Are individuals from group 2 heavier than the ones from group 1? Which group mean is higher than 7.5 g?

```
# data simulation
mu <- 12    # "true" mean of group 1 (reference group)
sigma <- 2 # Residual standard deviation (=within group SD)
b1 <- 3    # difference between the "true" means of group 1 and group 2
b2 <- -5   # difference between the "true" means of group 1 and group 3
n <- 90    # sample size
group <- factor(rep(c(1,2,3), each=30))

# simulate the y variable
simresid <- rnorm(n, mean=0, sd=sigma)
y <- mu + as.numeric(group=="2")*b1 + as.numeric(group=="3")*b2 +
   simresid
```

An ANOVA is a linear regression with a categorical predictor variable instead of a continuous one. The categorical predictor variable with k levels is (as a default in R) transformed to $k - 1$ indicator variables. An indicator variable is a binary variable containing 0 and 1 where 1 indicates a specific level (a category of a nominal variable). Often, one indicator variable is constructed for every level except for the reference level. In our example, the categorical variable is group (g) with the three levels 1, 2, and 3 ($k = 3$). Group 1 is taken as the reference level, and for each of the other two groups an indicator variable is constructed, $I(g_i = 2)$ and $I(g_i = 3)$. We can write the model as a formula:

$$\begin{aligned} \hat{y}_i &= \beta_0 + \beta_1 I(g_i = 2) + \beta_2 I(g_i = 3) \\ y_i &\sim Norm(\hat{y}_i, \sigma^2) \end{aligned} \tag{4-5}$$

where y_i is the i-th observation (weight measurement for individual i), and $\beta_{0,1,2}$ are the model coefficients. The residual variance is σ^2. The model coefficients $\beta_{0,1,2}$ constitute the deterministic part of the model. From the model formula it follows that the group means, m_g, are:

$$\begin{aligned} m_1 &= \beta_0 \\ m_2 &= \beta_0 + \beta_1 \\ m_3 &= \beta_0 + \beta_2 \end{aligned} \tag{4-6}$$

There are other possibilities to describe three group means with three parameters, for example:

$$\begin{aligned} m_1 &= \beta_1 \\ m_2 &= \beta_2 \\ m_3 &= \beta_3 \end{aligned} \tag{4-7}$$

In this case, the model formula would be:

$$\begin{aligned} &\widehat{y}_i = \beta_1 I(g_i = 1) + \beta_2 I(g_i = 2) + \beta_3 I(g_i = 3) \\ &y_i \sim Norm(\widehat{y}_i, \sigma^2) \end{aligned}$$

The way the group means are described is called the parameterization of the model. Different statistical softwares use different parameterizations. The parameterization used by R by default is the one shown in Equation 4-6. R automatically takes the first level as the reference (the first level is the first one alphabetically unless the user defines a different order for the levels). The mean of the first group (i.e., of the first factor level) is the intercept, β_0, of the model. The mean of another factor level is obtained by adding, to the intercept, the estimate of the corresponding parameter (which is also the difference from the reference group mean). R calls this parameterization "treatment contrasts".

The parameterization of the model is defined by the model matrix. In the case of a one-way ANOVA, there are as many columns in the model matrix as there are factor levels (i.e., groups); thus there are k factor levels and k model coefficients. Recall from Equation 4-3 that for each observation, the entry in the j-th column of the model matrix is multiplied by the j-th element of the model coefficients and the k products are summed to obtain the fitted values. For a data set with $n = 5$ observations of which the first two are from group 1, the third from group 2, and the last two from group 3, the model matrix used for the parameterization described in Equation 4-6 is

$$X = \begin{pmatrix} 1 & 0 & 0 \\ 1 & 0 & 0 \\ 1 & 1 & 0 \\ 1 & 0 & 1 \\ 1 & 0 & 1 \end{pmatrix}$$

If parameterization of Equation 4-7 were used,

$$X = \begin{pmatrix} 1 & 0 & 0 \\ 1 & 0 & 0 \\ 0 & 1 & 0 \\ 0 & 0 & 1 \\ 0 & 0 & 1 \end{pmatrix}$$

Other possibilities of model parameterizations, particularly for ordered factors, are introduced in Section 4.2.8.

Defining Factors and Factor Levels in R

We would like to draw your attention to some default features of R when reading data into the workspace (e.g., when using the function `read.table`). A column in the data table containing not only numbers but also text is, automatically, converted to a factor. A variable with only numbers (and, possibly, missing values, coded "NA") is converted to a numeric or integer variable. Thus, if we code our factor with numbers, we must tell R that it is a factor, for example, by using `dat$group <- factor(dat$group)`; it may be wise to always use letters to code factors (unlike our example above!). Using `str(dat)` in R tells us which variable is defined as a factor.

When converting a variable to a factor, by default, the order of the levels is alphabetic. If we want to change the order of the factor levels, for example, for the factor "age" with the levels "adult", "immature", and "juvenile", the code in R is: `dat$age <- factor(dat$age, levels = c("juvenile","immature", "adult"))`. The order of factor levels determines which one will be used as the reference level. It also determines the order in plots and in the output of various functions such as `table`. Using `levels(dat$group)` lists all factor levels in the order that R treats them.

Be very careful when changing a variable defined as a factor and coded with numbers into a numeric vector: the function `as.numeric` will return the number of the factor level, not the number itself! Thus, applying `as.numeric` to a factor with the levels "2", "33", and "36" will return a "1", "2", and "3" instead of "2", "33", and "36", respectively. If we want the 2s, 33s, and 36s, we have to do `as.numeric(as.character(..))`.

Factor levels are retained, even if we delete all cases of one level from the data set. R remembers this factor level even though there is no observation left. As a consequence, cross-tables obtained by `table`, for example, will show this level with a count of 0. To get rid of "empty" factor levels, we can do: `dat$age <- dat$age[drop=T]`, or `dat$age <- factor(dat$age)`, or, if we want to drop all unused levels of all variables, `dat <- droplevels(dat)`.

To obtain the parameter estimates for model parameterized according to Equation 4-6 we fit the model in R:

```
group <- factor(group) # define group as a factor!
mod <- lm(y ~ group)   # fit the model
```

```
mod
Call:
lm(formula = y ~ group)

Coefficients:
(Intercept)    group2    group3
     12.367     2.215    -5.430

summary(mod)$sigma
[1] 1.684949
```

The "Intercept" is β_0. The other coefficients are named with the factor name ("group") and the factor level (either 2 or 3). These are β_1 and β_2, respectively. Before drawing conclusions from an R output we need to examine whether the model assumptions are met, that is, we need to do a residual analysis as described in Chapter 6.

Different questions can be answered using the above ANOVA: What are the group means? What is the difference in the means between group 1 and group 2? What is the difference between the means of the heaviest and lightest group? In a Bayesian framework we can directly assess how strongly the data support the hypothesis that the mean of the group 2 is larger than the mean of group 1.

We first simulate from the posterior distribution of the model parameters.

```
bsim <- sim(mod, n.sim=1000)
```

Then we obtain the posterior distributions for the group means according to the parameterization of the model formula (Eq. 4-6).

```
m.g1 <- coef(bsim)[,1]
m.g2 <- coef(bsim)[,1] + coef(bsim)[,2]
m.g3 <- coef(bsim)[,1] + coef(bsim)[,3]
```

The histograms of the simulated values from the posterior distributions of the three means are given in Figure 4-6. The three means are well separated and, based on our data, we are confident that the group means differ. From these simulated posterior distributions we obtain the means and use the 2.5% and 97.5% quantiles as limits of the 95% credible intervals (Figure 4-6, right).

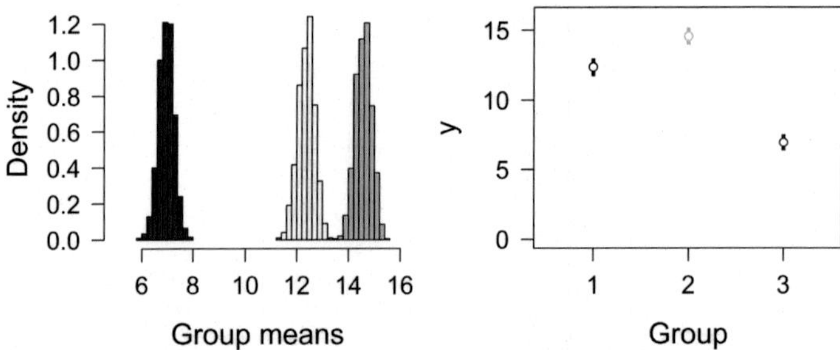

FIGURE 4-6 Distribution of the simulated values from the posterior distributions of the group means (*left*); group means with 95% credible intervals obtained from the simulated distributions (*right*). The color identifies the groups.

To obtain the posterior distribution of the difference between the means of group 1 and group 2, we simply calculate this difference for each draw from the joint posterior distribution of the group means:

```
d.g1.2 <- m.g1 - m.g2
mean(d.g1.2)
[1] -2.211714
quantile(d.g1.2, prob=c(0.025,0.975))
      2.5%       97.5%
-3.127215  -1.352867
```

The estimated difference is −2.2. We are 95% sure that the difference between the means of group 1 and 2 is between −3.1 and −1.4.

How strongly do the data support the hypothesis that the mean of group 2 is larger than the mean of group 1? To answer this question we calculate the proportion of the draws from the joint posterior distribution for which the mean of group 2 is larger than the mean of group 1.

```
sum(m.g2 > m.g1) / nsim   # probability that m2 > m1
[1] 1
```

This means that in all of the 1000 simulations from the joint posterior distribution, the mean of group 2 was larger than the mean of group 1. Therefore, there is a very high probability (i.e., it is close to 1; because probabilities are never exactly 1, we write >0.999) that the mean of group 2 is larger than the mean of group 1.

4.2.2 Frequentist Results from a One-Way ANOVA

Three quantities are important: total sum of squares SST $= \sum_{i=1}^{n}(y_i - \bar{y})^2$, sum of squares between groups SSB $= \sum_{i=1}^{n}(\bar{y}_{g_i} - \bar{y})^2$, and sum of squares within groups SSW $= \sum_{i=1}^{n}(y_i - \bar{y}_{g_i})^2$. The sum of SSB and SSW equals SST (Figure 4-7).

SSB represents the variance that is explained by the model, whereas SSW measures the unexplained or residual variance. How big must SSB be compared to SSW to conclude that the between-group variance is larger than expected by chance alone? The answer is given by an F-test. To that end, we first construct a null hypothesis H_0: "There is no difference between the group means": $\theta_1 = \theta_2 = \theta_3$. Given H_0 we expect that the variance between the groups equals the variance within the groups. To obtain comparable variances, the mean sum of squares between groups (MSB) and within groups (MSW) are calculated by dividing the sums of squares, SSB and SSW, by their respective degrees of freedoms (df).

The df for SSB is the number of groups minus one ($3 - 1 = 2$ in our case), because only two of the group means are free to vary. If two group means are fixed to a value, the third mean is defined by the first two means and the overall

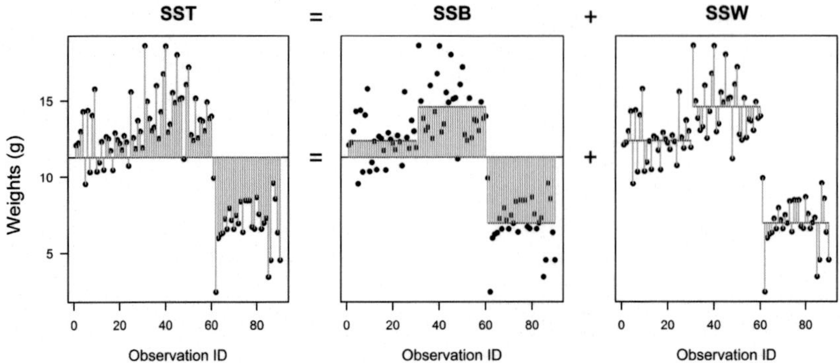

FIGURE 4-7 Total sum of squares (SST), between group sum of squares (SSB), and within group sum of squares (SSW) for the simulated data.

mean that is given by the data. Similarly, for SSW the degrees of freedom is the number of observations minus the number of model parameters (90 – 3 in our case), because, if 29 observations are free to vary in each group, the 30th observation can be obtained by the 29 other observations and the group mean, thus this observation is no longer "free". The ratio MSB/MSW gives the test statistic, *F*, which is close to 1 if the between and within group variances are equal. Typically, these results are presented in an ANOVA table (Table 4-1).

The *F*-distribution for a given numerator df and denominator df determines the probability distribution of the *F*-value given H_0 (Figure 4-8). From this distribution, the probability of getting a data set with the observed or a more extreme F-value under H_0 is obtained. In our example, the probability of observing an *F*-value ≥ 3.1, given H_0 is true, is 0.05. We reject H_0 if the

TABLE 4-1 ANOVA Table of the Simulated Data Example

Source	df	SS	MS	*F*
Between (= group)	2	928	464.1	163.5
Within (= residual)	87	247	2.8	
Total	89	1175		

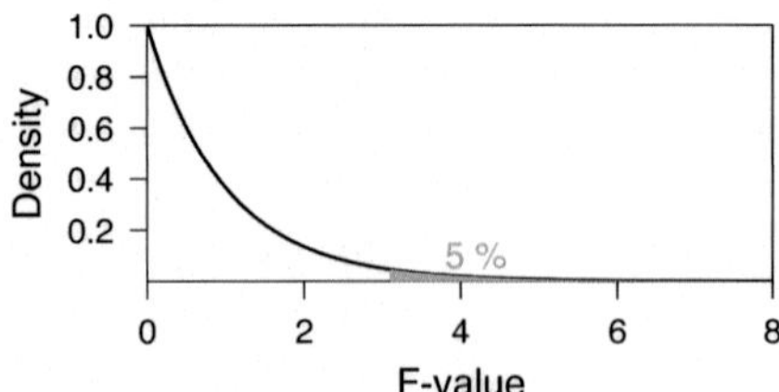

FIGURE 4-8 $F_{2,87}$-distribution with the 5% rejection region highlighted.

observed F-value is higher than 3.1 and conclude that at least one group's mean is significantly different from another group's mean.

In the summary output of the model (`summary(mod)`) standard errors of the coefficients, a t-value, and a p-value are given. This p-value is the outcome of a t-test for the null hypothesis that the coefficient, β_k, equals zero. The last part of the summary output gives the parameter σ and the residual degrees of freedom. The R^2 and the adjusted R^2 are explained in Section 10.2. We also find the F-value and the corresponding p-value in the summary output.

4.2.3 Two-Way ANOVA

We often want to model the effect of more than one grouping variable (factor) and quantify the differences between the group means for each factor separately. We illustrate the two-way ANOVA using the data set "periparusater": How big are the differences in wing length between the sexes and age classes in coal tits *Periparus ater*? Wing lengths were measured on 19 coal tit museum skins with known sex and age class (Figure 4-9). Sex has two levels (1 = male, 2 = female) and age class also has two levels (3 = juvenile, 4 = adult; according to the code defined by the EURING database, www.euring.org).

```
data(periparusater)
dat <- periparusater # give the data frame a shorter name
```

To describe differences in wing length between the age classes or between the sexes, a two-way ANOVA is fitted to the data. The two explanatory factors are specified on the right side of the model formula separated by the "+" sign, which means that the model is an additive combination of the two effects (as opposed to an interaction, see following).

```
mod <- lm(wing ~ sex + age, data=dat)
```

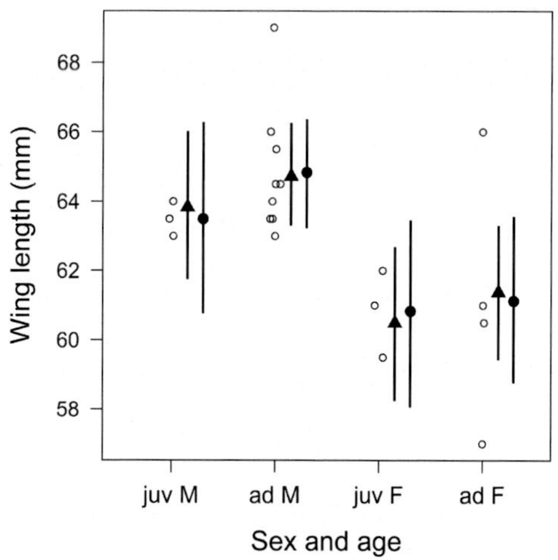

FIGURE 4-9 Wing length measurements (circles) on 19 museum skins of coal tits per age class and sex. Fitted values obtained are from the additive model (triangle) and from the model including an interaction (dot). Vertical bars = 95% credible intervals.

After having seen that the residual distribution does not appear to violate the model assumptions (as assessed with the methods presented in Chapter 6), we can draw inferences. We first have a look at the model parameter estimates:

```
mod
Call:
lm(formula = wing ~ sex + age, data = dat)
Coefficients:
(Intercept)       sex2      age4
    63.8378    -3.3423    0.8829
summary(mod)$sigma
[1] 2.134682
```

Note that R has taken the first level of sex ("1") and age ("3") as reference levels and provides estimates of what is to be added to get the fitted values of the other levels, that is, "sex2" and "age4."

To get the fitted values, we insert the parameter estimates into the model formula:

$$\widehat{y}_i = \widehat{\beta}_0 + \widehat{\beta}_1 I(sex_i = 2) + \widehat{\beta}_2 I(age_i = 4) \text{ which yields}$$
$$\widehat{y}_i = 63.8 - 3.3 I(sex_i = 2) + 0.9 I(age_i = 4).$$

Alternatively, we could use matrix notation. We construct a new data set that contains one "virtual" individual for each age and sex class. For this new data set the model matrix contains four rows (corresponding to juvenile male, adult male, juvenile female, and adult female) and three columns. The first column contains only ones because the values of this column are multiplied by the intercept (β_0) in the matrix multiplication. The second column contains an indicator variable for females (so only the rows corresponding to females contain a one) and the third column has ones for adults.

$$\widehat{\mathbf{y}} = \mathbf{X}\widehat{\boldsymbol{\beta}} = \begin{pmatrix} 1 & 0 & 0 \\ 1 & 0 & 1 \\ 1 & 1 & 0 \\ 1 & 1 & 1 \end{pmatrix} \times \begin{pmatrix} 63.8 \\ -3.3 \\ 0.9 \end{pmatrix} = \begin{pmatrix} 63.8 \\ 64.7 \\ 60.5 \\ 61.4 \end{pmatrix}$$

The result of the matrix multiplication is a vector containing the estimated wing lengths for juvenile males, adult males, juvenile females, and adult females, respectively, that is, the fitted values. The function `predict` does the calculations for us. Alternatively, we can use matrix multiplication.

```
newdat <- expand.grid(sex=factor(c(1,2)), age=factor(c(3,4)))
newdat$fit <- predict(mod, newdata=newdat) # or:
newdat$fit <- model.matrix(~ sex + age, data=newdat) %*% coef(mod)
```

The function `expand.grid` creates a data frame with all combinations of the variable values given, and `model.matrix` creates a model matrix (like the one shown) for a specific model formula and a data set. When creating the model matrix, care has to be taken that exactly the same model formula is used, for example, using "~ age + sex" in the previous example would produce the wrong results! We normally copy and paste the model formula from the lmcode to the model.matrixcode. Alternatively, the code `formula(mod)[c(1,3)]` extracts the model formula from a fitted model object.

To describe the uncertainty about these fitted values, we simulate 2000 sets of model parameters from the joint posterior distribution:

```
nsim <- 2000
bsim <- sim(mod, n.sim=nsim)
```

We then calculate the fitted values for each set of model parameters by hand using the matrix multiplication just described.

```
fitmat <- matrix(ncol=nsim, nrow=nrow(newdat))
Xmat <- model.matrix(formula(mod)[c(1,3)], data=newdat)
for(i in 1:nsim) fitmat[,i] <- Xmat %*% bsim@coef[i,]
```

The matrix "fitmat" now contains 2000 draws from the posterior distributions of the estimated average wing length for each of the four combinations of the predictor levels. The 2.5% and 97.5% quantiles of these 2000 values define the 95% credible interval. We extract these quantiles row-wise from "fitmat" by the function `apply`:

```
newdat$lower <- apply(fitmat, 1, quantile, prob=0.025)
newdat$upper <- apply(fitmat, 1, quantile, prob=0.975)
```

We can see that the fitted values are not equal to the arithmetic means of the groups; this is especially clear for juvenile males (Figure 4-9). The fitted values are constrained because only three parameters were used to estimate four means. In other words, this model assumes that the age difference is equal in both sexes and, vice versa, that the difference between the sexes does not change with age. If the effect of sex changes with age, we would include an interaction between sex and age in our model. Including an interaction adds a fourth parameter enabling us to estimate the group means exactly. In R, an interaction is indicated with the "`:`" sign. In printed text, an interaction is usually coded with an "×" like sex × age, or, in words, "sex by age."

```
mod2 <- lm(wing ~ sex + age + sex:age, data=dat)
```

Alternatively we can use the "*" sign, which tells R to include the interaction and the main effects `mod2 <- lm(wing ~ sex * age, data=dat)`, or the `^2` sign `mod2 <- lm(wing ~ (sex + age)^2, data=dat)`. The formula for this model is:

$$\hat{y}_i = \beta_0 + \beta_1 I(sex_i = 2) + \beta_2 I(age_i = 4) + \beta_3 I(sex_i = 2) I(age_i = 4)$$
$$y_i \sim Norm(\hat{y}_i, \sigma^2)$$

To obtain the fitted values for each sex and age class, the R code just mentioned can be recycled without any adaptation, except that we replace "mod" by "mod2". These fitted values are exactly equal to the arithmetic means of each group (Figure 4-9). We can also see that the uncertainty of the fitted values is larger for the model with an interaction than for the additive model. This is because, in the model including the interaction, an additional parameter has to be estimated based on the same amount of data. Therefore, the information available per parameter is smaller than in the additive model. In the additive model, some information is pooled between the groups.

How do we decide whether an interaction is needed or not? This is a delicate question that we discuss in more detail in Chapter 11. If it is of interest, we look at the estimated strength of the interaction which is measured, in the coal tit example, by the parameter β_3 (the 4th parameter) to decide whether it is important or not.

```
bsim2 <- sim(mod2, n.sim=nsim)
quantile(bsim2@coef[,4], prob=c(0.025, 0.5, 0.975))
      2.5%         50%      97.5%
-6.035500  -1.066639  3.680728
```

The degree to which a difference in wing length is "important" depends on the context of the study. Here, for example, we could consider effects of wing length on flight energetics and maneuverability or methodological aspects such as measurement error. Mean between-observer difference in wing length measurement is around 0.3 mm (Jenni & Winkler, 1989). Therefore, we may consider that the interaction is important when the data support the hypothesis that the interaction parameter is larger than 0.3 mm. The degree of support for this is 90%. Further, we think a difference of 1 mm in wing length may be relevant compared to the between-individual variation, which is around 2 mm (`summary(mod2)$sigma`).

```
mean(abs(bsim2@coef[,4])>0.3)
[1] 0.9
```

Therefore, we report the parameter estimates of the model including the interaction together with their credible intervals.

```
coef(mod2)
(Intercept)        sex2      age4   sex2:age4
  63.500000 -2.666667 1.333333   -1.041667
apply(bsim2@coef, 2, quantile, prob=c(0.025, 0.975))
      (Intercept)          sex2         age4  sex2:age4
2.5%    60.79368   -6.275990 -1.740102 -6.035500
97.5%   66.07708    1.219414   4.671309   3.680728
```

From these parameters we obtain the estimated differences in wing length between the sexes for juveniles: –2.6 mm (95% CrI: –6.3 to 1.2 mm);

```
quantile(bsim2@coef[,2], prob=c(0.025, 0.5, 0.975))
      2.5%          50%      97.5%
-6.275990 -2.608274 1.219414
```

and for adults: –3.7 mm (–6.5 to –0.9 mm).

```
quantile(bsim2@coef[,2]+bsim2@coef[,4], prob=c(0.025,0.5,0.975))
      2.5%          50%        97.5%
-6.423355 -3.726922 -1.155056
```

And, maybe we would also like to report the posterior probability for the hypothesis that male coal tits have longer wings than females for both adults and juveniles:

```
sum(bsim2@coef[,2]<0)/nsim   # for juveniles (reference level)
[1] 0.924
sum(bsim2@coef[,2]+bsim2@coef[,4]<0)/nsim   # for adults
[1] 0.9935
```

The data support the hypothesis of sexual size dimorphism in coal tits.

4.2.4 Frequentist Results from a Two-Way ANOVA

F-tests are used to test whether the residual variance is significantly reduced when a predictor is added to the model. In ANOVAs with more than one predictor, two different F-tests exist: the sequential and the marginal F-tests. In the sequential tests, the terms in the model are added sequentially in the order they appear in the model formula, whereas in the marginal F-tests each term is tested as if it was the last entering the model. When the data are balanced (i.e., the same numbers of samples are in each combination of factor levels) the two tests produce the same results. However, balanced data sets are rare; in practice, the two tests usually produce different results, and it is important to understand the different interpretations (Table 4-2).

In sequential tests, the F-test for the first term in the model (sex in our example) tests whether this term can explain a significant portion of the total

TABLE 4-2 Model Comparisons Done When Using Sequential and Marginal *F*-Tests for Wing Length Example: wing ~ sex + age + sex × age

Term	Sequential *F*-test	Marginal *F*-test excluding interaction for testing main effects	Marginal *F*-test
Sex	wing ~ 1 vs. wing ~ sex	wing ~ age vs. wing ~ sex + age	Not possible
Age	wing ~ sex vs. wing ~ sex + age	wing ~ sex vs. wing ~ sex + age	Not possible
Sex × age	wing ~ sex + age vs. wing ~ sex + age + sex × age	Not applicable	wing ~ sex + age vs. wing ~ sex + age + sex × age

Note: Marginal *F*-tests usually are not conducted for main effects that are involved in an interaction because such tests are in most cases not interpretable. Therefore, sometimes, the interaction is deleted first to test the main effects.

variance, whereas the marginal test for the same term tests whether this term can explain a significant amount of the remaining variance after having corrected for all the other terms in the model. The R function `drop1` returns marginal frequentist *F*-tests while `anova` returns sequential frequentist *F*-tests.

One characteristic of the sequential test is that the test result depends on the order of the terms in the model. This is because the proportion of variance that is explained by each of the two predictors is identified differently depending on the order of the terms in the model. Figure 4-10 shows two different ways to estimate the variance explained by each of the predictors. In the left panel, the sum of squares (SS) of the differences between the overall mean and the means for each sex (fitted values of the model wing ~ sex) is considered first; this can be interpreted as the variance in wing length that is explained by sex alone.

The SS of the differences between the means by sex (ignoring age) and the means by sex and age classes (fitted values of the model wing ~ sex + age) is then interpreted as the variance explained by age after correcting for the effect of sex. Alternatively, the variance that is explained by age alone can be calculated first and then the additional variance that is explained by sex can be determined (Figure 4-10, right panel). Because the data set is unbalanced, the SS depends on the order of the predictors in the model: the SS for age is larger in the right panel than in the left.

To conclude, when reading an ANOVA table, we need to determine whether the *F*-values presented are from sequential or marginal tests and we need to be aware of the different interpretations of the two test types.

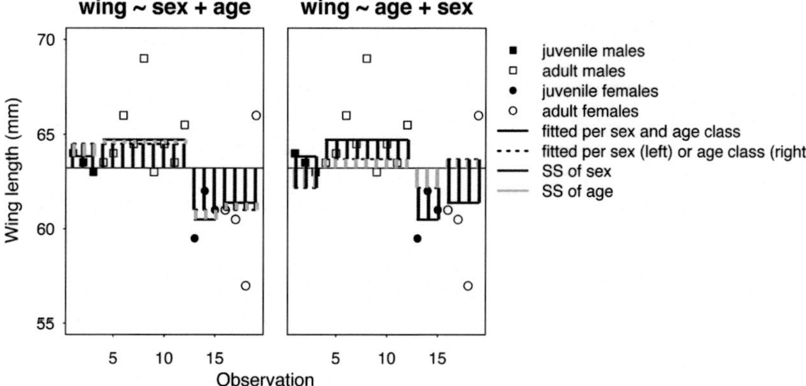

FIGURE 4-10 Variances (measured as sum of squares, SS) explained by each of the predictor variables sex and age depending on the order of the predictors in the model. Thin horizontal line = overall mean (null model), solid horizontal lines = fitted values of the model containing both predictors (sex and age), dotted horizontal lines = fitted values of the model containing only sex (left panel), or only age (right panel) as predictor. The SS of the vertical blue lines is interpreted as the variance in wing length that is explained by sex. The sum of the squares of the orange lines is interpreted as the variance in wing length that is explained by age. The SS of age is larger when age is entered as the first term in the model (right panel) than when it is entered as the second term (left panel).

4.2.5 Multiple Comparisons and Post Hoc Tests

In the null hypothesis testing framework, the issue of multiple testing arises when many different tests are done on the same set of data, for example, when comparing all possible pairs of means. By chance alone, 5% of the statistical tests will be considered significant even though the null hypothesis is in fact true. This is called type I error. The problem also pertains to Bayesian statistics: The true parameter will lie outside the 95% credible interval with a probability of 0.05, and, therefore, the larger the number of parameters the more credible intervals will not contain the true parameter value. However, Bayesian statisticians often treat a factor, for which many pairwise comparisons among levels are of interest, as a random factor (see Chapter 7). In such a model, the estimated group means are shrunk toward the overall mean depending on how precisely the group mean can be estimated.

Imprecise group mean estimates are shrunk to a higher degree than precise group mean estimates. This prevents one from overestimating specific differences that are large just because of random variation (Gelman et al., 2012). In frequentist statistics, different attempts have been made to adjust the significance level by the number of tests that have been made (e.g., Bonferroni or Dunn-Šidák method: Rice, 1989; Sokal & Rohlf, 2003; or the Tukey honest significant difference: Miller, 1980; Yandell, 1997). These corrections are usually applied only if an overall test has produced a significant result. Such tests are called post hoc tests. However, we do not use such post hoc tests but follow the guidelines regarding multiple testing of Gelman and Hill (2007):

- Formulate scientific hypotheses prior to data collection and stick to these during data analysis.
- Use, whenever possible, hierarchical models (mixed models).
- Accept that the true parameter value is outside the 95% CrI in 5% of the cases.
- Be honest—that is, describe the search effort that has been spent to find significant results, and also publish the nonsignificant results.

4.2.6 Analysis of Covariance

An analysis of covariance, ANCOVA, is a normal linear model that contains at least one factor and one continuous variable as explanatory variables. The continuous variable is also called a covariate, hence the name analysis of covariance. An ANCOVA can be used, for example, when we are interested in how the biomass of grass depends on the distance from the surface of the soil to the ground water in two different species (*Alopecurus pratensis*, *Dactylis glomerata*; Figure 4-11). The two species were grown by Ellenberg (1953, 1954) in tanks that showed a gradient in distance from the soil surface to the ground water. The distance from the soil surface to the ground water is used as a covariate ("water"). The model formula is then

$$\hat{y}_i = \beta_0 + \beta_1 I(species_i = Dg) + \beta_2 water_i$$
$$y_i \sim Norm\left(\hat{y}_i, \sigma^2\right)$$

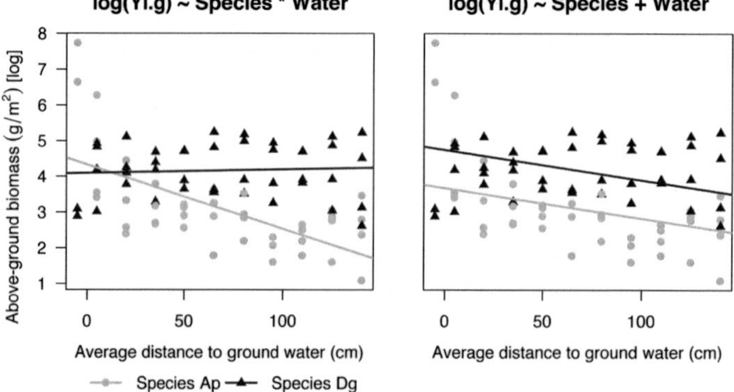

FIGURE 4-11 Aboveground biomass (*g*, log-transformed) in relation to distance to ground water and species (two grass species). *Left*: two separate (independent) regression lines per species, that is, from the model including an interaction. *Right*: the two regression lines from the ANCOVA without interaction.

To fit the model, it is important to first check whether the factor is indeed defined as a factor and the continuous variable contains numbers (i.e., numeric or integer values) in the data frame.

```
data(ellenberg)
index <- is.element(ellenberg$Species, c("Ap", "Dg"))
dat <- ellenberg[index,] # select two species
dat <- droplevels(dat) # drop unused factor levels
str(dat) # print the definitions of the variables
'data.frame': 88 obs. of 29 variables:
$ Year       : int 1952 1952 1952 1952 1952 1952 1952 1952 1952 1952 ...
$ Soil       : Factor w/ 2 levels "Loam","Sand": 1 1 1 1 1 1 1 1 1 1 ...
$ Water      : int -5 5 20 35 50 65 80 95 110 125 ...
$ Species    : Factor w/ 2 levels "Ap","Dg": 1 1 1 1 1 1 1 1 1 1 ...
$ Mi.g       : num NA 112.6 66.1 42.3 38.4 ...
$ Yi.g       : num NA 34.8 28 44.5 24.8 ...
[... and 23 more variables]
```

Species is a factor with two levels and Water is an integer variable, so we are fine and we can fit the model:

```
mod <- lm(log(Yi.g) ~ Species + Water, data=dat)
```

We log-transform the biomass to make the residuals closer to normally distributed (Chapter 6). Let's have a look at the model matrix (first six rows):

```
head(model.matrix(mod)) # print the first 6 rows of the matrix
   (Intercept) SpeciesDg Water
24           1         0     5
25           1         0    20
26           1         0    35
27           1         0    50
28           1         0    65
29           1         0    80
```

The first column of the model matrix contains only ones. These are multiplied by the intercept in the matrix multiplication that yields the fitted values. The second column contains the indicator variable for species *Dactylis glomerata* (Dg). Species *Alopecurus pratensis* (Ap) is the reference level. The third column contains the values for the covariate.

To have a first quick look at the parameter estimates, we often use the function `summary`, although it applies frequentist methods. We just ignore the standard errors, *t*-, *F*-, and *p*-values. But the summary output gives more information, which is also useful in a Bayesian framework, such as the note "4 observations deleted due to missingness," the degrees of freedom, and the R^2-value.

```
summary(mod)
Call:
lm(formula = log(Yi.g) ~ Species + Water, data = dat)
Residuals:
     Min        1Q  Median      3Q     Max
-1.8967 -0.6418 -0.1263 0.4482 4.0191
Coefficients:
              Estimate  Std. Error   t value    Pr(>|t|)
(Intercept) 3.678937    0.225652    16.304    < 2e-16 ***
SpeciesDg    1.065942    0.217117     4.910  4.66e-06 ***
Water       -0.008443    0.002403    -3.513   0.000728 ***
---
Signif. codes: 0 '***' 0.001 '**' 0.01 '*' 0.05 '.' 0.1 ' ' 1
Residual standard error: 0.995 on 81 degrees of freedom
   (4 observations deleted due to missingness)
Multiple R-squared: 0.3103, Adjusted R-squared: 0.2933
F-statistic: 18.22 on 2 and 81 DF, p-value: 2.919e-07
```

The parameters are the intercept ($\widehat{\beta}_0$), the difference between the grass species ($\widehat{\beta}_1$), and the slope for water ($\widehat{\beta}_2$). The regression lines are $\widehat{y}_i = \widehat{\beta}_0 + \widehat{\beta}_2\ water_i$ for species Ap and $\widehat{y}_i = \widehat{\beta}_0 + \widehat{\beta}_1 + \widehat{\beta}_2 water_i$ for species Dg. The model we just chose includes only one slope for water (i.e., the same slope for both species). Therefore, two parallel regression lines are fitted (Figure 4-11, right panel). However, it is pretty obvious that the two species react differently to water. Therefore, we need to include an interaction in the model. The model formula, then, is:

$$\widehat{y}_i = \beta_0 + \beta_1 I(Species_i = Dg) + \beta_2 Water_i + \beta_3 I(Species_i = Dg) Water_i$$

$$y_i \sim Norm\left(\widehat{y}_i, \sigma^2\right)$$

The two regression lines are: $\widehat{y}_i = \widehat{\beta}_0 + \widehat{\beta}_2 Water_i$ for species Ap and $\widehat{y}_i = \widehat{\beta}_0 + \widehat{\beta}_1 + (\widehat{\beta}_2 + \widehat{\beta}_3)\ Water_i$ for species Dg. β_1 is the difference in the intercept between the species and β_3 is the difference in the slope. When we fit the model including an interaction, we get exactly the same two regression lines as if we had fit two separate regressions, one for each species (Figure 4-11, left panel).

```
mod2 <- lm(log(Yi.g) ~ Species*Water, data=dat)
summary(mod2)
Call:
lm(formula = log(Yi.g) ~ (Species + Water)^2, data = dat)
Residuals:
     Min       1Q    Median      3Q     Max
-1.6158  -0.6135  -0.1063  0.5876  3.3203
```

```
Coefficients:
                 Estimate  Std. Error  t value    Pr(>|t|)
(Intercept)      4.330406    0.253108   17.109    < 2e-16 ***
SpeciesDg       -0.236995    0.357948   -0.662       0.51
Water           -0.017911    0.003075   -5.825   1.15e-07 ***
SpeciesDg:Water  0.018935    0.004349    4.354   3.92e-05 ***
-
Signif. codes: 0 '***' 0.001 '**' 0.01 '*' 0.05 '.' 0.1 ' ' 1
```

You may have noticed that the main effect of species turns negative ($\widehat{\beta}_1 = -0.2$) when we include an interaction in the model while, when no interaction is included, it was positive ($\widehat{\beta}_1 = 1.1$). This is because the main effect is the difference at the intercept, and the intercept is the value of the dependent variable when the covariate is zero. If zero is at the edge or far away from the range of the covariate values (here: water), and if we have a model with nonparallel regression lines (i.e., because we have included an interaction), the difference in the intercept is very different compared to the difference between the means of the species in the center of the data. Therefore, we do not interpret the main effects of terms that are involved in an interaction unless we have centered the covariate (i.e., subtracted the mean) before fitting the model. By centering, zero moves into the middle of the data, so that the intercept corresponds to the difference in fitted values measured in the center of the covariate.

Before drawing conclusions from the model, of course, we have to check how well the model fits to the data as described in Chapter 6. In the case of an ANCOVA, it makes sense to also plot the residuals against the covariate for each group separately. In our example, residual analyses showed a slight deviation from a linear relationship and substantial autocorrelation because the grass biomass was measured in different tanks. Measurements from the same tank were more similar than measurements from different tanks after correcting for the distance to water. Thus, the analysis we have done here suffers from pseudoreplication. We will reanalyze the example data in a more appropriate way in Chapter 7. Let's see what we would conclude from the current analysis (and compare this conclusion to the one from a mixed model later).

An ecologically important question could be: "At what distance to water does species Dg outperform species Ap?" We say that species Dg outperforms species Ap when its average biomass is larger than the one of Ap, thus to the right of the crosspoint of the two regression lines. We can get the distance to water for the crosspoint numerically by finding the difference between the two intercepts divided by the negative difference between the two slopes, or by the R function `crosspoint`. This crosspoint is a quantity that we derive from the parameter estimates from our model. Therefore, we

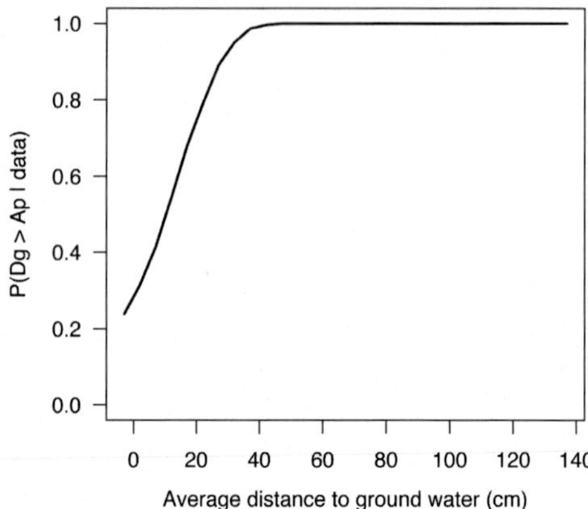

FIGURE 4-12 Posterior probability that the average biomass of the grass species Dg is larger than the one of grass species Ap in relation to distance to ground water.

have to propagate the uncertainty of the parameter estimates to the crosspoint. This is done by calculating the crosspoint for 2000 sets of model parameters from the joint posterior distribution. Then, the cumulative density distribution of the water-level values at the crosspoint gives the posterior probability that Dg outperforms Ap in relation to distance to water (Figure 4-12).

```
nsim <- 2000
bsim <- sim(mod2, n.sim=nsim)
xatcross <- crosspoint(coef(bsim)[,1], coef(bsim)[,3],
  coef(bsim)[,1]+coef(bsim)[,2], coef(bsim)[,3]+coef(bsim)[,4])[,1]
xatcross[xatcross< (-5)] <- -5
th <- hist(xatcross, breaks=seq(-5.5, 140.5, by=5))
plot(th$mids, cumsum(th$counts)/2000, type="l", lwd=2, las=1,
  ylim=c(0,1), ylab="P(Dg > Ap | data)", xlab="Average distance to
  ground water (cm)")
```

4.2.7 Multiple Regression and Collinearity

The term *multiple regression* was originally used for a normal linear model that contains more than one continuous, but no categorical, predictor variable. In practice, data containing many predictor variables often contain categorical and numerical predictor variables. Nowadays, "multiple regression" is often used for linear models containing many predictor variables independent of their data type.

Multiple regressions are fit in R in a similar way as simple regressions and ANOVA. We just add all the predictor variables in the formula notation. Here is an example with two numeric predictor variables:

```
data(mdat) # load a set of simulated data with correlated variables
mod <- lm(y~x1+x2, data=mdat)
summary(mod)
Call:
lm(formula = y ~ x1 + x2, data = dat)
Residuals:
    Min      1Q Median     3Q    Max
-4.9358 -1.4462 0.1653 1.3347 4.7456
Coefficients:
            Estimate Std. Error t value  Pr(>|t|)
(Intercept)  2.6120     0.2485  10.510  < 2e-16 ***
x1           1.1807     0.2289   5.158 1.32e-06 ***
x2          -0.2289     0.2009  -1.140    0.257
---
Signif. codes: 0 '***' 0.001 '**' 0.01 '*' 0.05 '.' 0.1 ' ' 1
Residual standard error: 1.804 on 97 degrees of freedom
Multiple R-squared: 0.2829, Adjusted R-squared: 0.2681
F-statistic: 19.14 on 2 and 97 DF, p-value: 9.884e-08
```

The model formula is

$$\widehat{y}_i = \beta_0 + \beta_1 x1_i + \beta_2 x2_i$$
$$y_i \sim Norm(\widehat{y}_i, \sigma^2)$$

The linear predictor describes a plane in 3D space and the data are a cloud in the same space (Figure 4-13). When more than two predictor variables are present, it is no longer possible to graphically represent the linear predictor. For graphical purposes, we can then reduce the dimensions of the linear predictor by fixing all but one or two predictor variables to a constant.

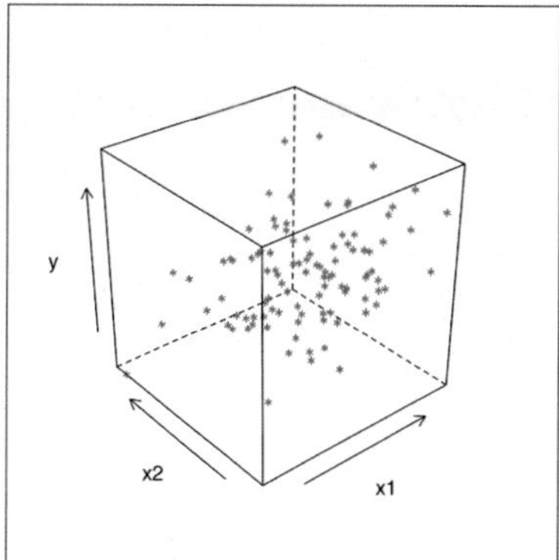

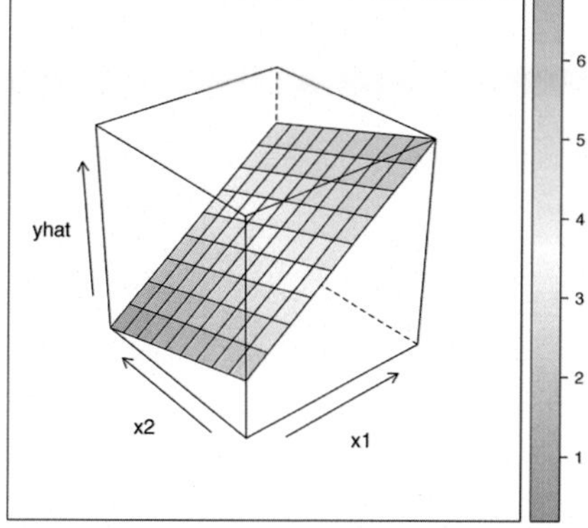

FIGURE 4-13 Three-dimensional graph of the data (*left*) and the linear predictor (*right*) in a multiple regression with two predictor variables.

Typically, multiple regression is used when the data originate from an observational study or from a field experiment with many influencing variables that could not be controlled for. Such data are mostly "unbalanced" or "collinear". Collinearity and unbalance can be visualized in a pairs plot (Figure 4-14), or measured by a correlation matrix in the case of continuous predictor variables.

```
cor(dat[,2:6]) # [,2:6] extracts column 2 to 6 from dat
           x1         x2          x3          x4         x5
x1  1.0000000 -0.3891033 -0.76776204 -0.48320754 -0.8610824
x2 -0.3891033  1.0000000  0.39080938  0.15687382  0.4462641
x3 -0.7677620  0.3908094  1.00000000  0.09897189  0.4021860
x4 -0.4832075  0.1568738  0.09897189  1.00000000  0.4858710
x5 -0.8610824  0.4462641  0.40218598  0.48587095  1.0000000
```

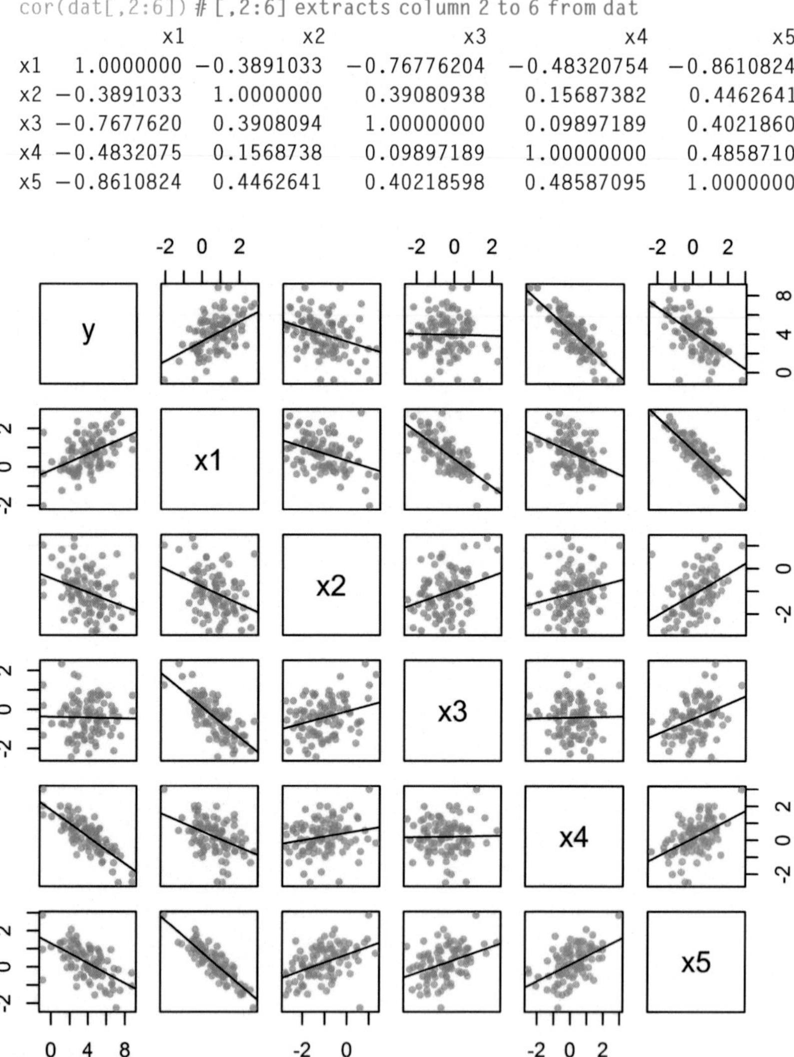

FIGURE 4-14 Pairs plot of a simulated data set with five numeric predictor variables ($x1$ to $x5$) and one outcome (y). To assess collinearity we only have to plot the x variables. However, if space allows, our habit is to also plot the outcome variable. This may help to detect nonlinear relationships in addition to assessing collinearity.

The function `pairs` plots each variable of a data frame against each other. In the argument "panel", we can give a function that specifies the graph we would like to draw.

```
own.graph <- function(x,y){
 points(x,y, pch=16, col=rgb(1,0.5,0,0.8))
 abline(lm(y~x))
 }
pairs(dat, panel=own.graph)
```

In the preceding example, many pairs of the five predictor variables strongly correlate, for example $x1$ and $x5$ (Figure 4-14). Collinearity has the same consequence for the regression analysis as when data are unbalanced in an ANOVA: the estimated effect of the one predictor depends on the presence of the other predictor in the model. Unbalance or collinearity can be due to two different causes: First, subjects are nonrandomly missing in the data. Second, the variables are naturally correlated even in a perfect random sample of the population. We normally speak of unbalance in the first case ("missing not at random") and of collinearity in the second.

For the data analyst, unbalance is more problematic than collinearity. Because of the nonrandom sample in the case of unbalance, naïvely estimated effects may be biased. In this case, it is only possible to obtain unbiased results when the missingness process is included in the model (see, e.g., Little & Rubin, 2002).

Collinearity (in a random sample) does not produce bias but it complicates the interpretation of the model coefficients. The estimated effect of predictor $x1$, $\widehat{\beta_1}$, in a model that also contains $x2$ is "corrected" for $x2$. This means that $\widehat{\beta_1}$ is the expected change in the outcome variable when $x1$ increases by 1 while $x2$ stays constant. However, if $x2$ is not in the model, $\widehat{\beta_1}$ is the expected change in the outcome variable when $x1$ increases by 1 and $x2$ changes by the average change of $x2$ when $x1$ increases by 1. Thus, in the latter case, $\widehat{\beta_1}$ is not a pure $x1$-effect; it is confounded by $x2$. Therefore, we often try to include as many variables as possible in the model. Unfortunately, sometimes, correlated predictors can cause numeric problems in the model fitting algorithm. And, the more parameters there are in the model, the larger is the variance in the parameter estimates simply because the model contains more parameters to be estimated. A predictor that is strongly correlated with another predictor only adds another parameter without adding much additional (i.e., independent) information. Thus, the variance of the parameter estimates increases overproportionally. The inflation of the variance of the parameter estimates due to collinearity can be measured by the variance inflation factor (VIF; Mansfield & Helms, 1982).

Collinearity often implicates the question of how many and which predictors should be added in a model. This is a model selection problem that we discuss in Chapter 11. In many cases, however, it is helpful to analyze the relationships between all the variables using multivariate statistics such as principal component analysis (Pearson, 1901; Manly, 1994), or using path analysis, structural equations, or a similar technique (Wright, 1921; Pearl, 2000) to model the relationships between all the variables. Sometimes, hierarchical partitioning methods help in analyzing complex data (Chevan & Sutherland, 1991).

4.2.8 Ordered Factors and Contrasts

In Sections 4.1 and 4.2.1 we became familiar with the model matrix for linear regression and ANOVA. Recall that the model matrix is an $n \times k$ matrix (with $n =$ sample size and $k =$ number of model coefficients) that is multiplied by the vector of the k model coefficients to obtain the fitted values of a normal linear model. The first column of the model matrix normally contains only 1s. This column is multiplied by the intercept. The other columns contain the observed values of the predictor variables if these are numeric variables, or indicator variables (= dummy variables) for factor levels if the predictors are categorical variables (= factors). For categorical variables the model matrix can be constructed in a number of ways (see also Section 4.2.1). Here we explain an additional possibility to construct a model matrix. This possibility is often used when the predictor variable is an ordered factor.

An ordered factor is a categorical variable with levels that have a natural order, for example, "low", "medium", and "high". How do we tell R that a factor is ordered? The swallow data contain a factor "nesting_aid" that contains the type of aid provided in a barn for the nesting swallows. The natural order of the levels is none < support (e.g., a wooden stick in the wall that helps support a nest built by the swallow) < artificial_nest < both (support and artificial nest). However, when we read in the data, R orders these levels alphabetically rather than according to the logical order:

```
data(swallows)
levels(swallows$nesting_aid)
[1] "artif_nest" "both" "none" "support"
```

And with the function contrasts we see how R will construct the model matrix:

```
contrasts(swallows$artnest)
            both  none  support
artif_nest    0     0         0
both          1     0         0
none          0     1         0
support       0     0         1
```

R will construct three dummy variables and call them "both", "none", and "support". The variable "both" will have an entry of 1 when the observation is "both" and 0 otherwise. Similarly, the other two dummy variables are indicator variables of the other two levels, and "artif_nest" is the reference level. The model coefficients can then be interpreted as the difference between "artif_nest" and each of the other levels (see 4.2.1). The instruction how to transform a factor into columns of a model matrix is called the contrasts.

Now, let's bring the levels into their natural order and define the factor as an ordered factor. Again, look at the levels and the contrasts.

```
swallows$nesting_aid <- factor(swallows$nesting_aid, levels=
  c("none", "support", "artif_nest", "both"), ordered=TRUE)
levels(swallows$nesting_aid)
[1] "none" "support" "aritf_nest" "both"
```

The levels are now in the natural order. R will, from now on, use this order for analyses, tables, and plots, and because we defined the factor to be an ordered factor, R will use polynomial contrasts:

```
contrasts(dat$nesting_aid)
             .L    .Q         .C
[1,] -0.6708204   0.5 -0.2236068
[2,] -0.2236068  -0.5  0.6708204
[3,]  0.2236068  -0.5 -0.6708204
[4,]  0.6708204   0.5  0.2236068
```

When using polynomial contrasts, R will construct three (= number of levels minus 1) variables that are called ".L", ".Q", and ".C" for linear, quadratic, and cubic effects. The contrast matrix defines which numeric value will be inserted in each of the three corresponding columns in the model matrix for each observation, for example, an observation with "support" in the factor "nesting_aid" will get the values −0.224, −0.5, and 0.671 in the columns L, Q, and C of the model matrix. These contrasts define yet another way to get four different group means (compare to 4.2.1):

$$m_1 = \beta_0 - 0.671 * \beta_1 + 0.5 * \beta_2 - 0.224 * \beta_3$$
$$m_2 = \beta_0 - 0.224 * \beta_1 - 0.5 * \beta_2 + 0.671 * \beta_3$$
$$m_3 = \beta_0 + 0.224 * \beta_1 - 0.5 * \beta_2 - 0.671 * \beta_3$$
$$m_4 = \beta_0 + 0.671 * \beta_1 + 0.5 * \beta_2 + 0.224 * \beta_3$$

The group means are the same, independent of whether a factor is defined as ordered or not. The ordering also has no effect on the variance that is explained by the factor "nesting_aid" or the overall model fit. Only the model coefficients and their interpretation depend on whether a factor is defined as ordered or not. When we define a factor as ordered, the coefficients can be interpreted as linear, quadratic, cubic, or higher order polynomial effects. The number of the polynomials will always be the number of factor levels minus 1 (unless the intercept is omitted from the model, in which case it is the number of factor levels). Linear, quadratic, and further polynomial effects are normally more interesting for ordered factors than single differences from a reference level because linear and polynomial trends tell us something about consistent changes in the outcome along the ordered factor levels. Therefore, an ordered factor with k levels is treated like a covariate consisting of the centered level numbers (-1.5, -0.5, 0.5, 1.5 in our case with four levels) and $k - 1$ orthogonal polynomials of this covariate are included in the model (see Section 4.2.9). Thus, if we have an ordered factor A with three levels, $y \sim A$ is equivalent to $y \sim x + x^2$, with $x = -1$ for the lowest, $x = 0$ for the middle, and $x = 1$ for the highest level.

Note that it is also possible to define own contrasts if we are interested in specific differences or trends. However, it is not trivial to find meaningful and orthogonal (= uncorrelated) contrasts.

4.2.9 Quadratic and Higher Polynomial Terms

The straight regression line for the biomass of grass species Ap *Alopecurus pratensis* dependent on the distance to the ground water does not fit well (refer to Figure 4-11). The residuals at low and high values of water tend to be positive, and intermediate water levels are associated with negative residuals. This clearly points out a violation of the model assumptions.

The problem is that the relationship between distance to water and biomass of species Ap is not linear. In real life, we often find nonlinear relationships, but if the shape of the relationship is quadratic (plus, potentially, a few more polynomials) we can still use "linear modeling" (the term *linear* refers to the linear function used to describe the relationship between the outcome and the predictor variables: $f(x) = \beta_0 + \beta_1 x + \beta_2 x^2$ is a linear function compared to, for example, $f(x) = \beta^x$, which is not a linear function). We simply add the quadratic term of the explanatory variable, that is, water in our example, as a further explanatory variable in the linear predictor:

$$\hat{y}_i = \beta_0 + \beta_1 \text{water}_i + \beta_1 water_i^2$$

A quadratic term can be fitted in R using the function `I()` which tells R that we want the squared values of distance to water. If we do not use `I()`, the `^2` indicates a two-way interaction (see Section 4.2.3).

```
lm(log(Yi.g) ~ Water + I(Water^2), data=dat[dat$species=="Ap",])
```

The cubic term would be added by `+I(Water^3)`.

As with interactions, a polynomial term changes the interpretation of lower level polynomials. Therefore, we normally include all polynomials up to a specific degree. Furthermore, polynomials are normally correlated (if no special transformation is used, see following) and problems associated with collinearity arise (see Section 4.2.7). To avoid collinearity among polynomials, so-called orthogonal polynomials can be used. These are polynomials that are uncorrelated. To that end, we can use the function `poly` which creates as many orthogonal polynomials of the variable as we want: `poly(dat$Water, 2)` creates two columns, the first one can be used to model the linear effect of water, the second one to model the quadratic effect of water:

```
t.poly <- poly(dat$Water, 2)
dat$Water.l <- t.poly[,1] # linear term for water
dat$Water.q <- t.poly[,2] # quadratic term for water
mod <- lm(log(Yi.g) ~ Water.l + Water.q, data=dat)
```

When orthogonal polynomials are used, the estimated linear and quadratic effects can be interpreted as purely linear and purely quadratic influences of the predictor on the outcome. The function `poly` applies a specific transformation to the original variables. To reproduce the transformation (e.g., for getting the corresponding orthogonal polynomials for new data used to draw an effect plot as shown in Section 4.1.3), the function `predict` can be used with the poly-object created based on the original data:

```
newdat <- data.frame(Water = seq(0,130))
# transformation analogous to the one used to fit the model:
newdat$Water.l <- predict(t.poly, newdat$Water)[,1]
newdat$Water.q <- predict(t.poly, newdat$Water)[,2]
```

These transformed variables can then be used to calculate fitted values that correspond to the water values specified in the new data.

FURTHER READING

Gelman et al. (2014) describe the theoretical background of a linear regression in the Bayesian framework. A more applied introduction to linear models in a Bayesian framework is given by Gelman and Hill (2007).

Easy-to-understand introductions to linear regression in a frequentist's framework, including R tutorials, are in Dalgaard (2008), Crawley (2007), and Faraway (2005). The technical background is described in Sokal and Rohlf (2003).

Engqvist (2005) reviews the results of ANCOVAs and found that many researchers misinterpret main effects in the presence of interaction terms.

The Ellenberg's data example is taken from Ellenberg (1953, 1954). This data set has been re-analysed using mixed models by Hector et al. (2012).

Hector et al. (2010) demonstrate the effect of collinearity on the variances explained by the different predictors and show a possibility to deal with collinearity. In simulation studies, Smith et al. (2009) and Dormann et al. (2013) compare different methods to deal with collinearity. Graham (2003) discusses different methods used to analyze correlated data.

Davis (2010) explains how to construct user-specified, orthogonal and nonorthogonal contrasts, and discusses the advantages of orthogonal contrasts above the nonorthogonal ones.
The Ellenberg data set has been thoroughly analyzed by Hector et al. (2012).

Chapter 5

Likelihood

Chapter Outline

5.1 THEORY

As described in Section 3.2, in Bayesian statistics the likelihood is the probability distribution of the data given the model $p(\mathbf{y}|\boldsymbol{\theta})$, also called the predictive density. In contrast, frequentists use the likelihood as a relative measure of the probability of the data given a specific model (i.e., a model with specified parameter values). Often, we see the notation $L(\boldsymbol{\theta}|\mathbf{y}) = p(\mathbf{y}|\boldsymbol{\theta})$ for the likelihood of a model. Let's look at an example. According to values that we found on the internet, black-tailed prairie dogs, *Cynomys ludovicianus*, weigh on average 1 kg with a standard deviation of 0.2 kg.

Using these values, we construct a model for the weight of prairie dogs $y_i \sim Norm(\mu, \sigma)$ with $\mu = 1$ kg and $\sigma = 0.2$ kg. We then go to the prairie and catch three prairie dogs. Their weights are 0.8, 1.2, and 1.1 kg. What is the likelihood of our model given this data? The likelihood is the product of the density function (the Gaussian function with $\mu = 1$ and $\sigma = 0.2$, see box) defined by the model and evaluated for each data point (Figure 5-1).

```
dnorm(x=0.8, mean=1, sd=0.2)*dnorm(x=1.2, mean=1, sd=0.2)*
   dnorm(x=1.1, mean=1, sd=0.2)
[1] 2.576671
```

This number (2.6) does not tell us much. But, another web page says that prairie dogs weigh on average 1.2 kg with a standard deviation of 0.4 kg. The likelihood for this alternative model is 0.6.

```
dnorm(x=0.8, mean=1.2, sd=0.4)*dnorm(x=1.2, mean=1.2, sd=0.4)*
   dnorm(x=1.1, mean=1.2, sd=0.4)
[1] 0.5832185
```

Bayesian Data Analysis in Ecology Using Linear Models with R, BUGS, and Stan
http://dx.doi.org/10.1016/B978-0-12-801370-0.00005-8.

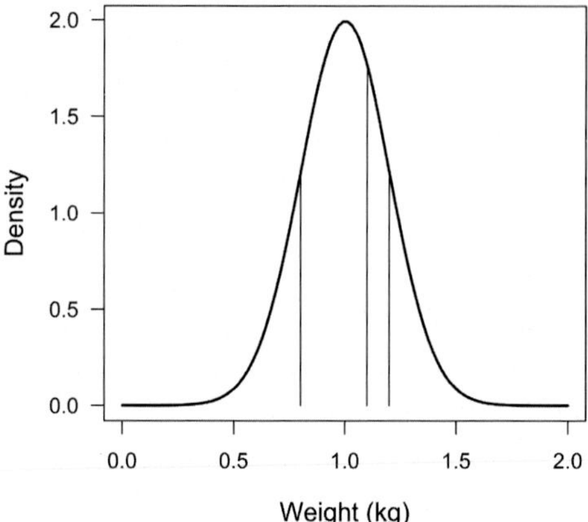

FIGURE 5-1 Density function of a Gaussian (= normal) distribution with mean 1 kg and standard deviation 0.2 kg. The vertical lines indicate the three observations in the data. The curve has been produced using the function `dnorm`.

Thus, the probability of observing the three weights (0.8, 1.2, and 1.1 kg) is four times larger given the first model than given the second model.

The Normal Distribution

Up to now, we have not had a close look at the normal distribution, also called the Gaussian distribution, even though we used it many times (e.g., to fit a linear regression). It is high time to do this now! The normal distribution is a bell-shaped function that describes the distribution of quantities with random (measurement) errors. It is an exponential function and is not extremely complicated:

$$p(y|\mu,\sigma) = 1\Big/\sqrt{2\pi\sigma}e^{\{-1/[\![2\sigma^2]\!][\![y-\mu]\!]^2\}}$$

The parameters μ (the mean) and σ (the standard deviation) define the shape of the function and y is the measurement (i.e., the x-axis in Figure 5-1). A short notation of the formula just shown is $y \sim Norm(\mu,\sigma)$. Note that we sometimes (as in Chapter 3) use $y \sim Norm(\mu, \sigma^2)$; that is, the variance instead of the standard deviation is given. The two formulas unambiguously specify the same normal distribution when parameter names are given (i.e., σ VS. σ^2). When numbers are given, such as $y \sim Norm(1.0, 0.2)$, we mean standard deviation by default.

For a biologist, it is not extremely important to know the mathematical formula of the normal distribution by heart, but it is helpful to remember that $y \sim Norm(\mu, \sigma)$ describes the bell-shaped distribution of a continuous numeric variable subject to random variance.

The normal density function is implemented in R by the function `dnorm`. The following code produces the bell-shaped curve in Figure 5-1.

```
x <- seq(0, 2, length=100)
dx <- dnorm(x, mean=1, sd=0.2)
plot(x,dx, type="l", xlab="Weight (kg)", ylab="Density", lwd=2, las=1)
```

`dnorm` gives the density of a normal distribution. `rnorm` draws random numbers from a normal distribution. We specify how many random numbers we would like to draw and the mean and the standard deviation of the normal distribution.

```
rnorm(5, 1, 0.2) # draws 5 random numbers from Norm(1, 0.2)
[1] 1.0226757 1.1376444 1.3970646 0.8193328 1.1983347
```

`pnorm` gives the proportion of the distribution that is to the left of a specific value, q.

```
pnorm(q = 0.8, 1, 0.2)
[1] 0.1586553
```

Thus, given the model *Norm*(1, 0.2) is realistic for the weights of prairie dogs, we expect around 16% of the prairie dogs to weigh less than 0.8 kg.

The function `qnorm` is the inverse function of `pnorm`, that is, it gives the value of y that is associated with a specific proportion of the distribution that is smaller than y. This is the quantile function for a theoretical normal distribution.

```
qnorm(0.1, 1, 0.2)
[1] 0.7436897
```

Thus, we expect 10% of the prairie dogs to weigh less than 0.74 kg.

The normal density function is a function of y with fixed parameters μ and σ. When we fix the data to value y (i.e., we have a data file with $n = 1$) and vary the parameter, we get the likelihood function given the data and we assume the normal distribution as our data model. When the data contain more than one observation, the likelihood function becomes a product of the normal density function for each observation.

$$L(\mu, \sigma|\mathrm{y}) = \Pi_{i=1}^{n} p\left(y_i, |\mu,\sigma\right)$$

We can insert the three observations into this function.

```
y <- c(0.8, 1.2, 1.1)
lf <- function(mu, sigma) prod(dnorm(y, mu, sigma))
```

Applying the likelihood function to the model from the first web page, we get the following likelihood.

```
lf(1, 0.2)
[1] 2.576671
```

5.2 THE MAXIMUM LIKELIHOOD METHOD

A common method of obtaining parameter estimates is the maximum likelihood method (ML). Specifically, we search for the parameter combination from the likelihood function for which the likelihood is maximized. Also in Bayesian statistics, ML is used to optimize algorithms that produce posterior distributions (e.g., in the function `sim`).

To obtain ML estimates for the mean and standard deviation of the prairie dog weights, we calculate the likelihood function for many possible combinations of means and standard deviations and look for the pair of parameter values that is associated with the largest likelihood value.

```
mu <- seq(0.6, 1.4, length=100)
sigma <- seq(0.05, 0.6, length=100)
lik <- matrix(nrow=length(mu), ncol=length(sigma))
for(i in 1:length(mu)){
  for(j in 1:length(sigma)){
   lik[i,j] <- lf(mu[i], sigma[j])
  }
}
contour(mu, sigma, lik, nlevels=20, xlab=expression(mu),
      ylab=expression(sigma), las=1, cex.lab=1.4) # Fig. 5-2
```

We can get the mu and sigma values that produce the largest likelihood by the function `optim`. This function finds the parameter values associated with the lowest value of a user defined function. Therefore, we define a function that calculates the negative likelihood.

```
neglf <- function(x) -prod(dnorm(y, x[1], x[2]))
MLest <- optim(c(1, 0.2), neglf)
MLest$par
[1] 1.0333272 0.1699747
points(MLest$par[1], MLest$par[2], pch=4)
```

The combination of mean = 1.03 and standard deviation = 0.17 is associated with the highest likelihood value (the "mountain top" in Figure 5-2). In the case of models with more than two parameters, the likelihood function cannot be visualized easily. Often, some parameters are fixed to one value, for example, their ML estimate, to show how the likelihood function changes in relation to one or two other parameters.

Likelihoods are often numbers very close to zero, especially when sample size is large. Then, likelihood values drop below the computing ability of computers (underflow). Therefore, the logarithm of the likelihood is often used. In addition, many standard optimizers are, in fact, minimizers such that the negative log-likelihood is used to find the ML estimates.

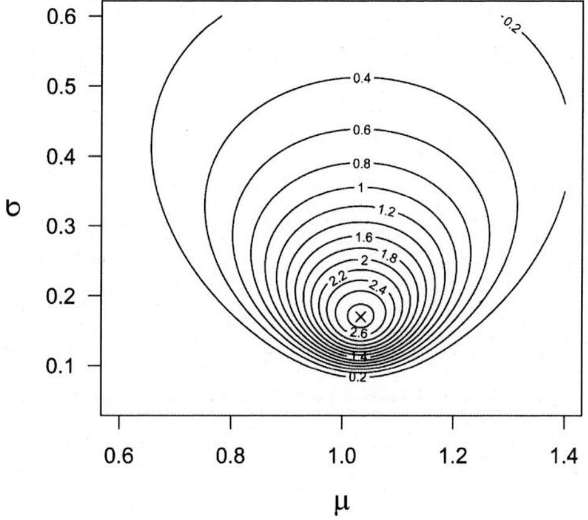

FIGURE 5-2 Likelihood function L(μ,σ|y) for $y = 0.8, 1.2, 1.1$ and the data model $y \sim Norm(\mu, \sigma)$. The cross indicates the highest likelihood value.

The ratio between the likelihoods of two different models—for example, $L(\mu = 1, \sigma = 0.2|y)/L(\mu = 1.2, \sigma = 0.4|y) = 2.6/0.6 = 4.4$—is called the likelihood ratio, and the difference in the logarithm of the likelihoods is called the log-likelihood ratio. It means, as just stated, that the probability of the data is 4.4 times higher given model 1 than given model 2 (note that with "model" we mean here a parameterized model—i.e., with specific values for the parameters μ and σ). From this, we may conclude that there is more support in the data for model 1 than for model 2.

The likelihood function can be used to obtain confidence intervals for the ML estimates. If the "mountain" in Figure 5-2 is very steep, the uncertainty about the ML estimate is low and the confidence intervals are small, whereas when the mountain has shallow slopes, the intervals will be larger. Such confidence intervals are called profile likelihood confidence intervals (e.g., Venzon & Moolgavkar, 1988).

5.3 THE LOG POINTWISE PREDICTIVE DENSITY

Up to now, we have discussed likelihood in a frequentist framework because it is so important in statistics. Let's now turn to the Bayesian framework. When Bayesians use the word likelihood, they mean the probability distribution of the data given the model, $p(y|\theta)$. However, the log pointwise predictive density (lppd) could be seen as a Bayesian analog to the "frequentist" log-likelihood. The lppd is the log of the posterior density integrated over the posterior distribution of the model parameters and summed over all observations in the data:

$$\text{lppd} = \sum\nolimits_{i=1}^{n} log \int p(y_i|\theta)p(\theta|y)d\theta$$

Based on simulated values from the joint posterior distribution of the model parameters, the log pointwise predictive density can be calculated in R as follows.

```
library(arm)
mod <- lm(y~1) # fit model by LS method
nsim <- 2000
bsim <- sim(mod, n.sim=nsim) # simulate from posterior dist. of
parameters
```

We prepare a matrix "pyi" to be filled up with the posterior densities for every observation y_i and for each of the simulated sets of model parameters.

```
pyi <- matrix(nrow=length(y), ncol=nsim)
for(i in 1:nsim) pyi[,i] <- dnorm(y, mean=bsim@coef[i,1],
sd=bsim@sigma[i])
```

Then, we average the posterior density values over all simulations (this corresponds to the integral in the previous formula).

```
mpyi <- apply(pyi, 1, mean)
```

Finally, we calculate the sum of the logarithm of the averaged density values.

```
sum(log(mpyi))
[1] 0.2401178
```

The log posterior density can be used as a measure of model fit. We will come back to this measurement in Chapter 11 where we discuss model selection and multimodel inference.

FURTHER READING

Link and Barker (2010) provide a very understandable chapter on the likelihood. Also, Burnham and Anderson (2002) give a thorough introduction to the likelihood and how it is used in the multimodel inference framework. An introduction to likelihood for mathematicians, but also quite helpful for biologists, is contained in Dekking et al. (2005).

Davidson and Solomon (1974) discuss the relationship between LS and ML methods. The difference between the ML- and LS-estimates of the variance is explained in Royle and Dorazio (2008).

Gelman et al. (2014) introduce the log predictive density and explain how it is computed (p. 167).

Chapter 6

Assessing Model Assumptions

Residual Analysis

Chapter Outline

6.1 MODEL ASSUMPTIONS

Every statistical model makes assumptions. We try to build models that reflect the data-generating process as realistically as possible. However, a model never is the truth. Yet, all inferences drawn from a model, such as estimates of effect size or derived quantities with credible intervals, are based on the assumption that the model is true. However, if a model captures the data-generating process poorly, for example, because it misses important structures (predictors, interactions, polynomials), inferences drawn from the model are probably biased and results become unreliable. In a (hypothetical) model that captures all important structures of the data generating process, the stochastic part, the difference between the observation and the fitted value (the residuals), should only show random variation. Analyzing residuals is a very important part of the data analysis process.

Residual analysis can be very exciting, because the residuals show what remains unexplained by the present model. Residuals can sometimes show surprising patterns and, thereby, provide deeper insight into the system. However, at this step of the analysis it is important not to forget the original research questions that motivated the study. Because these questions have been asked without knowledge of the data, they protect against data dredging. Of course, residual analysis may raise interesting new questions. Nonetheless, these new questions have emerged from patterns in the data, which might just be random, not systematic, patterns. The search for a model with good fit should be guided by thinking about the process that generated the data, not by

Bayesian Data Analysis in Ecology Using Linear Models with R, BUGS, and Stan
http://dx.doi.org/10.1016/B978-0-12-801370-0.00006-X.

trial and error (i.e., do not try all possible variable combinations until the residuals look good; that is data dredging). All changes done to the model should be scientifically justified. Usually, model complexity increases, rather than decreases, during the analysis.

6.2 INDEPENDENT AND IDENTICALLY DISTRIBUTED

Usually, we model an outcome variable as independent and identically distributed (iid) given the model parameters. This means that all observations with the same predictor values behave like independent random numbers from the identical distribution. As a consequence, residuals should look iid. Independent means that:

- The residuals do not correlate with other variables (those that are included in the model as well as any other variable not included in the model).
- The residuals are not grouped (i.e., the means of any set of residuals should all be equal).
- The residuals are not autocorrelated (i.e., no temporal or spatial autocorrelation exist; Sections 6.4 and 6.5).

Identically distributed means that:

- All residuals come from the same distribution.

In the case of a linear model with normal error distribution (Chapter 4) the residuals are assumed to come from the same normal distribution. Particularly:

- The residual variance is homogeneous (homoscedasticity), that is, it does not depend on any predictor variable, and it does not change with the fitted value.
- The mean of the residuals is zero over the whole range of predictor values. When numeric predictors (covariates) are present, this implies that the relationship between x and y can be adequately described by a straight line.

Residual analysis is mainly done graphically. R makes it very easy to plot residuals to look at the different aspects just listed. As a first example, we use the coal tit example from Chapter 4:

```
data(periparusater)
dat <- periparusater                  # re-name to shorter name
mod <- lm(wing ~ sex + age, data=dat) # fit an additive 2-way ANOVA
par(mfrow=c(2,2))                     # divide the graphic window
into 4 sub-windows
plot(mod)                             # produce  Fig. 6-1
```

The first panel of Figure 6-1 shows the residuals against the fitted values together with a smoother (red line). This plot is called a Tukey–Anscombe plot. The mean of the residuals should be around zero along the whole range of

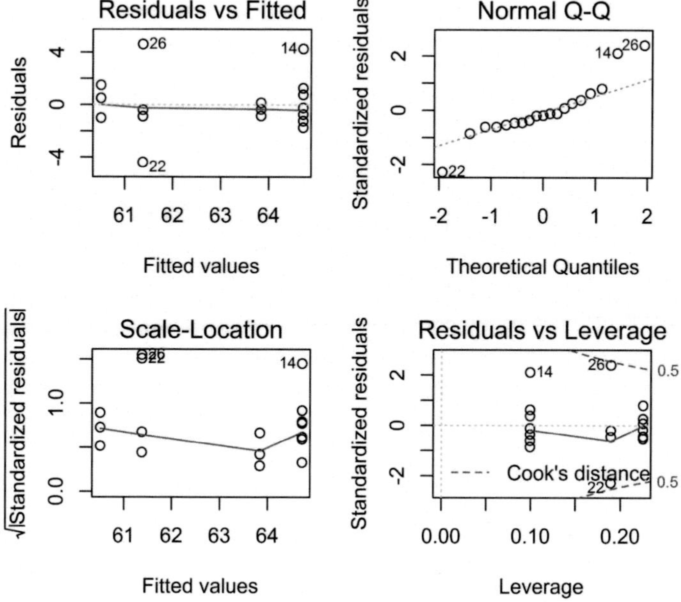

FIGURE 6-1 Residual plots for the two-way ANOVA fitted to the coal tit data set. See text for the interpretation of each panel.

fitted values. Note that smoothers are very sensitive to random structures in the data, especially for low sample sizes and toward the edges of the data range. Often, curves at the edges of the data do not worry us because the edges of smoothers are based on very small sample sizes.

The second panel shows a normal quantile-quantile (QQ) plot of the residuals. When the residuals are normally distributed, the points lie along the diagonal line. This plot is explained in more detail in Section 6.2.

The third panel shows the square root of the absolute values of the standardized residuals, a measure of residual variance versus the fitted values, together with a smoother. When the residual variance is homogeneous along the range of fitted values, the smoother is horizontal. Because of the low sample size, we are not worried about the slightly lower variance in the third group.

The fourth panel shows the residuals against the leverage. An observation with a measurement of a predictor variable far from the others has a large leverage. When all predictors are factors, observations with a rare combination of factor levels have higher leverage than observations with a common combination of factor levels. Thus, the observations can only have different leverage values when the data set is unbalanced. A high leverage does not necessarily mean that this observation has a big influence on the model. If that observation fits well to the pattern of all other data points, the observation does not have an unduly large influence on the model estimates, despite its large leverage. However, if it does not fit into the picture, this observation has a strong influence on the parameter estimates.

The influence of one observation on the parameter estimates is measured by Cook's distance. Observations with a large Cook's distance lie beyond the red dashed lines in the fourth of the residual plots (the 0.5 and 1 iso lines for Cook's distances are given as dashed lines). Observations with a Cook's distance larger than 1 are usually considered to be overly influential and should be checked.

The four plots produced by `plot(mod)` show the most important aspects of the model fit. However, often these four plots are not sufficient. In addition, we recommend plotting the residuals against all variables in the data set (including those not used in the current model). The function `names` extracts all variable names of a data frame, thus it reminds us what plots we need to produce.

```
names(dat)
[1] "country" "age"      "sex"      "weight"  "P8"      "wing"
```

We do not need to plot the residuals against the outcome variable (wing in this case). And, also a plot against the length of primary P8 is not needed, because this is also a measurement for wing length (P8 is the third outermost wing feather often forming the wing tip).

```
plot(resid(mod) ~ country, data=dat)
plot(resid(mod) ~ age, data=dat)
plot(resid(mod) ~ sex, data=dat)
plot(resid(mod) ~ weight, data=dat)
```

The residuals seem to be quite different between individuals from different countries (Figure 6-2). This may seriously affect the inferences we have drawn from the analysis. We could include country as a further predictor in the model to account for the between-country variance. However, this data set is so unbalanced and sample sizes extremely small for three of the four countries that we may want to restrict our analysis to the Swiss samples here.

```
table(dat$country)
Algeria   Bulgaria   Russia   Switzerland
      1          2        2            14
```

The plots of the residuals against sex and age do not show serious violation of the model assumptions: the medians of the residuals are around zero in all groups and the boxes indicate approximately equal variance in all groups. The correlation of the residuals with weight may be slightly positive, indicating, as expected, that heavier individuals have longer wings within the same age and sex class.

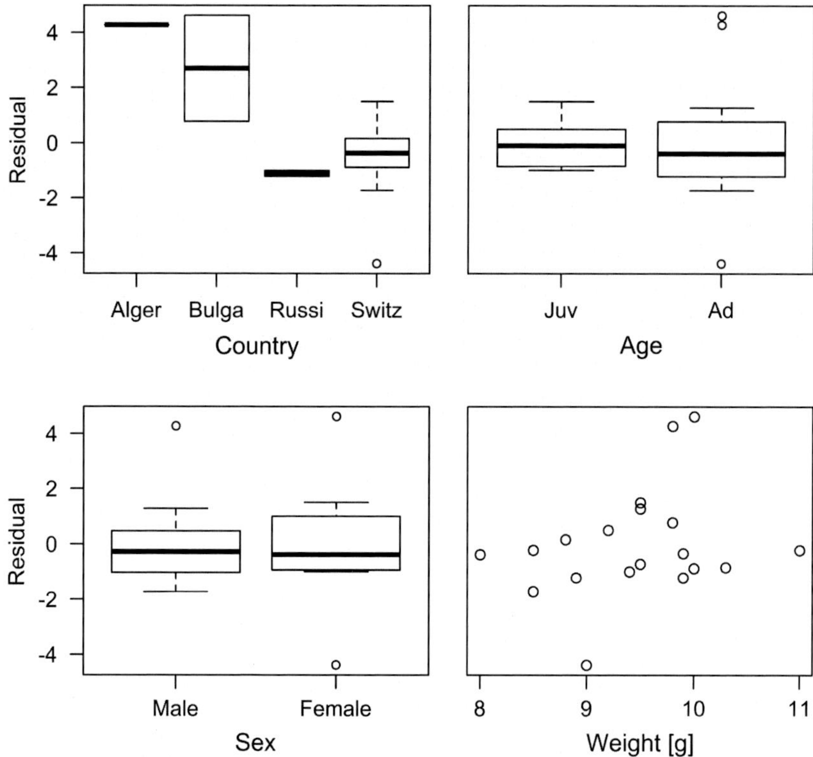

FIGURE 6-2 Residuals plotted against country, sex, age, and weight.

6.3 THE QQ PLOT

Each residual represents a quantile of the sample of n residuals. These quantiles are defined by n. A useful choice is the $((1,...,n) - 0.5)/n$-th quantiles. A QQ plot shows the residuals on the y-axis and the values of the $((1,...,\text{n}) - 0.5)/n$-th quantiles of a theoretical normal distribution on the x-axis. A QQ plot could also be used to compare the distribution of whatever variable with any distribution, but we want to use the normal distribution here because that is the assumed distribution of the residuals in the model. If the residuals are normally distributed, the points are expected to lie along the diagonal line in the QQ plot.

It is often rather difficult to decide whether a deviation from the line is tolerable or not. The function `compareqqnorm` may help. It draws, eight times, a random sample of n values from a normal distribution with a mean of zero and a standard deviation equal to the residual standard deviation of the model (Figure 6-3). It then creates a QQ plot for all eight random samples and for the

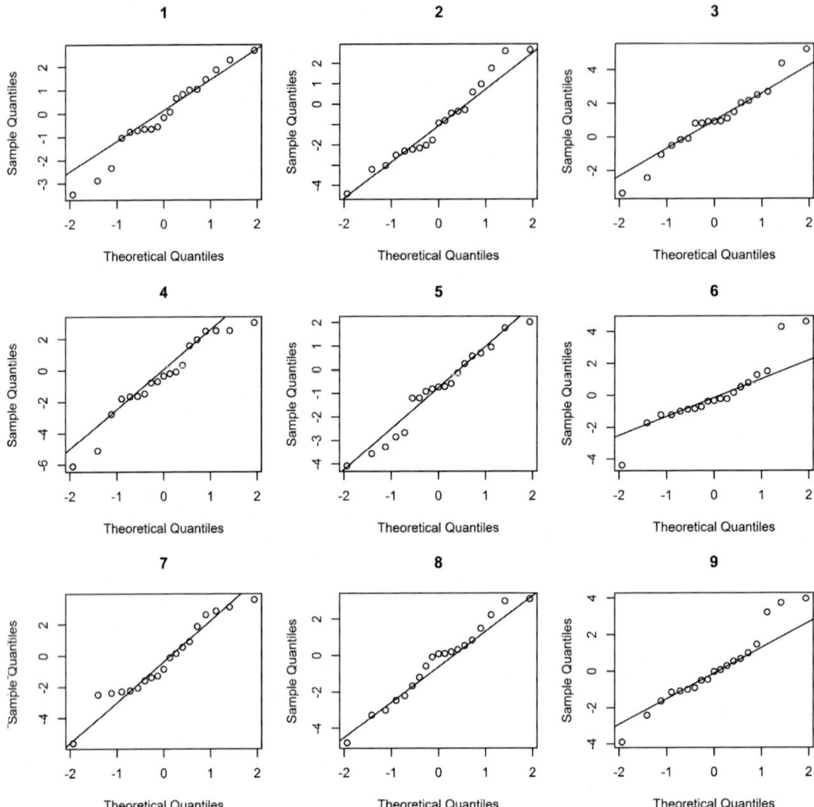

FIGURE 6-3 QQ plots of eight random samples from a normal distribution, and one QQ plot of the residuals of the linear model for the coal tit wing lengths. The model residual plot, in this case, was number 6 (the last plot in the second row).

observed residuals in random order. If the QQ plot of the observed residuals can easily be identified among the nine QQ plots, there is reason to think the distribution of the residuals deviates from normal. Otherwise, there is no indication to suspect violation of the normality assumption. The position of the residual plot of the model in the nine panels is printed to the R console.

```
compareqqnorm(mod)
[1] 6
```

6.4 TEMPORAL AUTOCORRELATION

Data are often collected over an extended time period (i.e., the study period). Often, conditions change over time, possibly without being noticed by the researcher. Such gradual changes in conditions can affect the measurements, so that measurements taken in the same time period are more similar to each other than to measurements from other time periods. This is called temporal autocorrelation. It is a sort of pseudoreplication because two (or more) adjacent data points are not independent and, thus, do not provide as much information as two independent data points. Ignoring temporal autocorrelation can result in overconfident and possibly incorrect conclusions.

We can detect temporal autocorrelation when looking at the autocorrelation plots of the residuals of a model. To illustrate this, we use counts of great tits *Parus major* observed at the mountain pass Ulmethöchi (Switzerland) between 1982 and 2007 during fall migration (Korner-Nievergelt et al., 2007).

```
data(parusmajor)
dat <- parusmajor                                    # rename
str(dat)
'data.frame':    434 obs. of   3 variables:
$ year   : int   1996 1994 1996 2000 1988 1996 2000 2001 2005 1988 ...
$ julian: int    266 267 267 267 268 268 268 268 268 269 ...
$ count : int    7 1 1 3 1 3 2 1 2 2 ...
```

The variable "count" contains the number of great tits observed per day at Ulmethöchi, "julian" is the day of the year (from 1 to 365 or 366), and "year" is the year. We can estimate the expected number of great tits per day and year using a normal linear model including a quadratic effect of day (the use of the function `poly` is explained in Section 4.2.9).

```
dat <- dat[order(dat$year, dat$julian),]  # order data acc. time
t.poly.jul <- poly(dat$julian, degree=2)
dat$julian.l <- t.poly.jul[,1]                # orthogonal linear trend
dat$julian.q <- t.poly.jul[,2]              # orthogonal quadratic trend

mod <- lm(count~year + julian.l + julian.q, data=dat)
```

We include the quadratic effect of "julian" because we expect the number of great tits to peak somewhere in the middle of the migration season. This is a very crude model that has much room for improvement. For example, a quadratic relationship assumes a regular curvature, which might not fit biological data well. Furthermore, the great tit is an irruptive species, migrating in spectacular numbers in some years while, in other years, migration is hardly noticeable. However, for illustrative purposes, we'll fit this model.

We detect temporal autocorrelation when looking at the autocorrelation plot of the residuals. Attention: this only works if the observations in the data set are in chronological order! Otherwise they need to be ordered before fitting the model (preceding code). The function `acf` plots two types of autocorrelations: the raw autocorrelations and the partial autocorrelations (Figure 6-4).

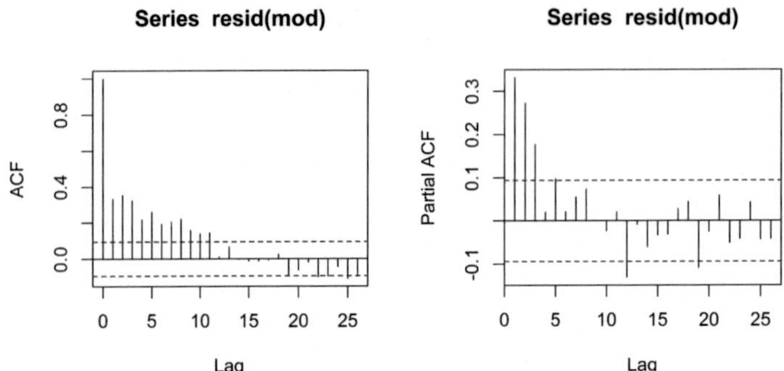

FIGURE 6-4 Autocorrelation and partial autocorrelation plot of the residuals of the great tit model.

```
par(mfrow=c(1,2))
acf(resid(mod))
acf(resid(mod), type="p")
```

Autocorrelation is a correlation of something, here the residuals, with itself. Of course, the correlation coefficient is 1 (hence, the first segment in a regular `acf` plot is always 1; see left panel of Figure 6-4). However, if the correlation coefficient is calculated between the residuals at time t and the residuals at time $t - 1$ (i.e., for time lag 1), or for another time lag, the correlation coefficient is expected to be 0, if the residuals are independent. The function `acf` plots the correlation coefficients of the residuals at time t with the residuals at time $t - k$ in relation to the lag k (x-axis).

For convenience, two blue dashed lines are added to indicate an approximated significance threshold at the 5% level, which is at $\pm \frac{1.96}{\sqrt{n}}$ according to the (frequentist) Pearson's correlation test. If a correlation coefficient crosses the blue dashed line, the correlation is statistically significant. If more than 5% of the correlation coefficients are beyond the blue lines, or, maybe more importantly, if the correlation coefficients are related to the lag (e.g., if they gradually decrease with the lag, or if they show a cyclic pattern), then the residuals are autocorrelated, and we cannot assume that they are independent. The autocorrelations for the great tit model start with values around 0.3 for lag 1 to lag 3 and then slightly decrease, but they are all positive until lag 11 (Figure 6-4, left panel). This is a clear indication that the residuals are autocorrelated.

If there is a correlation between residual t and $t - 1$, this same correlation also holds for the residuals of $t - 1$ and $t - 2$, and so on. Thus, residual t can be correlated with residual $t - 2$ via pseudocorrelation (i.e., we see a correlation between t and $t - 2$ only because both are correlated with t − 1). In contrast, partial correlation is the correlation that is left after having corrected for all the lower order correlations. The partial correlations for our example are shown in the right panel of Figure 6-4 (argument `type="p"`). They help us assess what process may have produced the correlations seen in the left panel.

Thus, this plot indicates up to what order an autocorrelation needs to be accounted for in the model when the model fit cannot be improved otherwise. A lag-1 process is called an autoregressive process of order 1 (AR1 process). In our case, there may be at least an AR3 process since we see substantial partial autocorrelation up to lag 3.

Before we start accounting for autocorrelation in a linear model we again think about the process that generated our data. Autocorrelation can indicate a lack of fit, for example, when a quadratic function does not describe the seasonal pattern of great tit migration well. The residuals will show a nonlinear pattern with time, resulting in autocorrelation in the residuals. In many cases, when the reason for the temporal autocorrelation is found and included (as a new predictor or specific structure) in the model, the autocorrelation disappears. For example, if the residuals in the great tit example are autocorrelated because the number of migrating great tits is very low in some years (thus all residuals of these years will be negative) whereas in other years this number is high, we may be able to substantially reduce temporal autocorrelation by including "year" as a random factor.

Sometimes, a look at the autocorrelation plot itself helps in finding the reason for temporal autocorrelation. For example, there may be a weekend effect because more observers were present on weekends. We would then expect a positive temporal correlation at lag 7. Such an effect could be accounted for by adding an indicator variable for the weekend days as a predictor in the model.

Only when we are not able to eliminate temporal autocorrelation by adjusting the systematic part of the model, do we need to account for temporal autocorrelation in the model. To do so, the residual variance is exchanged by an $n \times n$ matrix $\mathbf{V}$ that defines for each pair of observations (i,j) how they covary: the covariance matrix. Thus, the model formula becomes $\mathbf{y} \sim Norm(\mathbf{X}\boldsymbol{\beta}, \mathbf{V})$. The diagonal elements of the covariance matrix are the residual variances σ^2; the off-diagonal elements are the covariances between observations i and j. Of course, it is not possible to estimate separate (independent) covariances for each pair of observations because a covariance is not defined when sample size is only one. But if these covariances are parameterized so that only a few parameters define the covariances for all pairs of observations, these parameters become estimable. For example, we can define the covariance to be a function of the temporal (or spatial, see next section) distance between observation i and j, $f(dist(i,j))$. Then, the covariance matrix becomes (example for a data set with sample size n = 6)

$$\mathrm{V} = \begin{vmatrix} \sigma^2 & f(dist(1,2)) & f(dist(1,3)) & f(dist(1,4)) & f(dist(1,5)) & f(dist(1,6)) \\ f(dist(2,1)) & \sigma^2 & f(dist(2,3)) & f(dist(2,4)) & f(dist(2,5)) & f(dist(2,6)) \\ f(dist(3,1)) & f(dist(3,2)) & \sigma^2 & f(dist(3,4)) & f(dist(3,5)) & f(dist(3,6)) \\ f(dist(4,1)) & f(dist(4,2)) & f(dist(4,3)) & \sigma^2 & f(dist(4,5)) & f(dist(4,6)) \\ f(dist(5,1)) & f(dist(5,2)) & f(dist(5,3)) & f(dist(5,4)) & \sigma^2 & f(dist(5,6)) \\ f(dist(6,1)) & f(dist(6,2)) & f(dist(6,3)) & f(dist(6,4)) & f(dist(6,5)) & \sigma^2 \end{vmatrix}.$$

When we scale the variance-covariance matrix so that the diagonal elements are all 1, we obtain the correlation matrix with entries (i,j): $\text{cor}(\varepsilon_i, \varepsilon_j) = f(dist(i,j))/\sigma^2$. One of the simplest temporal correlation structures is the AR1 process. It models the residuals at time t as linearly dependent on the residuals at time $t-1$: $\varepsilon_t = \rho\varepsilon_{t-1} + \eta_t$, with $\eta_t \sim Norm(0,\sigma_\varepsilon)$. From this process it follows that the correlation matrix is

$$\text{cor}(\varepsilon_i, \varepsilon_j) = \begin{cases} 1 & \text{if } i = j \\ \rho^{|j-i|} & \text{else} \end{cases}$$

The parameter ρ has to be estimated from the data. We can do this, for example, by using the function `gls` from the package nlme (Pinheiro et al., 2011).

```
library(nlme)
mod <- gls(count ~ year + julian.l + julian.q, data=dat,
           correlation = corAR1())
summary(mod)
Generalized least squares fit by REML
   Model: count ~ year + julian.l + julian.q
   Data: dat
         AIC         BIC       logLik
   3607.234 3631.617 -1797.617

Correlation Structure: AR(1)
 Formula: ~1
 Parameter estimate(s):
        Phi
0.3422066

Coefficients:
                  Value   Std.Error    t-value   p-value
(Intercept)   -326.8006    419.4594 -0.7790995    0.4363
year             0.1693      0.2100  0.8059450    0.4207
julian.l        48.5121     20.1315  2.4097555    0.0164
julian.q       -55.7998     22.1730 -2.5165677    0.0122

 Correlation:
          (Intr)   year   juln.L
year      -1.000
julian.l  -0.012  0.012
julian.q   0.062 -0.062    0.003

Standardized residuals:
       Min          Q1         Med         Q3        Max
-0.8619025 -0.5815432 -0.3272695  0.1234870  6.0001067

Residual standard error: 16.59327
Degrees of freedom: 434 total; 430 residual
```

The estimate of the parameter ρ (for some reason called "phi" instead of "rho" in the R output) is 0.34, and we see that the standard errors are substantially larger compared to the model that does not take temporal autocorrelation into account (they are 0.1 for year and 16.5 for the linear and quadratic day effect). This makes sense because we are no longer "cheating" by boosting our sample size with pseudoreplication. While this model certainly improves the situation, it may still not be good enough since the `acf` plot with `type="p"` suggests dependence up to lag 3. For modeling more complex autocorrelation structures, methods specific for time series can be used; see the help files of the functions `ar`, `arma`, or `arima` for more.

The function `sim` cannot handle gls, ar, and similar objects. Therefore, it is not (yet) possible to simulate from the posterior distribution when using `gls`. The R function `MCMCglmm`, BUGS, or Stan all fit autoregressive models in a Bayesian framework. Table 6-1 gives an overview over the R functions we know of that can fit linear models including a temporal autocorrelation for the residuals.

6.5 SPATIAL AUTOCORRELATION

Data measured at different points in space are often spatially (auto-)correlated. For example, when we measure bird density using point counts (ornithologists stop at different points in the landscape and count all the birds they see within a given distance for a given time period) we have to decide on the density of the spatial arrangement of the points. The sampling points could be 100 m, 500 m, 1 km, or 10 km apart. The distances between the points should be small enough to reach the spatial resolution that is necessary. However, if distances are chosen to be too small, measurements at neighboring points may be very similar to each other. An observation in close proximity to another observation does not substantially increase the information in the data. Such measurements only artificially increase sample size without contributing independent information. This is, again, a typical case of pseudoreplication. In the ideal case, the points are as close as possible but far enough so that they are independent. The distance between independent sample points is determined by many factors such as the heterogeneity of the habitat, the presence of covariates that are not measured (e.g., a gradient in soil nutrients), dispersal distances, interaction with other individuals of the same or of other species, or spatial activity of the target species. We often do not know in advance how far the sampling points should be, and often, spatial autocorrelation cannot be avoided. Therefore, we need tools to measure spatial correlation and account for it in the model. In the following section, we show how to detect and measure spatial correlation. Models that account for spatial correlations are introduced in Chapter 13.

TABLE 6-1 Nonexhaustive List of R Functions That Allow Inclusion of Temporal Autocorrelation in a Linear Model

Function	Package	Description, Options, and Reference
`MCMCglmm`	`MCMCglmm`	Generalized linear mixed models using Markov chain Monte Carlo simulation; experienced users can specify any correlation structure, see `vignette(MCMCglmm)`; Hadfield (2010)
`gls`	`nlme`	Generalized least-squares method; works for all normal linear models. The following standard temporal correlation structures are available (Pinheiro & Bates, 2000): -`corAR1` autoregressive process of order 1 -`corARMA` autoregressive moving average process, with arbitrary orders for the autoregressive and moving average components -`corCAR1` continuous autoregressive process (AR(1) process for a continuous time covariate)
`lme`	`nlme`	REML or ML; works only for normal linear models that include at least one random factor. Same correlation structures as in gls; Pinheiro and Bates (2000)
`gam`	`mgcv`	Generalized additive model; gam allows inclusion of the same correlation structures as the function gls; Wood (2006)
`gamm`	`mgcv`	Generalized additive mixed model; gamm allows inclusion of the same correlation structures as the function gls; Wood (2006)
`glmmPQL`	`MASS`	Penalized quasilikelihood for generalized linear mixed models; the same correlation structures as for gls can be used; Venables and Ripley (2002)
`geeglm`	`geepack`	Generalized estimation equation: this is an alternative fitting algorithm for fitting a GLM that allows taking correlation structures into account. One of the following correlation structures can be chosen: "independence," "exchangeable," "ar1," "unstructured," and "userdefined." See Zuur et al. (2009) p. 304 ff

Note: In most functions, the autocorrelation structure can be given in the argument "correlation."

Our general approach to analyzing spatial data is:

(1) Fit a model including all important predictor variables while ignoring spatial correlation.
(2) Analyze the residuals and try to find predictor variables that can explain the spatial structure of the residuals.

(3) If, after having included all important predictors, spatial correlation is still present in the residuals, allow for spatial correlation in the model (Chapter 13).

The data set "frogs" contains the number of water frogs (of the genus *Pelophylax*) counted in ponds of the canton of Aargau in Switzerland. We model the water frog counts by elevation, year, presence of fish and vegetation, and the size of the pond. We also include the interaction between fish and vegetation since we think that predation by fish is less severe when the frogs can hide in vegetation. We z-transform year and elevation before the analysis so that their means are 0 and their standard deviations are 1.

Two measurements for the size of the ponds (in m^2) are zero. Because a pond without area would not exist, we assume that these zero values were rounded from an original measurement below 0.5 m^2. Therefore, we replace the zero values by 0.25. Alternatively, we could add 1 to all measurements before the log transformation to avoid losing the two observations. Before the log transformation, we "linearized" the area measurements by the square-root transformation. This two-step transformation (the logarithm of the square root) produces a size variable that is nicely symmetrically distributed. Subsequently, it also is z-transformed (standardized, i.e., centered and scaled, so that the mean is zero and the standard deviation 1; see Chapter 16).

```
data(frogs)
frogs$year.z <- scale(frogs$year)
frogs$elevation.z <- scale(frogs$elevation)
frogs$waterarea[frogs$waterarea==0] <- 0.25
frogs$waterarea.sqrt.l <- log(sqrt(frogs$waterarea))
frogs$waterarea.sqrt.l.z <- scale(frogs$waterarea.sqrt.l)
```

We use the counts of the second visit (count2) as the outcome variable. We started with a Poisson model (Chapter 8), as it is often used for counts, and found a huge degree of overdispersion (Section 8.2.3), showing that the Poisson model does not fit. Next, we included an observation level random factor to account for the extra variance. Unfortunately, this model seems to fit so badly that neither `glmer` nor `glmmPQL` could fit the model. The normal linear model (`lm`) for the logarithm of the count increased by 1 could be a pragmatic and often robust alternative. However, because the mean count (12.8) is low compared to the variance (standard deviation was 37.2) this model would predict negative counts in many cases. Therefore, we decided to use a negative binomial model that seems to fit quite well but not perfectly. However, for the purpose of showing how to detect spatial correlation, the negative binomial model is good enough.

The Negative Binomial Model and the Frog Count Data

The negative binomial distribution describes the natural distribution of the number of failures counted before a given number of successes have occurred for a constant success probability. It often describes count data with a large variance quite

Continued

The Negative Binomial Model and the Frog Count Data—cont'd

well. In this book, we will not introduce the negative binomial model in detail for two reasons. First, only very rarely will biological count data be expressed by the numbers of events that happened until a predefined number of anti-events have happened. Thus, the negative binomial model is, in most cases, not an explicit model of the data-generating process. For this reason, we prefer to use Poisson models including explicit parameter(s) for the extra variance. Second, the function `sim` cannot (yet) handle negative binomial models. To fit a negative binomial model in a Bayesian framework, the glmmADMB package, BUGS, or Stan can be used.

Nevertheless, we pragmatically use a negative binomial model for the water frogs in this chapter, to keep the model as simple as possible. More appropriate, but also more complex models, would include an observation model that accounts for the fact that some frogs are overlooked, such as *N*-mixture models (Royle, 2004). Variants of *N*-mixture models are introduced in Chapter 14. Further, we pragmatically use frequentist methods to fit the model, because we have no prior information and we are only interested in the spatial distribution of the residuals. In this case, the spatial distribution of the residuals does not depend on whether the model was fitted in a frequentist or Bayesian framework.

```
library(MASS)        # provides the function glm.nb
mod <- glm.nb(count2~elevation.z + year.z + fish + vegetation +
              waterarea.sqrt.l.z + fish:vegetation, data=frogs)
```

Of course, we always look at the standard residual plots generated by `plot(mod)` first, but for now, let's head directly to analyzing the spatial distribution of the residuals. The bubble plot (obtained by the function `bubble` from the package sp; Pebesma & Bivand, 2005) is a nice way to look at the spatial distribution of the residuals. The function `bubble` is applied to objects of the class "SpatialPointsDataFrame." We create such an object by defining the *x*- and *y*-coordinates in a `data.frame` together with the residuals. We use the function `resid` to obtain deviance residuals. Such residuals should be comparable between the different observations and therefore can be used to assess spatial correlation. More commonly, standardized (*z*-transformed residuals, `resid(mod, type="pearson")`) are used. But these do not seem to be provided for the negative binomial model, at least we did not found documentation on this.

```
library(sp)          # function bubble
spdata <- data.frame(resid=resid(mod), x=frogs$x, y=frogs$y)
coordinates(spdata) <- c("x", "y")
bubble(spdata, "resid", col=c("blue", "orange"), main="Residuals",
     xlab="X-coordinates", ylab="Y-coordinates")
```

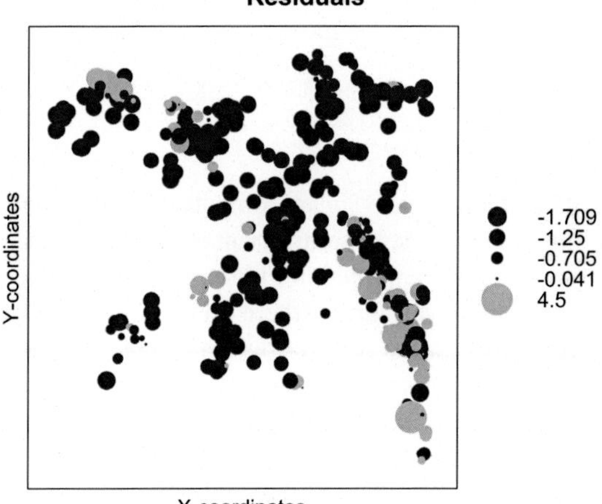

FIGURE 6-5 Spatial distribution of residuals of the model for the frog counts. Orange points indicate positive and blue points indicate negative residuals. The size of the points corresponds to the absolute values of the residuals.

The bubble plot (Figure 6-5) shows three clusters of positive residuals: in the southeastern part of the canton, the northwestern part, and just southwest of the center of the canton. From this plot we learn that the residuals are not randomly distributed across the study area. We may have missed important predictor variable(s) to explain why the frog populations of the three subareas were larger than expected by the model.

Another reason could be that there is natural spatial correlation between the frog populations of neighboring ponds because frogs migrate between different ponds. Thus, a pond in close proximity to a very good pond for frogs (with a high number of frogs) may have large numbers too because frogs emigrate from the good pond. In the first case, spatial correlation will disappear if the important predictors are included in the model. In the second case, spatial correlation will not disappear because there is a natural connection between the ponds. In this case, the strength and shape of the spatial correlation may be of biological interest.

A more formal way to look at spatial correlation is by using a semivariogram (often, but incorrectly, called a "variogram"). A semivariogram shows 0.5 times the mean squared differences between measurements (= semivariance) dependent on the distance between the two measurements. Given independence, the expected value of the semivariogram is the variance of the measurement, which is 1 if the measurement is standardized. Values smaller than the variance (or smaller than 1) mean that the measurements are more similar to each other than expected by chance, thus they are correlated. We apply the function `variogram` (from the gstat package; Pebesma, 2004) to the standardized residuals. The standardized residuals are the ones that are extracted from a glm.nb-object by the function `resid`.

```
library(gstat)
vario.mod <- variogram(resid(mod)~1, spdata)
plot(vario.mod)
```

By default, the function `variogram` computes the semivariances between all pairs of points. Then, these semivariances are averaged over distance classes and these averages are plotted in Figure 6-6 (left panel), which is called the "sample" or "residual" variogram (there are other types of variograms that we do not consider here). In the semivariogram (Figure 6-6, left panel) we see spatial correlations up to a distance of around 7500 m (7.5 km). This is called the range. We would expect the semivariogram to reach zero as the distance approaches zero. However, the semivariogram often starts at some value higher than zero for distances close to zero due to unexplained variance in the data such as measurement errors. This is called the nugget effect.

The semivariogram shown in the right panel of Figure 6-6 shows anisotropy, that is, the strength of spatial correlation is different in different directions. The different directions along which the semivariances should be calculated are given in the argument alpha within the variogram function.

```
vario.mod.6dir <- variogram(resid(mod)~1, spdata, alpha=seq(0,
150, by=30))
plot(vario.mod.6dir)
```

Sometimes spatial correlation varies with covariates. We may then try to add covariates to model anisotropy. To account for spatial correlation in a

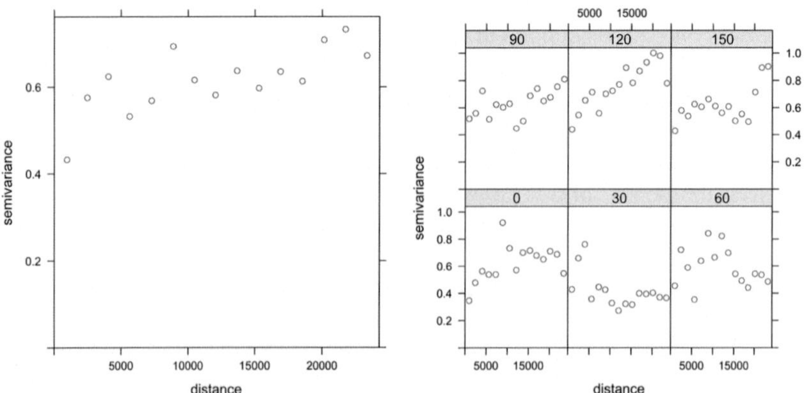

FIGURE 6-6 Semivariograms of the residuals of the frog count model. *Left panel*: averaged over all observations independent of the direction. *Right panel*: semivariograms for six different directions indicating anisotropy.

model, we need to find a parametric function that describes the relationship between correlation and distance between pairs of observations, $f(\text{dist}(i,j))$. As for models with temporal correlation, this function is then used in the covariance matrix **V** to describe the correlations among the residuals (Section 6.4).

In Chapter 13, we introduce linear models with spatial correlation using Bayesian methods in R. None of the types of models that can be handled by `sim` allow inclusion of spatial autocorrelation. The commonly used R functions that allow inclusion of spatial correlations in a linear model are given in Table 6-2; only two of them use Bayesian methods.

The Link to Mixed Models

Models that include temporal or spatial autocorrelation are closely related to mixed models and, therefore, many procedures written to fit mixed models also fit models including temporal or spatial autocorrelation. A mixed model is a model that accounts for the nonindependence of observations that belong to the same group. For example, when repeated measurements were taken on the same individual, the grouping factor is the individual. The residual covariance matrix for a data set with three measurements on each of two individuals would be

$$\mathbf{V} = \begin{vmatrix} \sigma^2 & \varphi & \varphi & 0 & 0 & 0 \\ \varphi & \sigma^2 & \varphi & 0 & 0 & 0 \\ \varphi & \varphi & \sigma^2 & 0 & 0 & 0 \\ 0 & 0 & 0 & \sigma^2 & \varphi & \varphi \\ 0 & 0 & 0 & \varphi & \sigma^2 & \varphi \\ 0 & 0 & 0 & \varphi & \varphi & \sigma^2 \end{vmatrix}$$

The (within-individual) residual variance σ^2 is on the diagonal and φ is the covariance of the observations of the same individual. It is assumed that observations on different individuals are independent because all covariances between observations from different individuals are zero. In Chapter 7 we will use a different (but synonymous) notation for mixed models.

6.6 HETEROSCEDASTICITY

Heteroscedasticity means nonhomogeneity of the residual variance. This lack of homogeneity in residual variance may occur when residual variance differs between groups or when it depends on a covariate; for example, when the measurement error decreases over time because the person taking the measurements becomes more experienced.

As with other violations of model assumptions, heteroscedasticity can be graphically detected by plotting the residuals. The third plot in Figure 6-1 (Scale-Location) shows the variance along the range of fitted values. When the

TABLE 6-2 Nonexhaustive List of R Functions That Allow Inclusion of Spatial Correlation in a Linear Model

Function	Package	Estimation Method, Options, and Reference
`spGLM`	`spBayes`	See chapter 13; Finley and Banerjee (2013)
`MCMCglmm`	`MCMCglmm`	Generalized linear mixed models using Markov chain Monte Carlo simulation; experienced users can specify any correlation structure, see `vignette(MCMCglmm)`; Hadfield (2010)
`gls`	`nlme`	Generalized least-squares method; works for all normal linear models. The following standard spatial correlation structures are available (Pinheiro & Bates, 2000): -`corExp` exponential spatial correlation -`corGaus` Gaussian spatial correlation -`corLin` linear spatial correlation -`corRatio` rational quadratics spatial correlation -`corSpher` spherical spatial correlation
`lme`	`nlme`	REML or ML; works only for normal linear models that include at least one random factor. Same correlation structures as in gls; Pinheiro and Bates (2000)
`gam`	`mgcv`	Generalized additive model; it allows inclusion of the same correlation structures as the function gls; Wood (2006)
`gamm`	`mgcv`	Generalized additive mixed model; it allows inclusion of the same correlation structures as the function gls; Wood (2006)
`glmmPQL`	`MASS`	Penalized quasilikelihood for generalized linear mixed models; the same correlation structures as for gls can be used; Venables and Ripley (2002)
`spautolm`	`spdep`	Fits spatial regression models using a mixture of ML and generalized least-square methods; Bivand et al. (2008)

Note: In most functions, the autocorrelation structure can be given in the argument "correlation."

variance is increasing or decreasing with the fitted value, we may try to use an error distribution with a variance that depends on the mean (such as the Poisson for count data), or we can try to transform the outcome variable. In many cases, the variance does not depend on the fitted value per se but on one of the predictors. To detect heterogeneity, the residuals, or the square root of the absolute values of the residuals, need to be plotted against all variables that may influence the variance in the data. Here, we show two examples from the Ellenberg data set introduced in Chapter 4.

```
data(ellenberg)
dat <- ellenberg[complete.cases(ellenberg[c("Yi.g", "Water",
"Species")]),]
mod <- lm(log(Yi.g) ~ Water + Species + Water:Species, dat)
par(mfrow=c(1,2))
plot(resid(mod) ~ Species, dat)
scatter.smooth(dat$Water, sqrt(abs(resid(mod))), xlab="Water")
```

The boxplot of the residuals versus the species (Figure 6-7, left panel) shows that the residual variance for species *A. elatius* (Ae) may be larger than for species *F. pratensis* or *P. palustris* (Fp, Pp). When plotting the square root of the absolute values of the residuals versus distance to water (Figure 6-7, right panel), we see that the variance seems to be quite homogeneous except for the distance to the water level of -5 cm. This is the situation where the grasses were actually overflown by water. Thus, this might have provided completely different growing conditions.

If heteroscedasticity needs to be accounted for in the model, we "simply" include a function for the variance:

$$\widehat{y}_i = \beta_o$$
$$y_i \sim Norm\left(\widehat{y}_i, \sigma_i^2\right)$$
$$\sigma_i = f(x_i)$$

Most of the functions listed earlier in Table 6.1 also allow one to include such a function for the variance. Often, the argument is called "weights." Pinheiro and Bates (2000) explain in detail how such variance functions are built when using the function `lme` from the package nlme. However, as with temporal correlation, the R functions used to fit models that include a variance function cannot (yet) be handled by `sim`, and thus, it is not so easy to obtain simulations from the posterior distribution. We recommend using BUGS or Stan to fit such models.

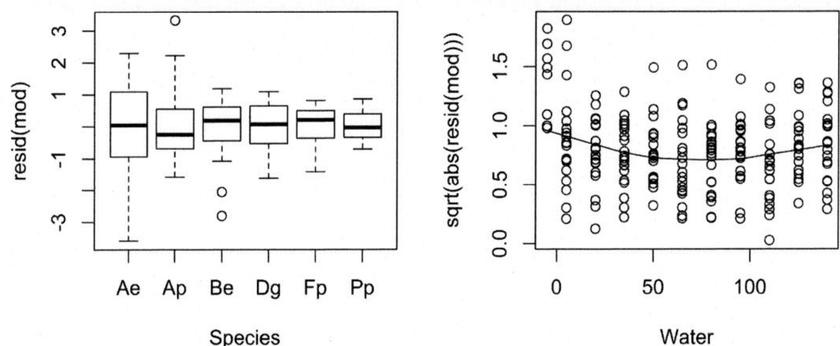

FIGURE 6-7 Residuals of a linear model fitted to the Ellenberg data versus species, and the square root of the absolute value of the residuals versus distance to water, with a smoother.

FURTHER READING

The book by Stahel (2002) contains a very good and comprehensive introduction to residual analysis (the book is in German).

Hoeting (2009) explains why it is important to account for spatial and temporal correlation in data analysis. She recommends a couple of books and additional literature on the topic.

Pinheiro and Bates (2000) is the reference for linear models with correlation structures in R.

Zuur et al. (2009, p. 168) explain, very understandably, how to fit models to data with heterogeneity, such as temporal or spatial correlation and heteroscedasticity.

Bivand et al. (2008) describe how to do spatial data analyses and how to import GIS data into R and analyze and plot them within R.

Chapter 7

Linear Mixed Effects Models

Chapter Outline

7.1 BACKGROUND

7.1.1 Why Mixed Effects Models?

Mixed effects models (or hierarchical models; see Gelman & Hill, 2007, for a discussion on the terminology) are used to analyze nonindependent, grouped, or hierarchical data. For example, when we measure growth rates of nestlings in different nests by taking mass measurements of each nestling several times during the nestling phase, the measurements are grouped within nestlings (because there are repeated measurements of each) and the nestlings are grouped within nests. Measurements from the same individual are likely to be more similar than measurements from different individuals, and individuals from the same nest are likely to be more similar than nestlings from different nests. Measurements of the same group (here, the "groups" are individuals or nests) are not independent. If the grouping structure of the data is ignored in the model, the residuals do not fulfill the independence assumption.

Predictor variables can be measured on different hierarchical levels. For example, in each nest some nestlings were treated with a hormone implant whereas others received a placebo. Thus, the treatment is measured at the level of the individual, while clutch size is measured at the level of the nest. Clutch size was measured only once per nest but entered in the data file more than once (namely for each individual from the same nest). Similarly, all

http://dx.doi.org/10.1016/B978-0-12-801370-0.00007-1.

observations of one individual have the same value for treatment (but different values for individual measures such as weight). This results in pseudoreplication if we do not account for the hierarchical data structure in the model. Mixed models allow modeling of the hierarchical structure of the data and, therefore, account for pseudoreplication.

Mixed models are further used to analyze variance components. For example, when the nestlings were cross-fostered so that they were not raised by their genetic parents, we would like to estimate the proportions of the variance (in a measurement, e.g., wing length) that can be assigned to genetic versus to environmental differences.

Mixed models contain fixed and random effects. Note that, by definition, fixed and random effects are factors. Fixed effects have a finite ("fixed") number of levels; for example, the factor "sex" has the levels male and female and (in many studies) nothing more. In contrast, random effects have a theoretically infinite number of levels of which we have measured a random sample. For example, we have measured 10 nests, but there are many more nests in the world that we have not measured. Normally, fixed effects have a low number of levels whereas random effects have a large number of levels (at least 3!). For fixed effects we are interested in the specific differences between levels (e.g., between males and females), whereas for random effects we are only interested in the between-level (= between-group, e.g., between-nest) variance rather than in differences between specific levels (e.g., nest A versus nest B).

Typical fixed effects are: treatment, sex, age classes, or season. Typical random effects are: nest, individual, field, school, or study plot. It depends sometimes on the aim of the study whether a factor should be treated as fixed or random. When we would like to compare the average size of a corn cob between specific regions, then we include region as a fixed factor. However, when we would like to measure the size of a corn cob for a larger area within which we have measurements from a random sample of regions, then we treat region as a random factor.

7.1.2 Random Factors and Partial Pooling

In a model with fixed factors, the differences of the group means to the mean of the reference group are separately estimated as model parameters. This produces $k-1$ (independent) model parameters, where $k =$ number of groups (or number of factor levels). In contrast, for a random factor, only one parameter, namely the between-group variance, is estimated. To estimate this variance, we look at the differences of the group means to the population mean; that is, we look at k differences from the population mean. These k differences are not independent. They are assumed to be realizations of the same (in most cases normal) distribution with mean zero. They are like residuals, and we usually call them b_g; each is the difference, b, between the mean of group, g, and the mean of all groups. The variance of the b_g values is the between-group variance.

Treating a factor as a random factor is equivalent to partial pooling of the data. There are three different ways to obtain means for grouped data. First, the grouping structure of the data can be ignored. This is called complete pooling (left panel in Figure 7-1). Second, group means may be estimated separately for each group. In this case, the data from all other groups are ignored when estimating a group mean. No pooling occurs in this case (right panel in Figure 7-1). Third, the data of the different groups can be partially pooled (i.e., treated as a random effect). Thereby, the group means are weighted averages of the population mean and the unpooled group means. The weights are proportional to sample size and the inverse of the variance (see Gelman & Hill, 2007, p. 252).

complete pooling	partial pooling	no pooling
$\widehat{y}_i = \beta_o$ $y_i \sim Norm(\widehat{y}_i, \sigma^2)$	$\widehat{y}_i = \beta_o + b_{g_i}$ $y_i \sim Norm(\widehat{y}_i, \sigma^2)$ $b_g \sim Norm(0, \sigma_b^2)$	$\widehat{y}_i = \beta_{g_i}$ $y_i \sim Norm(\widehat{y}_i, \sigma_{gi}^2)$

What is the advantage of analyses using partial pooling (i.e., mixed, hierarchical, or multilevel modeling) compared to the complete or no pooling analyses? Complete pooling ignores the grouping structure of the data. As a result, the credible interval of the population mean may be too narrow. We are too confident in the result because we assume that all observations are independent when they are not. This is a typical case of pseudoreplication. On the other hand, the no pooling method (which is equivalent to treating the factor as fixed) has the danger of overestimation of the between-group variance because the group means are estimated independently of each other.

The danger of overestimating the between-group variance is particularly large when sample sizes per group are low and within-group variance large. In

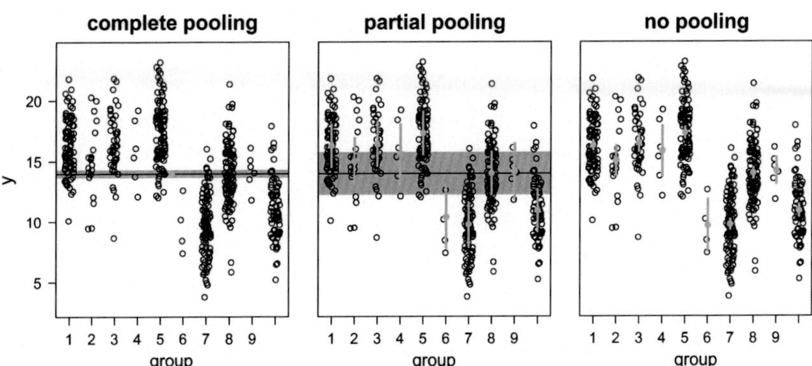

FIGURE 7-1 Three possibilities to obtain group means for grouped data $\widehat{y}_i$: complete pooling, partial pooling, and no pooling. Open symbols = data, orange dots with vertical bars = group means with 95% credible intervals, horizontal black line with shaded interval = population mean with 95% credible interval.

contrast, the partial pooling method assumes that the group means are a random sample from a common distribution. Therefore, information is exchanged between groups. Estimated means for groups with low sample sizes, large variances, and means far away from the population mean are shrunk toward the population mean. Thus, group means that are estimated with a lot of imprecision (because of low sample size and high variance) are shrunk toward the population mean. How strongly they are shrunk depends on the precision of the estimates for the group specific means and the population mean.

An example will help make this clear. Imagine that we measured 60 nestling birds from 10 nests (6 nestlings per nest) and found that the average nestling mass at day 10 was around 20 g with a between-nest standard deviation of 2 g. Then, we measure only one nestling from one additional nest (from the same population) whose mass was 12 g. What do we know about the average mass of this new nest? The mean of the measurements for this nest is 12 g, but with $n = 1$ uncertainty is high. Because we know that the average mass of the other nests was 20 g, and because the new nest belonged to the same population, a value higher than 12 g is a better estimate for an average nestling mass of the new nest than the 12 g measurement of one single nestling, which could, by chance, have been an exceptionally light individual. This is the shrinkage that partial pooling allows in a mixed model. Because of this shrinkage, the estimates for group means from a mixed model are sometimes called shrinkage estimators. A consequence of the shrinkage is that the residuals are positively correlated with the fitted values.

To summarize, mixed models are used to appropriately estimate between-group variance and to account for nonindependency among data points.

7.2 FITTING A LINEAR MIXED MODEL IN R

To introduce the linear mixed model (LMM), we use repeated hormone measures of nestling barn owls *Tyto alba*. The cortbowl data set contains stress hormone data (corticosterone, variable "totCort") of nestling barn owls that were either treated with a corticosterone implant or with a placebo implant as the control group. The aim of the study was to quantify the corticosterone increase due to the corticosterone implants (Almasi et al., 2009). In each brood, one or two nestlings were implanted with a corticosterone implant and one or two nestlings with a placebo implant (variable "Implant"). Blood samples were taken just before implantation, and at days 2 and 20 after implantation. In total, there are 287 measurements of 151 individuals (variable "Ring") of 54 broods.

Because the measurements from the same individual are nonindependent, we use a mixed model to analyze these data: Two additional arguments for a mixed model are: (1) the mixed model allows prediction of corticosterone levels for an "average" individual, whereas the fixed effect model allows prediction of corticosterone levels only for the 151 individuals that were

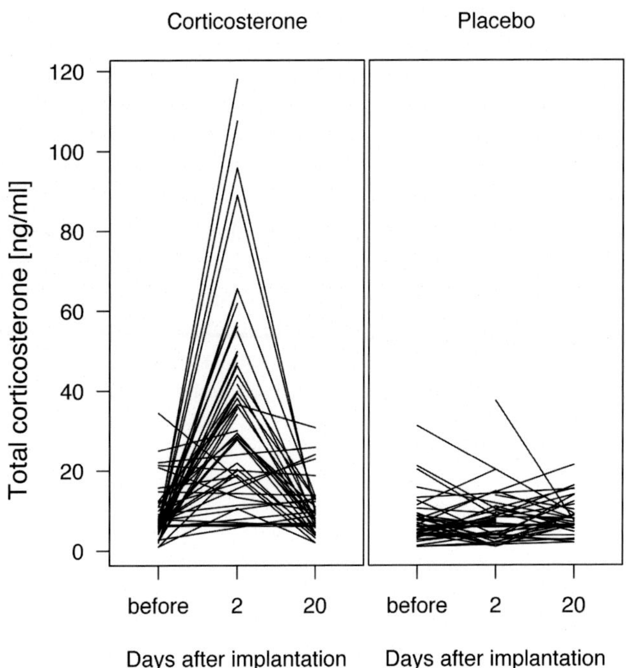

FIGURE 7-2 Total corticosterone at day 0 ("before"), and at day 2 and day 20 after implantation of a corticosterone implant or a placebo. Lines connect measurements of the same individual.

sampled; and (2) fewer degrees of freedom are needed. If we include individual as a fixed factor, we would use up 150 degrees of freedom, while just one degree of freedom is needed for the random factor.

We first create a graphic to show the development for each individual, separately for owls receiving corticosterone versus owls receiving a placebo (Figure 7-2).

We fit a normal linear model with "Ring" as a random factor, and "Implant", "days", and the interaction of "Implant" × "days" as fixed effects. Note that both "Implant" and "days" are defined as factors, thus R creates indicator variables for all levels except the reference level. Later, we will also include "Brood" as a grouping level; for now, we ignore this level and start with a simpler (less perfect) model for illustrative purposes.

$$\begin{aligned}\widehat{y}_i &= \beta_o + b_{Ring_i} + \beta_1 I(Implant_i = P) + \beta_2 I(days_i = 2) + \beta_3 I(days_i = 20) + \\ &\quad \beta_4 I(Implant_i = P) I(days_i = 2) + \beta_5 I(Implant_i = P) I(days_i = 20) \\ y_i &\sim Norm(\widehat{y}_i, \sigma^2) \\ b_{Ring} &\sim Norm(0, \sigma_b^2)\end{aligned}$$

Several different functions to fit a mixed model have been written in R: `lme`, `lmer`, `gls`, `gee`. We primarily use `lmer` from the package lme4 (which is automatically loaded to the R console when loading arm), because we find its usage intuitive and the function `sim` can treat `lmer` objects but none of the others.

```
data(cortbowl)
dat <- cortbowl
mod <- lmer(log(totCort) ~ Implant + days + Implant:days + (1|Ring),
            data=dat, REML=TRUE)
```

The function `lmer` is used similarly to the function `lm`. The only difference is that the random factors are added in the model formula within parentheses. The "1" stands for the intercept and the "|" means "grouped by". "(1|Ring)", therefore, adds the random deviations for each individual to the average intercept. These deviations are the b_{Ring} in the previous model formula. Corticosterone data are log transformed to achieve normally distributed residuals.

```
mod
Linear mixed model fit by REML ['lmerMod']
Formula: log(totCort) ~ Implant + days + Implant:days + (1 | Ring)
   Data: dat
REML criterion at convergence: 611.9053
Random effects:
 Groups   Name        Std.Dev.
 Ring     (Intercept) 0.3384
 Residual             0.6134
Number of obs: 287, groups: Ring, 151
Fixed Effects:
(Intercpt)    ImplP   days2     days20  ImplP:days2  ImplP:days20
     1.914   -0.085   1.653   0.263278       -1.720        -0.095
```

The output of the `lmer` object tells us that the model was fitted using the REML method (see 7.3). The "REML criterion" is the statistic describing the model fit for a model fitted by REML. The model output further contains the parameter estimates. These are grouped into a random effects and fixed effects section. The random effects section gives the estimates for the between-individual standard deviation ($\widehat{\sigma_{Ring}} = 0.34$) and the residual standard deviation ($\widehat{\sigma} = 0.61$). The fixed effects section gives the estimates for the intercept ($\widehat{\beta_0} = 1.9$), which is the mean for an "average" individual that received a corticosterone implant at the day of implantation.

The other model coefficients are defined as follows: the difference in log(totCort) between placebo- and corticosterone-treated individuals before implantation ($\widehat{\beta_1} = -0.09$), the difference between day 2 and before implantation for the corticosterone-treated individuals ($\widehat{\beta_2} = 1.65$), the difference between day 20 and before implantation for the corticosterone-treated individuals ($\widehat{\beta_3} = 0.26$), and the interaction parameters that tell us how the differences between day 2 and before implantation ($\widehat{\beta_4}$), and day 20 and before implantation ($\widehat{\beta_5}$), differ for the placebo-treated individuals compared to the corticosterone-treated individuals ($\widehat{\beta_4} = -1.72$, $\widehat{\beta_5} = -0.10$).

Neither the model output shown earlier nor the summary function (not shown) give any information about the proportion of variance explained by the model such as an R^2. This is because it is not straightforward to obtain a measure of model fit in a mixed model, and different definitions of R^2 exist (Chapter 10, Section 10.2).

The function fixef extracts the estimates for the fixed effects; the function ranef extracts the estimates for the random deviations from the population intercept for each individual. The random deviations measure, for each individual, how much its shrunken mean corticosterone value deviates from the mean of all individuals with the same treatment on the same day (the difference between the orange dot and the horizontal black line in the middle panel in Figure 7-1).

```
round(fixef(mod),3)
(Intercpt)    ImplP    days2  days20  ImplP:days2       ImplP:days20
     1.914   -0.085    1.653   0.263       -1.720             -0.095
ranef(mod)
$Ring
            (Intercept)
898054      0.248849794
898055      0.118458626
898057     -0.107882775
898058      0.069989589
898059     -0.080864976
898061     -0.083968388
.........
```

7.3 RESTRICTED MAXIMUM LIKELIHOOD ESTIMATION

For a mixed model, the restricted maximum likelihood estimation (REML) method is used by default instead of the maximum likelihood (ML) method (see Chapter 5). This is because the ML method underestimates the variance parameters because this method assumes that the fixed parameters are known without uncertainty when estimating the variance parameters. However, the parameters for the fixed effects are estimates with uncertainty. The REML method uses a mathematical trick to make the estimates for the variance parameters independent of the estimates for the fixed effects.

We recommend reading the very understandable description of the REML method in Zuur et al. (2009). For our purposes, the relevant difference between the two methods is that the ML estimates are unbiased for the fixed effects but biased for the random effects, whereas the REML estimates are biased for the fixed effects and unbiased for the random effects. However, when sample size is large compared to the number of model parameters, the differences between the ML and REML estimates become negligible. As a guideline, use REML if the interest is in the random effects (variance parameters) and ML if the interested is in the fixed effects. The estimation method can be chosen by setting the argument "REML" to "FALSE" (default is "TRUE").

```
mod <- lmer(log(totCort) ~ Implant + days + Implant:days + (1|Ring),
        data=dat, REML=FALSE) # using ML
```

7.4 ASSESSING MODEL ASSUMPTIONS

As with a simple linear model, the assumptions are carefully checked before inference is drawn from a mixed model. The assumptions are, as explained in Chapter 6, that the residuals are independent and identically distributed (iid). In principle, the same methods described in Chapter 6 are used to assess violation of model assumptions in mixed models. However, the function `plot` does not produce the standard diagnostic residual plots from an `lmer` object. Therefore, these plots have to be coded by hand.

```
par(mfrow=c(2,2))
scatter.smooth(fitted(mod),resid(mod)); abline(h=0, lty=2)
title("Tukey-Anscombe Plot") # residuals vs. fitted
```

This code produces the first plot of Figure 7-3. We see that the residuals scatter around zero with a few exceptions in the lower left part of the panel. A positive correlation between the residuals and the fitted values (not present in the example data) would not bother us statistically, but it indicates strong shrinkage and may have biological meaning.

```
qqnorm(resid(mod), main="normal QQ-plot, residuals") #qq of residuals
qqline(resid(mod))
scatter.smooth(fitted(mod), sqrt(abs(resid(mod)))) #res. var vs. fitted
```

The few small measurements that do not fit well to the model are also recognizable in the QQ plot of the residuals (a banana-like deviation from the straight line at the left edge of the data) and when plotting the square root of the absolute values of the residuals against the fitted values. Because the number of such cases is low and all other observations seem to fulfill the model assumptions well, we accept the slight lack of fit. But we are aware of the fact that model predictions for small corticosterone levels are unreliable.

In addition to the checks presented in Chapter 6, the assumption that the random effects, that is the values of b_{Ring}, are normally distributed, needs to be checked. In the following, we do this using a QQ plot (Figure 7-3). Note that we need to extract the b_{Ring} using "[,1]" because the `ranef` object is two-dimensional.

```
qqnorm(ranef(mod)$Ring[,1], main="normal QQ-plot, random effects")
qqline(ranef(mod)$Ring[,1])   # qq of random effects
```

We do not see a serious deviation in the distribution of the random effects from the normal distribution.

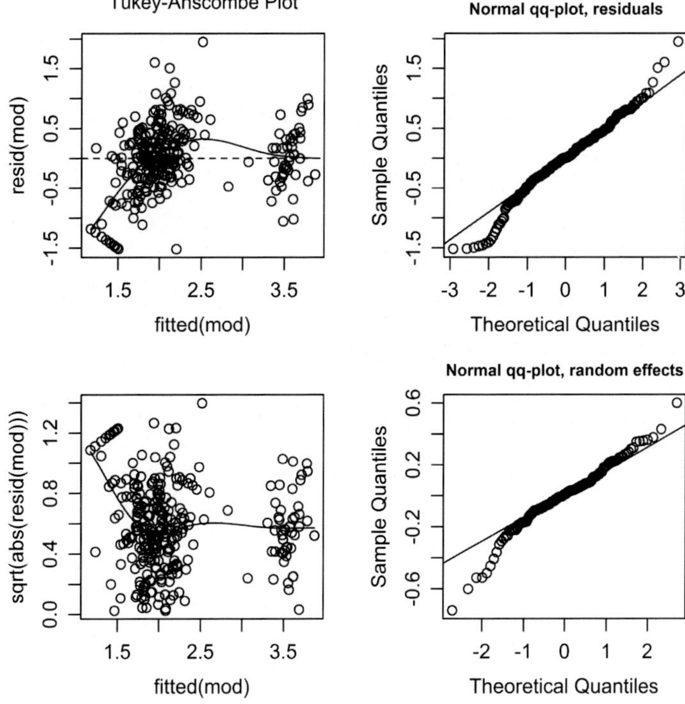

FIGURE 7-3 Diagnostic residual and random effect plots to assess model assumptions of the corticosterone model. *Upper left*: residuals versus fitted values. *Upper right*: Normal QQ plot of the residuals. *Lower left*: square-root of the absolute values of the residuals versus fitted values. *Lower right*: Normal QQ plot of the random effects.

7.5 DRAWING CONCLUSIONS

We use the function `sim` to draw 2000 random values from the joint posterior distribution of the model parameters; that is, we draw 2000 values for each parameter while taking the correlation between the parameters into account.

```
nsim <- 2000
bsim <- sim(mod, n.sim=nsim)
str(bsim)
Formal class 'sim.merMod' [package "arm"] with 3 slots
  ..@ fixef: num [1:2000, 1:6] 2.11 1.82 1.99 2.04 1.88 ...
  .. ..- attr(*, "dimnames")=List of 2
  .. .. ..$ : NULL
  .. .. ..$ : chr [1:6] "(Intercept)" "ImplantP" "days2" "days20" ...
  ..@ ranef:List of 1
  .. ..$ Ring: num [1:2000, 1:151, 1] 0.3545 0.3593 0.3137 ...
  .. .. ..- attr(*, "dimnames")=List of 3
  .. .. .. ..$ : NULL
  .. .. .. ..$ : chr [1:151] "898054" "898055" "898057" "898058" ...
  .. .. .. ..$ : chr "(Intercept)"
  ..@ sigma: num [1:2000] 0.603 0.629 0.629 0.61 0.595 ...
```

The object produced by the `sim` function when applied to a mer object (= objects produced by the functions `lmer` or `glmer`) contains three slots (remember that slots are addressed using the `@` sign in R). The slot "fixef" is a matrix with as many columns as there are parameters in the fixed part of the model. The number of rows corresponds to the number of simulations (we have saved this number in the object "nsim"). The slot "ranef" is a list that contains one element for each random factor. In the previous example, there is only one random factor ("Ring"). Therefore, the list contains only one element.

Each element is a three-dimensional array where the first dimension represents the number of simulations (2000 in our case), the second dimension is the number of groups (factor levels; here this is the number of individuals), and the third dimension corresponds to the number of parameters that are grouped by the specific factor. In the prevoius example, we have included only a random intercept in the model. Therefore, this dimension has a length of one. In the following we will fit models with more parameters per random effect.

From the simulated values, the 2.5% and 97.5% quantiles can be used for the 95% credible interval:

```
round(apply(bsim@fixef, 2,quantile, prob=c(0.025,0.5,0.975)),3)
```

```
      (Intercept) ImplantP days2 days20 ImplantP:days2 ImplantP:days20
2.5%        1.740   -0.351 1.407  0.007         -2.069           0.454
50%         1.912   -0.089 1.657  0.262         -1.711           0.089
97.5%       2.082    0.181 1.894  0.499         -1.358           0.269
```

Some effort is needed to interpret models with interactions. We see that in the corticosterone-treated nestlings, the logarithm of the corticosterone measure increases from before implantation to day 2 by 1.7 (95% CrI: 1.4 – 1.9), which is quite substantial. To get this increase for the placebo-treated individuals, we have to add the interaction parameter to this increase, thus the increase in placebo nestlings is 1.7 − 1.7 = 0. The interaction parameter, −1.7 (95% CrI:−2.1 to −1.4), measures the difference between placebo and corticosterone nestlings in their response to the treatment; it is, therefore, an important result.

To better see what really happens, we often plot the fitted values with CrI and the observations in one plot. To this end, we first prepare a new data frame that contains a row for each factor level ("Implant" and "days"). Then, the fitted value for each of the factor-level combinations is calculated 2000 times (for each set of model parameters from the simulated posterior distribution) to obtain 2000 simulated values from the posterior distribution of the fitted values. The 2.5% and 97.5% quantiles of these fitted values are used as lower and upper bounds of the 95% credible interval.

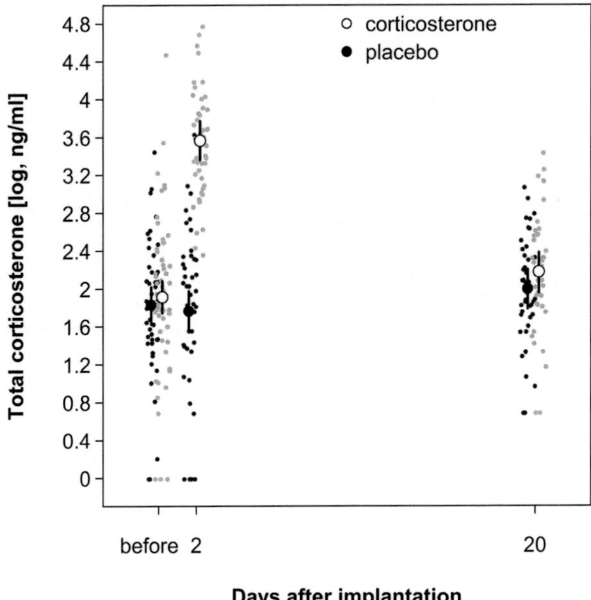

FIGURE 7-4 Predicted total corticosterone values with 95% CrI of placebo-implanted nestlings (closed symbol) and corticosterone-implanted nestlings (open symbol) in relation to days after implantation. Blue dots are raw data of placebo-implanted nestlings, and orange dots are raw data of corticosterone-implanted nestlings.

```
newdat<-expand.grid(Implant=factor(c("C","P"),
            levels=levels(dat$Implant)),
            days =factor(c(1,3,21),levels=levels(dat$days)))
Xmat <- model.matrix(~ Implant + days + Implant:days, data=newdat)
fitmat <- matrix(ncol=nsim, nrow=nrow(newdat))
for(i in 1:nsim) fitmat[,i] <- Xmat %*% bsim@fixef[i,] # fitted values
newdat$lower <- apply(fitmat, 1, quantile, prob=0.025)
newdat$upper <- apply(fitmat, 1, quantile, prob=0.975)
newdat$fit <- Xmat %*% fixef(mod)
```

The fitted values given in Figure 7.4, together with their uncertainty measures (CrI), take into account that we had repeated measures for each individual.

7.6 FREQUENTIST RESULTS

The estimation of standard errors for the model parameters in mixed models is difficult using frequentist methods (see Further Reading at the end of the chapter). Often, likelihood ratio tests are done to decide whether one model fits the data substantially better than another, which is suggested if the ratio between the likelihoods (see Chapter 5) of two models is much larger than one. If the likelihood ratio is significantly larger than one, the larger model is said to be significantly better than the smaller one. From such a result, it is concluded that the parameter, or the group of parameters, that are missing in the smaller model significantly increase the likelihood when they are added to the model, and thus these parameters are declared to be important in explaining the variance of the outcome variable.

The likelihood ratio test statistic (LRT) is two times the difference in the logarithms of the likelihoods of the two models. Given the null hypothesis that the two models fit the data equally well, and given that the observations are independent, this test statistic follows an χ^2 distribution. However, for mixed models, the independence assumption is not met.

In these cases, the assumption of the χ^2 distribution given the null hypothesis is an approximation only. The problem is that in mixed models it is difficult to obtain the degrees of freedom. However, for testing fixed effects, and when sample size is large, the approximate likelihood ratio test is reliable in practice. In contrast, when testing random effects, or when sample size is small, the approximate likelihood ratio test can be misleading. In such cases it is recommended to simulate the distribution of the LRT given the null hypothesis using a parametric bootstrap. Faraway (2006) presents how this parametric bootstrap is done in R.

7.7 RANDOM INTERCEPT AND RANDOM SLOPE

In the preceding model, only the intercept β_0 was modeled per individual (the model allowed for between-individual variance in β_0). But a random effect does not need to be restricted to the intercept. Any parameter can be modeled, if the data allow. For example, in the previous model we cannot include an individual-specific difference between corticosterone and placebo treatment because each individual obtained only one treatment. Therefore, the data do not contain information about between-individual differences in the treatment effect, and it does not make sense to include such a structure in the model.

In another study on barn owls, we were interested in the effect of corticosterone on growth rate. Here, we measured wing length (as a proxy for size) at different ages of individuals that had been treated either with corticosterone or with a placebo implant. We used the slope of the regression line for wing length on age as a growth rate measure, and we were interested in the difference in this slope between corticosterone- and placebo-implanted individuals.

We expected that growth rate differed between individuals due to between-individual differences in body condition and also because the age at which the implant was implanted differed between the individuals because barn owl nestlings hatch asynchronously (the implantation was done on the same day for all the nestlings in a nest). Therefore, we modeled the slope of the regression line for each individual. Otherwise, our confidence in the (population) slope parameter would be too high (the credible intervals would be too small). This is because individual-specific slopes in the data produce a kind of pseudoreplication when equal slopes between individuals are assumed in the model (Schielzeth & Forstmeier, 2009).

The data for the growth rate study are in the wingbowl data set. In total, we have 209 measurements of 86 individuals (variable "Ring") from 24 broods.

In the model, we include "age" (as a continuous covariate), "implant" (as a fixed effect), and their interaction in the fixed part; and "ring" (= id of individual) is included as a random effect. We model both the intercept and the slope for the covariate "age" dependent on "ring." We use this notation with b_{Ring} being an individual-specific deviation from the population mean parameter value, but we add a second numerical subscript, $b_{1,Ring}$, to indicate the random intercept and $b_{2,Ring}$ for the random slope. The two random parameters are assumed to follow a multivariate normal distribution (MVNorm); that is, they are both normally distributed and are assumed to be correlated and this correlation is estimated. Hence, the formula for the random intercept and random slope model is:

$$\begin{aligned}
\widehat{y}_i &= \beta_o + b_{1,Ring_i} + (\beta_1 + b_{2,Ring_i})age_i + \beta_2 I(Implant = P) \\
&\quad + \beta_3 age_i I(Implant = P) \\
y_i &\sim Norm(\widehat{y}_i, \sigma^2) \\
\mathbf{b}_{1:2,Ring} &\sim MVNorm(\mathbf{0}, \Sigma)
\end{aligned}$$

The vector $\mathbf{b}_{Ring}$ contains the two parameters $\mathrm{b}_{1,Ring}$ and $\mathrm{b}_{2,Ring}$. The matrix Σ contains the variances of and the covariances between the intercept and the slope. The notation `(Age|Ring)` means that both the intercept and the Age-effect are grouped by Ring. We find it advisable to center and scale covariates, especially for mixed models with some complexity because noncentered covariates lead to a stronger correlation between the estimated parameters, which may cause nonconvergence of the fitting algorithm. Hence, we center and scale the variable "Age":

```
data(wingbowl)
dat <- wingbowl
dat$Age.z <- scale(dat$Age)
mod <- lmer(Wing ~ Age.z + Implant + Age.z:Implant + (Age.z|Ring),
          data=dat, REML=FALSE)
mod
Linear mixed model fit by maximum likelihood ['lmerMod']
Formula: Wing ~ Age.z + Implant + Age.z:Implant + (Age.z | Ring)
   Data: dat
      AIC       BIC    logLik   deviance   df.resid
1280.4391 1307.1778 -632.2195  1264.4391        201
```

```
Random effects:
 Groups    Name        Std.Dev.  Corr
 Ring      (Intercept) 6.394
           Age.z       1.898     -0.12
 Residual              2.542
Number of obs: 209, groups: Ring, 86
Fixed Effects:
    (Intercept)        Age.z    ImplantP      Age.z:ImplantP
        155.442       24.954       4.554               2.185
```

The estimate of the between-nestling standard deviation for the intercept is 6.4 and for the age-effect (slope), this value is 1.9. We see that there is a negative correlation between the intercept and the slope (−0.12). This is not unusual, and it would be even stronger with noncentered predictors: this means we can find different regression lines that fit the data similarly well when we increase the intercept and simultaneously decrease the slope.

We can use the diagnostic residual plots as described in Chapter 6, and the R code of the previous section, to produce a QQ plot of the random effects. Because two parameters were modeled per individual, we have to produce two different QQ plots to assess whether the random effects are normally distributed. The different parameters for each random effect are extracted from the `ranef` object using the squared brackets because the random effects are given as a matrix containing one row per individual and one column per parameter (here: intercept and age-effect).

```
qqnorm(ranef(mod)$Ring[,1]) # intercept
qqline(ranef(mod)$Ring[,1])

qqnorm(ranef(mod)$Ring[,2]) # slope
qqline(ranef(mod)$Ring[,2]) # plots not shown
```

The diagnostic residual plots did not indicate strong violation of the model assumptions; therefore, we can start drawing inferences from the model. Our question was how strongly does corticosterone affect growth rate? From the model output we see that individuals with a placebo implant grow 2.2 mm more per standard deviation of age, that is, `2.2/sd(dat$Age)` = 0.4 mm per day compared to the individuals with a corticosterone implant. We can get the 95% CrI of the parameter estimates using `sim`.

```
nsim <- 2000
bsim <- sim(mod, n.sim=nsim)
apply(bsim@fixef, 2, quantile, prob=c(0.025, 0.975))
      (Intercept)       Age.z    ImplantP  Age.z:ImplantP
2.5%     153.4488    24.00212    1.595778       0.8697857
97.5%    157.4537    25.91693    7.448484       3.4909336
```

The CrI for the interaction effect of 0.4 mm, on the original Age-scale, is:

```
quantile(bsim@fixef[,"Age.z:ImplantP"]/sd(dat$Age), prob=c(0.025,
  0.975))
2.5%   97.5%
0.164  0.657
```

Thus, given the data and the model, we are 95% sure that the wings of placebo nestlings grow between 0.16 mm and 0.66 mm faster per day than the wings of corticosterone nestlings. To be better able to assess the biological relevance of this effect, we may want to plot the two regression lines (averaged over the individuals) as well as the individual-specific regression lines (Figure 7-5). To do so, we calculate the fitted values for each age and implant combination 2000 times, each with a different set of model parameters from their posterior distribution.

```
newdat <- expand.grid(Age=seq(23, 45, length=100),
                      Implant=levels(dat$Implant))
newdat$Age.z <- (newdat$Age - mean(dat$Age))/sd(dat$Age)
Xmat <- model.matrix(~Age.z+Implant + Age.z:Implant, data=newdat)
fitmat <- matrix(ncol=nsim, nrow=nrow(newdat))
for(i in 1:nsim) fitmat[,i] <- Xmat %*% bsim@fixef[i,]
newdat$lower <- apply(fitmat, 1, quantile, prob=0.025)
newdat$upper <- apply(fitmat, 1, quantile, prob=0.975)
```

We do not extract the mean of the posterior distribution of the fitted values because we use the ML estimates for drawing the mean regression lines. These estimates are not subject to simulation error. Note that we use Age.z on the *x*-axis, but we label the *x*-axis with back-transformed values so that we are able to use the `abline` function to draw the regression lines directly from the model parameters.

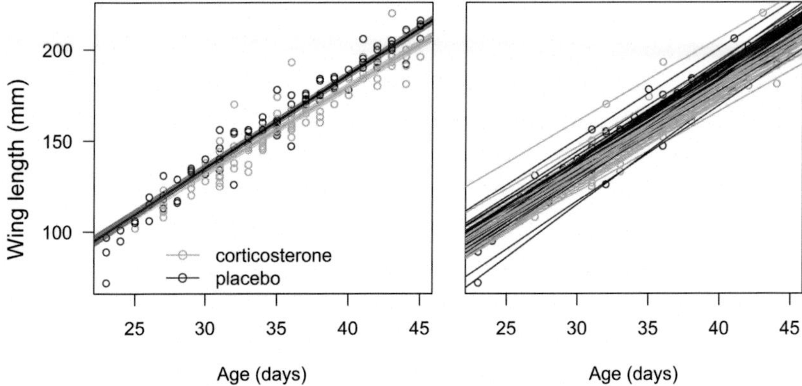

FIGURE 7-5 *Left*: Population regression lines (bold lines) with 95% credible intervals (semi-transparent color) for the corticosterone (orange) and placebo (blue) treated barn owl nestlings. *Right*: Individual-specific regression lines. Circles are the raw data.

```
par(mfrow=c(1,2), mar=c(5,1,1,0.1), oma=c(0,4,0,0))
plot(dat$Age.z, dat$Wing, pch=1, cex=0.8, las=1,
  col=c("orange", "blue")[as.numeric(dat$Implant)],
  xlab="Age (days)", ylab=NA, xaxt="n")
at.x_orig <- seq(25,45,by=5) # values on the x-axis, original scale
at.x <- (at.x_orig - mean(dat$Age))/sd(dat$Age) # transformed scale
axis(1, at=at.x, labels=at.x_orig) # original values at transformed
mtext("Wing length (mm)", side=2, outer=TRUE, line=2, cex=1.2,
  adj=0.6)
abline(fixef(mod)[1], fixef(mod)[2], col="orange", lwd=2) # for C
abline(fixef(mod)[1]+fixef(mod)[3], fixef(mod)[2]+fixef(mod)[4],
  col="blue", lwd=2)                                # for P

# add transparent polygons to visualize the 95% CrI
for(i in 1:2){
  index <- newdat$Implant==levels(newdat$Implant)[i]
  polygon(c(newdat$Age.z[index], rev(newdat$Age.z[index])),
    c(newdat$lower[index], rev(newdat$upper[index])),
    border=NA, col=c(rgb(1,0.65,0,0.5), rgb(0,0,1,0.5))[i])
}
legend(x=-1.5,y=100, c("corticosterone", "placebo"),
  pch=c(1,1),col=c("orange","blue"), bty = "n",lwd=2,cex=1,
  pt.cex=1.2)
```

To draw the individual-specific regression lines, we add the individual-specific deviations from the intercept and the slope, respectively, to the population intercept and slope. We do this in a separate plot so as not to overload the figure.

```
plot(dat$Age.z, dat$Wing, pch=1, cex=0.8, las=1,
  col=c("orange", "blue")[as.numeric(dat$Implant)],
  xlab="Age (days)", ylab=NA, yaxt="n", xaxt="n")
at.x_orig <- seq(25,45,by=5)
at.x <- (at.x_orig - mean(dat$Age))/sd(dat$Age)
axis(1, at=at.x, labels=at.x_orig)
indtreat <- tapply(dat$Implant, dat$Ring, function(x) as.
  character(x[1]))
for(i in 1:86){
  if(indtreat[i]=="C") abline(fixef(mod)[1] + ranef(mod)$Ring[i,1],
                fixef(mod)[2] + ranef(mod)$Ring[i,2],
                col="orange") else
                  abline(fixef(mod)[1] + fixef(mod)[3] +
                    ranef(mod)$Ring[i,1], fixef(mod)[2] +
                    fixef(mod)[4] + ranef(mod)$Ring[i,2],
                    col="blue")
}
```

We see a discernible effect of corticosterone on growth rate such that the wing length of corticosterone-treated nestlings is, on average, around 1 cm smaller than in placebo-treated individuals at the end of the nestling phase.

7.8 NESTED AND CROSSED RANDOM EFFECTS

Random factors can be nested or crossed. Each level of a factor that is nested within another factor occurs only in one level of the other factor (Figure 7-6, left). For example, the factor "nestling" is nested in the factor "nest" because the same nestling cannot be in two nests. In contrast, when two factors are crossed, all possible combinations of the factor levels occur in the data set (Figure 7-6, right). For example, the factors "month" and "year" are crossed, because all months occur in every year.

It is important to specify factors as nested and crossed random factors, because a falsely specified structure leads to indecipherable results. If unique level names are used (in contrast to starting with the number 1 in each group), the nested structure is defined in the data and there is no need to explicitly specify the nested design in the `lmer` (or `glmer`, see Chapter 9) function; if nestlings (or other levels nested in another factor) are not labeled uniquely (i.e., different nestlings from different nests have the same name), the nesting structure has to be specified in the model formula. The following is an example of a nested random effect and one of a crossed random effect.

In the two barn owl example data sets described earlier, the individuals were actually not independent because they were grouped in nests. Thus, we should have included nest as another random factor in the two models. Because each individual only appears in one nest, these two random factors are nested. There are two possible ways to specify nested random effects in the function `lmer`. The first is to add the factor nest ("Brood") as a second random effect in the model formula. This only fits nested random effects if the nested structure is differentiated from the names of the factor levels, because the nested structure is not explicitly defined in the model formula; all nestlings must have a unique name.

```
data(cortbowl),
dat <- cortbowl
mod <- lmer(log(totCort) ~ Implant + days + Implant:days + (1|Brood) +
  (1|Ring), data=dat, REML=FALSE)
```

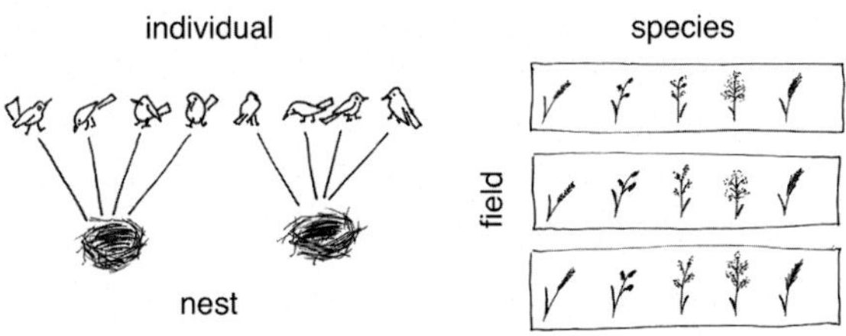

FIGURE 7-6 Nested and crossed structures of two factors. In the left panel, each level of the factor "individual" only occurs in one level of the factor "nest". These are called nested effects. In the right panel, each level of the factor "species" occurs in all levels of the factor "field". Therefore, these factors are called crossed.

```
mod
Linear mixed model fit by maximum likelihood ['lmerMod']
Formula: log(totCort) ~ Implant + days + Implant:days + (1|Brood) +
(1|Ring)
    Data: cortbowl
       AIC          BIC       logLik    deviance
  604.2934     637.2287    -293.1467    586.2934
Random effects:
  Groups    Name          Std.Dev.
  Ring      (Intercept)   0.1917
  Brood     (Intercept)   0.2486
  Residual                0.6117
Number of obs: 287, groups: Ring, 151; Brood, 54
Fixed Effects:
(Intercept)      ImplP   days2   days20   ImplP:days2   ImplP:days20
      1.953    -0.1034   1.639    0.256        -1.693         -0.076
```

The second possibility is to define explicitly the nested structure in the model formula using the "F1/F2" notation ("F2 is nested in F1").

```
mod <- lmer(log(totCort) ~ Implant + days + Implant:days +
  (1|Brood/Ring), data=dat, REML=FALSE)
mod
[..]
Groups              Name              Std.Dev.
Ring:Brood     (Intercept)             0.1917
Brood          (Intercept)             0.2486
Residual                               0.6117
[..]
```

These two models are the same, hence not all model output is repeated. The only difference is in the name of the individual random factor: it is called "Ring" in the first version, whereas in the second version it is called "Ring:-Brood," meaning "Ring nested in Brood."

To specify crossed random effects, we can only use the first specification in the previous example. For example, in Ellenberg's data (Chapter 4), six different grass species were grown in four different situations: two tanks in each of two years, and the two tanks contained different soil. This four-level factor has been called "gradient" by Hector et al. (2012). Four species were grown in all four gradients and two species were grown in only two of the gradients. Aboveground biomass was measured for 11 different water conditions within each species–gradient combination.

```
data(ellenberg)
ellenberg$gradient <- paste(ellenberg$Year, ellenberg$Soil)
table(ellenberg$Species, ellenberg$gradient)
```

	1952 Loam	1952 Sand	1953 Loam	1953 Sand
Ae	11	11	11	11
Ap	11	11	11	11
Be	11	11	11	11
Dg	11	11	11	11
Fp	11	11	0	0
Pp	11	11	0	0

We assume that the six grass species are a random sample from the family Poaceae (true grasses) and we may be interested in the general relationship between water condition (distance to ground water; using a linear and a quadratic term) and Poaceae biomass, as well as in species-specific reactions to water conditions. Therefore, we treat species as a random factor. We also include gradient as a random factor to correct for any between-gradient variance.

```
ellenberg$water.z <- as.numeric(scale(ellenberg$Water))
mod <- lmer(log(Yi.g) ~ water.z + I(water.z^2) +
            (water.z + I(water.z^2)|Species) + (1|gradient),
            data=ellenberg)
```

We did not find any suspicious pattern in the residuals and, therefore, plot the species-specific and the overall (for all Poaceae) relationship between water gradient and biomass in a figure. We see that the average biomass of Poaceae does not change with distance to ground water, but the different species react quite differently to the water condition (Figure 7-7).

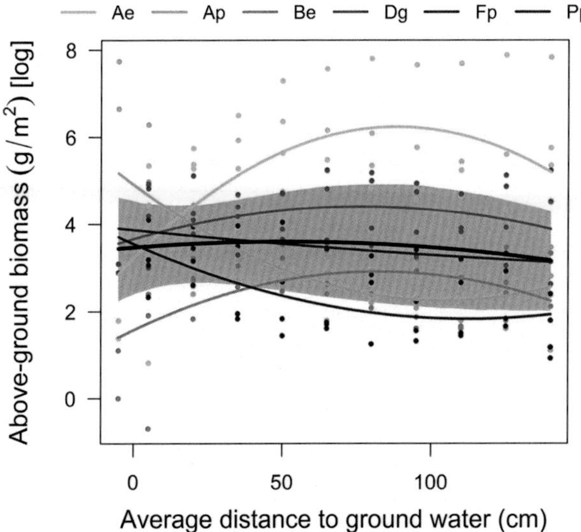

FIGURE 7-7 Average biomass of Poaceae plants (black line with 95% credible interval in gray), and the species-specific biomasses (colored lines), in relation to water condition. Dot are the raw data, each color is a different species.

7.9 MODEL SELECTION IN MIXED MODELS

Random effects are inexpensive in terms of degrees of freedom, because only one parameter per random effect is used. Further, natural processes vary on many different levels and, therefore, including random effects in a model leads to more realistic models in most cases. However, sometimes the model-fitting algorithms do not converge when the model is overloaded with random structures. Therefore, before adding a random effect to a model, be sure that the data contain some information about the specific variance parameter.

Sometimes, we would like to decide based on the data whether a random effect should be included in the model or not. This is model selection, and it is discussed in more detail in Chapter 11. However, when analyzing random factors, the following recommendations may be kept in mind: (1) As the random effects are estimated conditional on the fixed effects, model selection in the random part of the model should be done using a realistic fixed part of the model. This should include all possible predictors. (2) Random factors that are included because of the study design (e.g., subject of repeated measures, blocks) should, whenever possible, remain in the model. And (3) to get unbiased estimates for variance parameters (i.e., for the random effects) use REML.

FURTHER READING

The book by Gelman and Hill (2007) is all about hierarchical models. Pinheiro and Bates (2000) is the reference for fitting mixed models in S and R. General guidelines to build a mixed model are given in Verbeke and Molenberghs (2000). Zuur et al. (2009) give a detailed example on model selection in mixed models.

Sometimes, covariates have different effects within and between groups of measurements. Van de Pol and Wright (2009) present a simple method to distinguish such different effects using mixed models.

McCulloch and Neuhaus (2011) studied the effect of nonnormally distributed random effects in linear models and found that model predictions seem to be quite robust against the violation of the normal distribution assumption of random effects.

The wingbowl data set has been analyzed in detail in Almasi et al. (2012) taking into account that the corticosterone implant affects blood corticosterone concentration for only three days. The Ellenberg data set has been analyzed in detail, including also the effects of soil type and year, in Hector et al. (2012).

Chapter 8

Generalized Linear Models

Chapter Outline

8.1 BACKGROUND

Up to now, we have dealt with models that assume normally distributed residuals (first row in Figure 8-1). Sometimes the nature of the outcome variable makes it impossible to fulfill this assumption as might occur with binary variables (e.g., alive/dead, a specific behavior occurred/did not occur), proportions (which are confined to be between 0 and 1), or counts that cannot have negative values. For such cases, models for distributions other than the normal distribution are needed; such models are called generalized linear models (GLM). They consist of three elements: the linear predictor, the link function, f, and the error distribution.

The linear predictor is exactly the same as in normal linear models. It is a linear function that defines the relationship between the dependent and the explanatory variables. The link function transforms the expected values of the outcome variable into the range of the linear predictor, which ranges from $-\infty$ to $+\infty$. Or, perhaps more intuitively, the inverse link function transforms the

http://dx.doi.org/10.1016/B978-0-12-801370-0.00008-3.

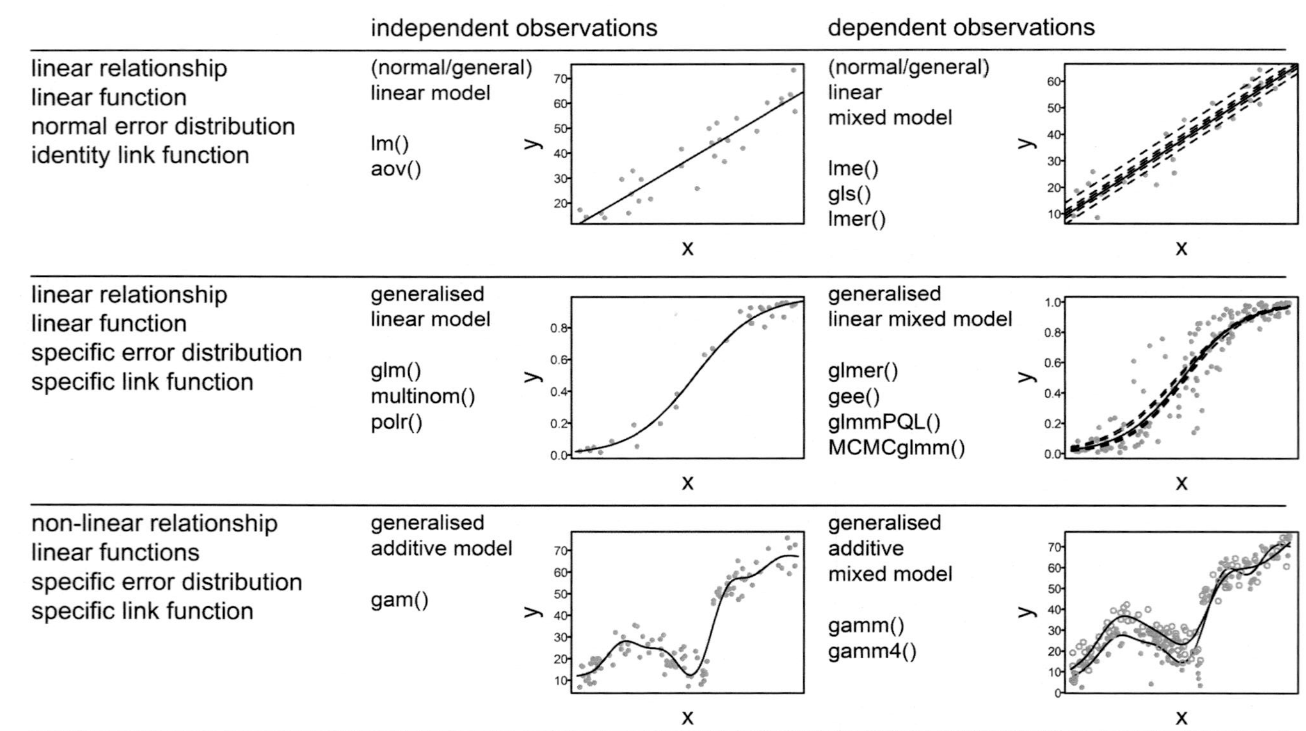

FIGURE 8-1 Overview of the six most often used types of linear models including some corresponding R functions. The first column gives the models used for independent data and the second column gives those used for dependent (e.g., hierarchical) data.

values of the linear predictor into the range of the outcome variable. Then, a specific error distribution, for example, binomial or Poisson, is used to describe how the observations scatter around the expected values. A general model formula for a generalized linear model is:

$$\mathbf{y} \sim ExpDist(\hat{\mathbf{y}}, \boldsymbol{\theta})$$

$$f(\hat{\mathbf{y}}) = \mathbf{X}\boldsymbol{\beta}$$

where *ExpDist* is a distribution of the exponential family and f is the link function. $\mathbf{y}$ are observed values of the outcome variable, $\boldsymbol{\beta}$ contains the model parameters in the linear predictor (also called the model coefficients), and $\mathbf{X}$ is the model matrix containing the values of the predictor variables. $\boldsymbol{\theta}$ is an optional vector of additional parameters needed to define the error distribution (e.g., the number of trials in the binomial distribution).

The normal linear model is a specific case of a generalized linear model, namely when *ExpDist* equals the normal distribution and f is the identity function ($f(x) = x$). Statistical distributions of the exponential family are normal, binomial, Poisson, inverse-normal, gamma, negative binomial, among others. The normal, binomial, and Poisson distributions are by far the most often used distributions. Most, but not all, data we gather in the life sciences can be analyzed assuming one of these three distributions. Thus, these are the ones we deal with in the book.

8.2 BINOMIAL MODEL

8.2.1 Background

The binomial distribution describes the distribution of the number of successful trials among a defined number of trials. For example, we define the observation of a specific behavior as a success and the lack of the behavior as a failure. Each animal is tested 10 times and the number of successful trials among the 10 tests is counted. This experiment is statistically equivalent to flipping a coin 10 times and counting the number of times the outcome is a tail. The binomial distribution is a discrete distribution and only nonnegative integers are possible. The probability of observing exactly y successful trials among n trials is:

$$p(y|n,p) = \binom{n}{y} p^y (1-p)^{n-y}$$

Figure 8-2 shows two examples of a binomial distribution. The expected value of a binomial distribution is $E(y) = np$ and the variance is $var(y) = np(1 - p)$. Note that the variance is defined by n and p, that is, there is no separate variance parameter. This is something we have to be careful about when fitting a binomial model to data. Often, real data show a higher

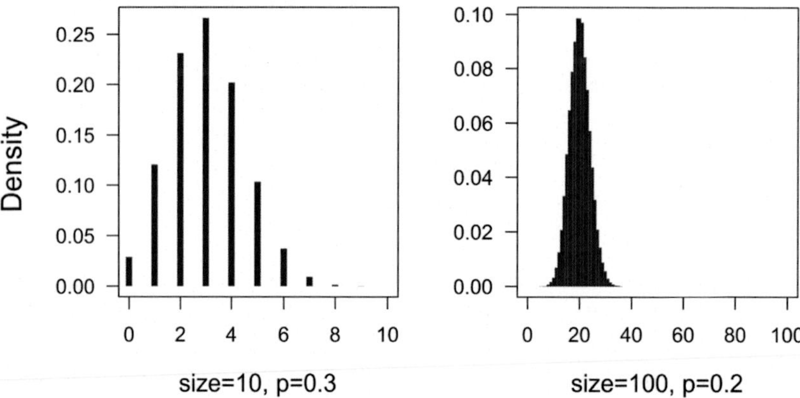

FIGURE 8-2 Two examples of a binomial distribution. Size: number of trials (the argument in the corresponding R functions, for example in rbinom, is called "size"), *p*: success probability.

variance than assumed by the binomial distribution. This is called overdispersion (see Section 8.2.3). Another characteristic of the binomial distribution is that its variance is highest when $p = 0.5$ and the variance reaches 0 when p is 0 or 1.

When fitting a binomial model to data, we have to estimate p. Often we are interested in correlations between p and one or several explanatory variables. Therefore, we model p as linearly dependent on the explanatory variables. Because the values of p are squeezed between 0 and 1 (because it is a probability), p is transformed by the link-function before the linear relationship is modeled.

$$\textit{link-function}(\widehat{\mathrm{p}}) = \mathbf{X}\widehat{\boldsymbol{\beta}}$$

Functions that can transform a probability into the scale of the linear predictor ($-\infty$ to $+\infty$) are, for example, logit, probit, cloglog, or cauchit. These link functions differ slightly in the way they link the outcome variable to the explanatory variables (Figure 8-3). The logit link function is the most often used link function in binomial models. However, sometimes another link function might fit the data better.

8.2.2 Fitting a Binomial Model in R

As an example, we use data from a study on the effects of anthropogenic fire regimes traditionally applied to savanna habitat in Gabon, Central Africa (Walters, 2012). Young trees survive fires better or worse depending, among other factors, on the fuel load, which, in turn, depends heavily on the time since the last fire happened. Thus, plots were burned after different lengths of time since the previous fire (4, 9, or 12 months ago). Trees that resprouted after the previous (first) fire were counted before and after the experimental

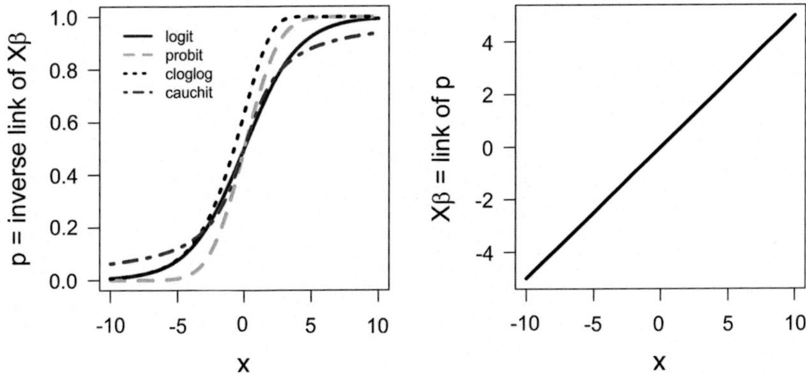

FIGURE 8-3 Left panel: Shape of different link functions commonly used in binomial models. Right panel: The relationship between the predictor X (x-axis) and p on the scale of the link function (y-axis) is assumed to be linear.

(second) fire to estimate their survival of the experimental fire depending on the time since the previous fire.

The outcome variable is the number of surviving trees among the total number of trees per plot. The explanatory variable is the time since the previous fire, a factor with three levels: "4m", "9m", and "12m". Assuming that the data follow a binomial distribution, the following model can be fitted to the data:

$$\mathrm{p}(y_i|p_i, n_i) \sim Binom(p_i, n_i)$$

$$\mathrm{logit}(p_i) = \beta_0 + \beta_1 I(treatment_i = 9\mathrm{m}) + \beta_2 I(treatment_i = 12\mathrm{m})$$

where $\mathrm{logit}(p_i) = \log(p_i/(1 - p_i))$, y_i = number of successes (i.e., survivors in our example) for plot i, p_i = probability of success for observation i (depending on the predictor values measured for observation i), n_i = number of trials in plot i (here, the number of trees on plot i before the experimental fire; not to be confused with the sample size of the data, here the number of plots: the latter is the number of rows in the data set, usually denoted "*n*" or "*N*"; hence, the index i makes a difference).

To fit the model, we use the function `glm`. It uses the "iteratively reweighted least-squares method" which is an adaptation of the LS method for fitting generalized linear models. With the argument `family` the error distribution and the link function can be specified. For a binomial model the logit function is used as the default link function. To change the link function we specify, for instance, `family=binomial(link=cloglog)`. The available families and their link functions are listed in Table 8-1.

The outcome variable has to be given as a matrix with two columns. The first column contains the number of successes (number of survivors) and the second column contains the number of failures (number of trees killed by the fire). We build this matrix using `cbind` ("column bind").

TABLE 8-1 Families for the glm-Function with Their Link Functions

	Family					
Link	**Binomial**	**Gamma**	**Gaussian**	**Inverse-Gaussian**	**Poisson**	**Negative-binomial**
logit	D					
probit	x					
cloglog	x					
identity		x	D		x	x
inverse		D				
log		x			D	D
$1/\mu^2$				D		
sqrt						x
cauchit	x				x	
exponent (μ^a)			x			

Note: R defaults are indicated with D; x = other commonly used link functions for the error distribution of the corresponding family. The negative-binomial family is in the library MASS. All others are in "stats" which is loaded automatically when R is started.
Source: Table modified from Venables and Ripley (2002).

```
data(resprouts)
dat <- resprouts
mod <- glm(cbind(succ, fail) ~ treatment, data=dat,
           family=binomial)

mod

Call:  glm(formula = cbind(succ, fail) ~ treatment,
           family = binomial, data = dat)

Coefficients:
(Intercept)  treatment9m  treatment12m
-1.241       1.159        -2.300

Degrees of Freedom: 40 Total (i.e. Null); 38 Residual
Null Deviance:      845.8
Residual Deviance: 395    AIC: 514.4
```

Experienced readers will be alarmed because the residual deviance is much larger than the residual degrees of freedom, which indicates overdispersion. We will soon discuss overdispersion, but, for now, we continue with the analysis for the sake of illustration.

The estimated model parameters are $\widehat{\beta}_0 = -1.24$, $\widehat{\beta}_1 = 1.16$, and $\widehat{\beta}_2 = -2.30$. These estimates tell us that tree survival was higher for the 9-month fire lag treatment compared to the 4-month treatment (which is the reference level), but lowest in the 12-month treatment. To obtain the mean survival probabilities per treatment, some math is needed because we have to back-transform the linear predictor to the scale of the outcome variable. The mean survival probability for the 4-month treatment is $\text{logit}^{-1}(-1.24) = \frac{e^{-1.24}}{1+e^{-1.24}} = 0.22$, for the 9-month treatment it is $\text{logit}^{-1}(-1.24 + 1.16) = 0.48$, and for the 12-month treatment it is $\text{logit}^{-1}(-1.24 - 2.30) = 0.03$. The function `plogis` gives the inverse of the logit function and can be used to estimate the survival probabilities, for example:

```
plogis(-1.24+1.16)
[1] 0.4800107
```

The direct interpretation of the model coefficients β_1 and β_2 is that they are the log of the ratio of the odds of two treatment levels (i.e., the log odds ratio). The odds for treatment "4 months" are 0.22/(1 – 0.22) = 0.29 (calculated using nonrounded values), which is the estimated ratio of survived to killed trees in this treatment. For treatment "9 months," the odds are 0.48/(1 – 0.48) = 0.92, and the log odds ratio is log(0.92/0.29) = 1.1590 = β_1.

The model output includes the null deviance and the residual deviance. Deviance is a measure of the difference between the data and a model. It corresponds to the sum of squares in the normal linear model. The smaller the residual deviance the better the model fits to the data. Adding a predictor reduces the deviance, even if the predictor does not have any relation to the outcome variable. The Akaike information criterion (AIC) value in the model output (last line) is a deviance measure that is penalized for the number of model parameters. It can be used for model comparison or model averaging (Chapter 11).

The residual deviance is defined as minus two times the difference of the log-likelihoods of the saturated model and our model. The saturated model is a model that uses the observed proportion of successes as the success probability for each observation $y_i \sim Binom(y_i/n_i, n_i)$. The saturated model has the highest possible likelihood (given the data set and the binomial model). This highest possible likelihood is compared to the likelihood of the model at hand, $y_i \sim Binom(p_i, n_i)$ with p_i dependent on some predictor variables. The null deviance is minus two times the difference of the log-likelihoods of the saturated model, and a model that contains only one overall mean success probability, the null model $y_i \sim Binom(p, n_i)$. The null deviance corresponds to the total sum of squares, that is, it is a measure of the total variance in the data.

8.2.3 Assessing Model Assumptions: Overdispersion and Zero-Inflation

As for the normal linear model, the binomial model assumes that the residuals are independent and identically distributed (iid). Independent means that every observation i is independent of the other observations. Particularly, there are no groups in the data and no temporal or spatial correlation. It is important to note that the number of successes is a single observation. In the preceding example, the number of surviving trees in one plot is a single observation and, thus, the trees in that plot need not to be independent (by definition they are not independent since they are all from the same plot). In binomial models, residuals are only identically distributed conditional on the fitted success probabilities and the number of trials. A critical violation of the binomial distribution assumption is overdispersion (see later). Further, a linear relationship between the outcome variable and the predictors is assumed on the scale of the link function. The assessment of the assumptions is more difficult than for a normal linear model. But it is equally important!

One difficulty is that different types of residuals exist. The standard residual plots obtained by `plot(mod)` produce the same four plots as for an `lm` object (refer to Figure 6-1), but it uses the deviance residuals for the first three plots (residuals versus fitted values, QQ plot, and residual variance versus fitted values) and the Pearson's residuals for the last (residuals versus leverage). The deviance residuals are the contribution of each observation to the deviance of the model. This is the default type when the residuals are extracted from the model using the function `resid`. The Pearson's residual for observation i is the difference between the observed and the fitted number of successes divided by the standard deviation given the number of trials and the fitted success probability: $\varepsilon_i = \frac{y_i - n_i\widehat{p_i}}{\sqrt{n_i\widehat{p_i}(1-\widehat{p_i})}}$. Other types of residuals are "working," "response," or "partial" (see Davison & Snell, 1991). For the residual plots, R chooses the type of residuals so that each plot should look roughly like the analogous plot for the normal linear model. However, in most cases the plots look awkward due to the discreteness of the data, especially when success probabilities are close to 0 or 1. We recommend thinking about why they do not look perfect; with experience, serious violations of model assumptions can be recognized. But often posterior predictive model checking or graphical comparison of fitted values to the data are better suited to assess model fit in GLMs (Chapter 10).

In our example data there are obviously a number of influential points (especially the data points with row numbers 7, 20, and 26; Figure 8-4). The corresponding data points may be inspected for errors, or additional predictors may be identified that help to explain why these points are extreme (Are they close/far from the village? Were they grazed? etc.). For whatever reason, the variance in the data is larger than assumed by the binomial distribution. We detect this higher variance in the mean of the absolute values of the standardized residuals that is clearly larger than one. This is overdispersion, which we mentioned earlier and deal with next.

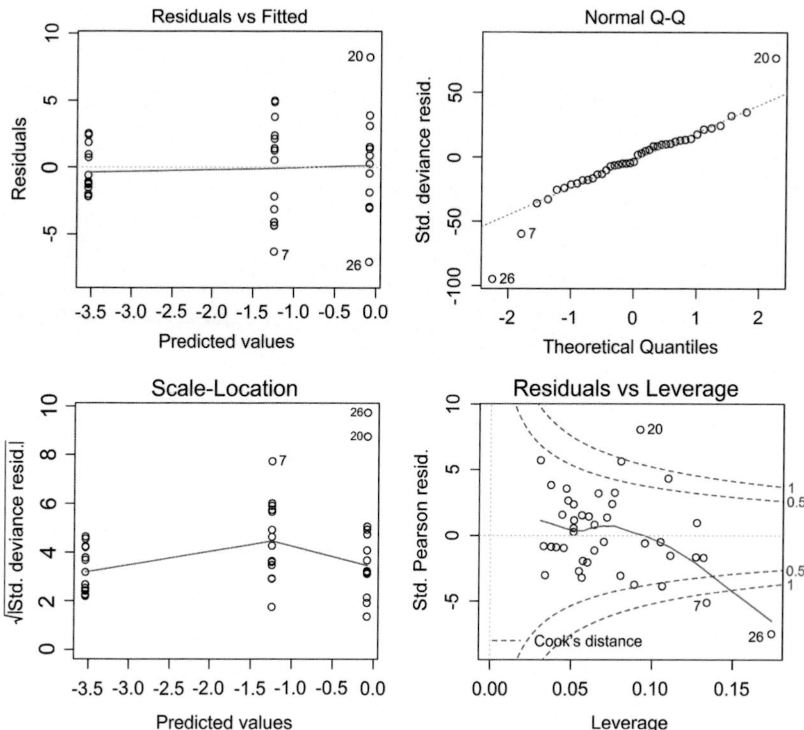

FIGURE 8-4 The four standard residual plots for a binomial generalized linear model obtained using the `plot` function.

The binomial model does not contain any variance parameter, because the variance is determined by the n_i and p_i; in our example the latter is fully defined by β_0, β_1, and β_2: $p_i = \text{logit}^{-1}(\beta_0 + \beta_1 \text{I}(\text{treatment}_\text{i} = \text{"9m"}) + \beta_2 \text{I}(\text{treatment}_\text{i} = \text{"12m"}))$, and n_i is part of the data. Similarly, in a Poisson model (which we will introduce in the next chapter) the variance is defined by the mean. Unfortunately, real data—as in our example—often show higher and sometimes lower variance than expected by a binomial (or a Poisson) distribution (Figure 8-5). When the variance in the data is higher than expected by the binomial (or the Poisson) distribution we have overdispersion. The uncertainties for the parameter estimates will be underestimated if we do not take overdispersion into account. Overdispersion is indicated when the residual deviance is substantially larger than the residual degrees of freedom. This always has to be checked in the output of a binomial or a Poisson model. In our example, the residual deviance is 10 times larger than the residual degrees of freedom, thus, we have strong overdispersion.

What can we do when we have overdispersion? The best way to deal with overdispersion is to find the reason for it. Overdispersion is common in

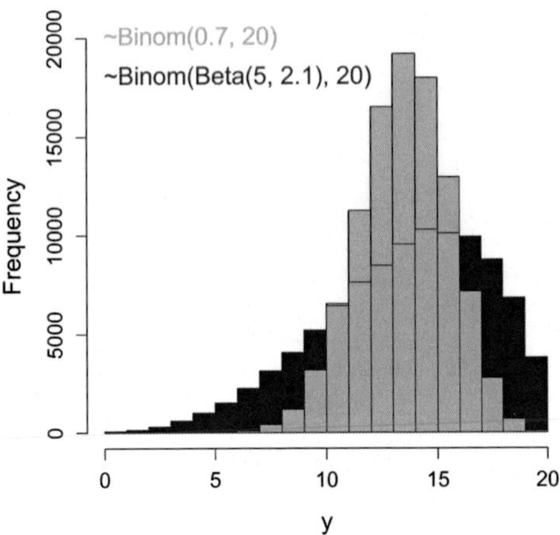

FIGURE 8-5 Histogram of a binomial distribution without overdispersion (orange) and one with the same total number of trials and average success probability, but with overdispersion (blue).

biological data because animals do not behave like random objects but their behavior is sensitive to many factors that we cannot always measure such as social relationships, weather, habitat, experience, and genetics. In most cases, overdispersion is caused by influential factors that were not included in the model. If we find them and can include them in the model (as fixed or as random variables) overdispersion may disappear. If we do not find such predictor variables, we have at least two options.

Option 1

Fit a quasibinomial or quasi-Poisson model by specifying "quasibinomial" or "quasipoisson" in the family-argument:

```
mod <- glm(cbind(succ,fail) ~ treatment, data=dat,
          family=quasibinomial)
```

This will fit a binomial model that estimates, in addition to the other model parameters, a dispersion parameter, ω, that is multiplied by the binomial or Poisson variance to obtain the residual variance: $\text{var}(y) = \omega np(1 - p)$, or $\text{var}(y) = \omega\lambda$, respectively.

This inflated variance is then used to obtain the standard errors of the parameter estimates, thus the posterior distributions produced by `sim` will have larger variances. However, the quasibinomial distribution is an unnatural distribution (there is no physical justification for it, such as number of coin flips that are tails among a defined number of coin flips). It is a binomial or a Poisson distribution that is stretched so that its variance parameter fits to the variance in the data without having a formal model for this variance. Therefore, we prefer to use the second option that explicitly models the additional variance.

Option 2

Add an observation-level random factor (i.e., a factor with the levels 1 to n, the sample size). This allows and accounts for extra variance in the data. To do that, we have to fit a generalized linear mixed model (GLMM) instead of a GLM (Chapter 9).

What do we have to do when the residual deviance is smaller than the residual degrees of freedom, that is, when we have "underdispersion"? Some statisticians do not bother about underdispersion, because, when the variance in the data is smaller than assumed by the model, uncertainty is overestimated. This means that conclusions will be conservative (i.e., on the "safe" side). However, we think that underdispersion should bother us as biologists (or other applied scientists). In most cases, underdispersion means that the variance in the data is smaller than expected by a random process, that is, the variance may be constrained by something. Thus, we should be interested in thinking about the factors that constrain the variance in the data. An example is the number of surviving young in some raptor species, (e.g., in the lesser spotted eagle *Aquila pomarina*). Most of the time two eggs are laid, but the first hatched young will usually kill the second (which was only a "backup" in case the first egg does not yield a healthy young). Because of this behavior, the number of survivors among the number of eggs laid will show much less variance than expected from n_i and p_i, leading to underdispersion. Clutch size is another example of data that often produces underdispersion (but it is a Poisson rather than a binomial process, because there is no n_i).

Sometimes, apparent under- or overdispersion can be caused by too many 0s in the data than assumed by the binomial or Poisson model. For example, the number of black stork *Ciconia nigra* nestlings that survived the nestling phase is very often 0, because the whole nest was depredated or fell from the tree (black storks nest in trees). If the nest survives, the number of survivors varies between 0 and 5 depending on other factors such as food availability or weather conditions. A histogram of these data shows a bimodal distribution with one peak at 0 and another peak around 2.5. It looks like a Poisson distribution, but with a lot of additional 0 values. This is called zero-inflation. Zero-inflation is often the result of two different processes being involved in producing the data.

The process that determines whether a nest survives differs from the process that determines how many nestlings survive, given the nest survives. When we analyze such data using a model that assumes only one single process it will be very hard to understand the system and the results are likely to be biased because the distributional assumptions are violated. In such cases, we will be more successful when our model explicitly models the two different processes. Such models are zero-inflated binomial or zero-inflated Poisson models. We will analyze the black stork nestling example using a zero-inflated Poisson mixed model in Chapter 14. When all predictors can be treated as fixed effects, the R functions from the package pscl can be used to fit zero-inflated binomial or Poisson models (Zeileis et al., 2008).

Zero-inflation typically occurs in count data. However, it can also occur in continuous measurements. For example, the amount of rain per day measured in mm is very often zero, and, when it is not zero, it is a number following a specific (possibly normal) continuous distribution. Such data may be analyzed using tobit models (Tobin, 1958). Several R packages provide tobit models, such as censReg (Henningsen, 2013), AER (Kleiber & Zeileis, 2008), and MCMCpack (Martin et al., 2011).

8.2.4 Drawing Conclusions

For the moment, we use the quasibinomial GLM to analyze the tree sprout data. To draw inferences, we simulate 2000 values from the joint posterior distribution of the model parameters.

```
nsim <- 2000
bsim <- sim(mod, n.sim=nsim)      # simulate from the posterior distr.
```

For each set of simulated model parameters, we derive the linear predictor by multiplying the model matrix with the corresponding set of model parameters. Then, the inverse logit function ($\text{logit}^{-1}(x) = e^x/(1 + e^x)$; R function `plogis`) is used to obtain the fitted value for each fire lag treatment.

```
newdat <- data.frame(treatment=
              factor(c("4m","9m","12m"),levels=c("4m","9m","12m")))
Xmat   <- model.matrix(~treatment, newdat)
fitmat <- matrix(nrow=nrow(newdat), ncol=nsim)
for(i in 1:nsim) fitmat[,i] <- plogis(Xmat %*% bsim@coef[i,])
```

Lastly, we extract, for each treatment level, the 2.5% and 97.5% quantile of the posterior distribution of the fitted values and plot it together with the estimates (the fitted values) per treatment and the raw data (Figure 8-6).

```
newdat$lwr <- apply(fitmat, 1, quantile, prob=0.025)
newdat$upr <- apply(fitmat, 1, quantile, prob=0.975)
newdat$fit  <- plogis(Xmat %*% coef(mod))  # fitted values
```

The posterior probability of the hypothesis that tree sprout survival is higher when the experimental fire takes place 9 months compared to only 4 months after the last fire can be obtained from the object that we called "fitmat". This 3×2000 matrix contains in each column a set of three reasonable survival probabilities (one for each fire lag treatment), given the data. The proportion of simulations in which the survival of treatment "9m" is higher than that of treatment "4m" is the posterior probability that "9m" has higher tree survival than "4m". Because "9m" had a higher survival probability than "4m" in 1998 out of 2000 simulations, we conclude that this hypothesis is very strongly supported. We can do the same calculations for any comparison.

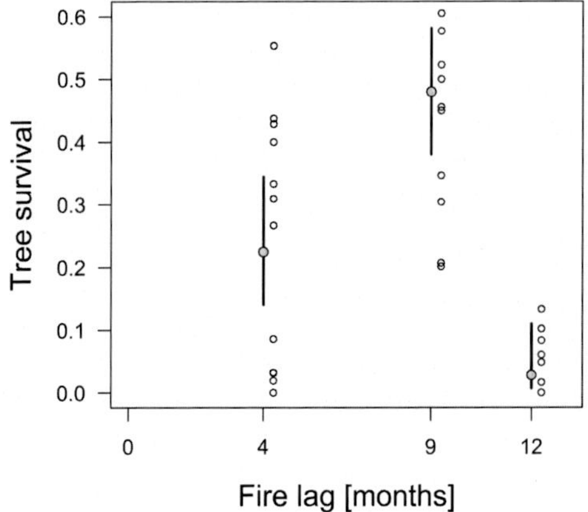

FIGURE 8-6 Proportion of surviving trees (circles) for three fire lag treatments with estimated mean proportion of survivors using a quasibinomial model. Gray dots = fitted values, black vertical bars = 95% credible intervals.

```
sum(fitmat[2,] > fitmat[1,])/nsim  # probability that survival 9m>4m
[1]   0.999
sum(fitmat[1,] > fitmat[3,])/nsim  # probability that survival 4m>12m
[1]   0.999
```

8.2.5 Frequentist Results

A frequentist likelihood ratio test can be based on the difference between deviances. To compare the null model with the model containing treatment as a predictor variable, the difference between the null deviance and the residual deviance can be compared to a χ^2-distribution with degrees of freedom equal to the difference in the number of parameters between the two models. This gives the p-value of the likelihood ratio test:

```
1 - pchisq(854.84-395.03, df=2)
[1] 0
```

Alternatively, the functions `anova` or `drop1` could be used to obtain sequential and marginal likelihood ratio tests, respectively. However, the likelihood ratio test is not reliable when overdispersion is present. See, for example, Crawley (2007) or Bolker et al. (2008) for recommendations on how to do likelihood ratio tests in quasibinomial or quasi-Poisson models. The functions `anova` and `drop1` produce similar ANOVA tables as for `lm` objects (see Chapter 4), however they contain deviances instead of sums of squares.

In cases with only one predictor (as in our example), the sequential and marginal likelihood ratio tests produce the same results. Up to now, we have not shown how to obtain fitted values with their 95% confidence intervals using frequentist methods. We should do this at least once in this book because it can be handy to very quickly assess parameter uncertainty. We obtain fitted values using the function `predict` and applying some math if a link function has been used. The latter is needed because the function `predict` applied to a glm object does not provide confidence intervals. We first predict in the scale of the linear predictor, that is logit($\widehat{p}_i$),

```
newdat$logitfit <- predict(mod, newdata=newdat)
```

and extract the standard error for this estimate.

```
newdat$logitfit.se <- predict(mod, newdata=newdat,
                              se.fit=TRUE)$se.fit
```

Second, the lower and upper limit of the 95% confidence interval of the logit($\widehat{p}_i$) are calculated by subtraction and addition, respectively, of 2 (or 1.96) times the standard error.

```
newdat$logitfit.lwr <- newdat$logitfit - 2*newdat$logitfit.se
newdat$logitfit.upr <- newdat$logitfit + 2*newdat$logitfit.se
```

At last, we back-transform the lower and upper limit of the 95% confidence interval to the proportion scale:

```
newdat$lwr <- plogis(newdat$logitfit.lwr)
newdat$upr <- plogis(newdat$logitfit.upr)
```

The 95% confidence intervals of proportions are asymmetric, especially when the mean is close to 0 or 1. The estimates and intervals will be the same as for the Bayesian inference when flat priors have been used as in `sim` (see Figure 8-6).

8.3 FITTING A BINARY LOGISTIC REGRESSION IN R

If the outcome variable can only take one of two values (e.g., a species is present or absent, or the individual survived or died; often coded as 1 or 0) we use a logistic regression. It corresponds to a generalized linear model with binomial error distribution of which the number of trials equals 1 and the probability of obtaining a 1 equals p. Now, the binomial distribution can be written as $p(y|p) = p^y(1-p)^{1-y}$. This is the Bernoulli distribution: *Binom*(p, 1) = *Bernoulli*(p). The expected value of a Bernoulli-distributed variable is $E(y) = p$ and its variance is $var(y) = p(1-p)$. As for the binomial distribution, the variance of the Bernoulli distribution is defined by p and no separate variance parameter exists. However, because the data can only take the values 0 and 1, there is no possibility that the data can show a higher variance than $p(1-p)$. Therefore, we do not have to worry about overdispersion when the outcome variable is binary.

A binary logistic regression is fitted in R in exactly the same way as a binomial model except that the outcome variable consists of the values 0 and 1 rather than a two-column matrix giving the number of successes and failures. As an example, we use presence-absence data of little owls *Athene noctua* in nest boxes during the breeding season. The original data are published in Gottschalk et al. (2011); here we use only parts of these data. The variable "PA" contains the presence of a little owl: 1 indicates a nestbox used by little owls, whereas 0 stands for an empty nestbox. The variable "elevation" has the elevation in meters above sea level. We are interested in how the presence of the little owl is associated with elevation within the study area, that is, how the probability of presence changes with elevation. Our primary interest, therefore, is the slope β_1 of the regression line.

$$y_i \sim Bernoulli(p_i)$$

$$\text{logit}(p_i) = \beta_0 + \beta_1 \text{ elevation}_i, \text{ where } \text{logit}(p_i) = \log(p_i/(1-p_i))$$

To fit this model, we use the `glm` function with family = binomial:

```
mod <- glm(PA~elevation, data=dat, family=binomial)
```

Note, if we forget the `family` argument, we fit a normal linear model, and there is no warning by R!

```
mod
Call:  glm(formula = PA ~ elevation, family = binomial, data = dat)
Coefficients:
(Intercept)   elevation
   0.579449   -0.006106

Degrees of Freedom: 360 Total (i.e. Null); 359 Residual
Null Deviance:   465.8
Residual Deviance: 445.6   AIC: 449.6
```

The residual plots normally look quite awful because the residual distribution very often has two peaks, a negative and a positive one resulting from the binary nature of the outcome variable. However, it is still good to have a look at these plots using `plot(mod)`. At least the average should roughly be around zero and not show a trend. An often more informative plot to judge model fit for a binary logistic regression is to compare the fitted values with the data. To better see the observations, we slightly jitter them in the vertical direction.

```
plot(fitted(mod), jitter(dat$PA, amount=0.05), xlab="Fitted values",
     ylab="Probability of presence", las=1, cex.lab=1.2, cex=0.8)
```

If the model would fit the data well, the data would be, on average, equal to the fitted values. Thus, we add the $y = x$-line to the plot using the `abline` function with intercept 0 and slope 1.

```
abline(0,1, lty=3)
```

Of course, binary data cannot lie on this line because they can only take on the two discrete values 0 or 1. However, the mean of the 0 and 1 values should lie on the line if the model fits well. Therefore, we calculate the mean for suitably selected classes of fitted values. In our example, we choose a class width of 0.1. Then, we calculate means per class and add these to the plot, together with a classical standard error that tells us how reliable the means are. This can be an indication whether our arbitrarily chosen class width is reasonable.

```
t.breaks <- cut(fitted(mod), seq(0,1, by=0.1))
means  <- tapply(dat$PA, t.breaks, mean)
semean <- function(x) sd(x)/sqrt(length(x))
means.se <- tapply(dat$PA, t.breaks, semean)
points(seq(0.05, 0.95, by=0.1), means, pch=16, col="orange")
segments(seq(0.05, 0.95, by=0.1), means-2*means.se,
    seq(0.05, 0.95,by=0.1), means+2*means.se,lwd=2, col="orange")
```

The means of the observed data do not fit well to the data (Figure 8-7, left panel). For low presence probabilities, the model overestimates presence probabilities whereas, for medium presence probabilities, the model underestimates presence probability. This indicates that the relationship between little owl presence and elevation may not be linear. After including polynomials up to the fourth degree, we obtained a reasonable fit (Figure 8-7, right panel).

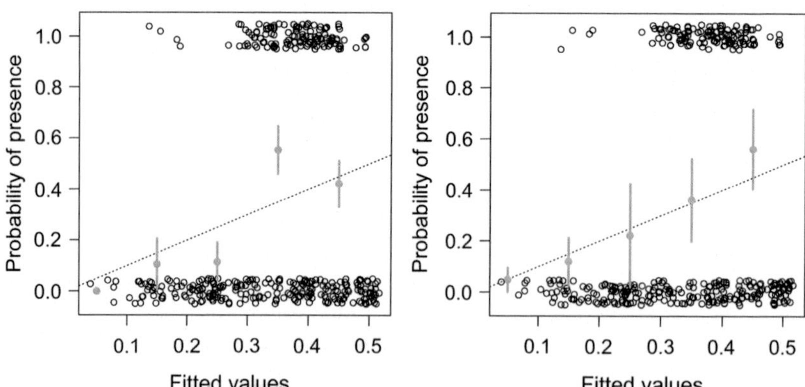

FIGURE 8-7 Goodness of fit plots. Observed little owl presence versus fitted values. Open circles = observed presence (1) or absence (0) jittered in the vertical direction; orange dots = mean (and 95% confidence intervals given as vertical bars) of the observations within classes of width 0.1 along the *x*-axis. The dotted line indicates perfect coincidence between observation and fitted values. *Left*: a model including only a linear trend for elevation; *right*: a model including polynomial effects of elevation.

Now, we are ready to draw inferences. We can simulate the posterior distribution of β_1 and obtain the 95% CrI. (The R output is slightly edited.)

```
mod <- glm(PA~elevation + I(elevation^2) + I(elevation^3) +
          I(elevation^4), data=dat, family=binomial)

bsim <- sim(mod, n.sim=2000)
apply(bsim@coef, 2, mean)
   (Intcpt)      elev.   I(elev.^2)  I(elev.^3)  I(elev.^4)
   -24.346      0.397       -0.002       0.000       0.000
apply(bsim@coef, 2, quantile, prob=c(0.025, 0.975))
        (Intcpt)    elev.  I(elev.^2) I(elev.^3)  I(elev.^4)
2.5%    -35.966    0.191      -0.004      0.000       0.000
97.5%   -13.054    0.607      -0.001      0.000       0.000
```

To interpret this polynomial function, an effect plot is helpful. To that end, and as we have done before, we calculate fitted values over the range of the covariate, together with credible intervals:

```
newdat <- data.frame(elevation = seq(80,600,by=1))
Xmat <- model.matrix(~elevation+ I(elevation^2) + I(elevation^3) +
  I(elevation^4), data=newdat) # the model matrix
fitmat <- matrix(nrow=nrow(newdat), ncol=nsim)
for(i in 1:nsim) fitmat[,i] <- plogis(Xmat %*% bsim@coef[i,])
newdat$lwr <- apply(fitmat,1,quantile,probs=0.025)
newdat$fit <- plogis(Xmat %*% coef(mod))
newdat$upr <- apply(fitmat,1,quantile,probs=0.975)
```

We now can plot the data together with the estimate and its credible interval. We, again, use the function `jitter` to slightly scatter the points along the y-axis to make overlaying points visible.

```
plot(dat$elevation, jitter(dat$PA, amount=0.05), las=1,
  cex.lab=1.4, cex.axis=1.2, xlab="Elevation",
  ylab="Probability of presence")
lines(newdat$elevation, newdat$fit, lwd=2)
lines(newdat$elevation, newdat$lower, lty=3)
lines(newdat$elevation, newdat$upper, lty=3)
```

We see that the little owl presence probability is highest at elevations just below 200 m. Presence probability sharply decreases for lower and higher elevations (Figure 8-8). Because we only looked at one predictor variable among many potentially relevant variables, the results may be confounded with other predictors and may have little general validity.

8.3.1 Some Final Remarks

Binary data do not contain a lot of information. Therefore, large sample sizes are needed to obtain robust results.

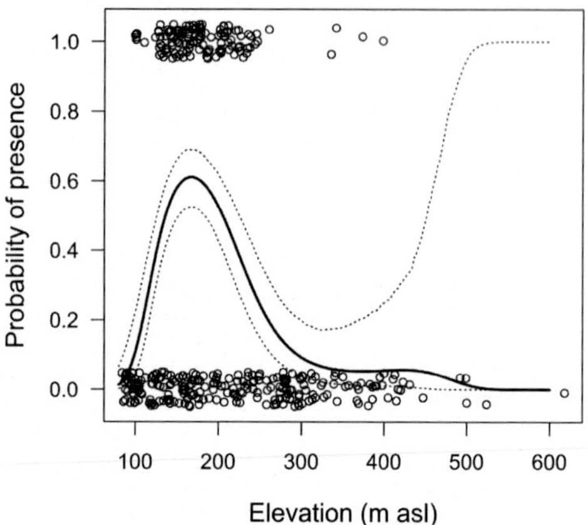

FIGURE 8-8 Little owl presence data versus elevation with regression line and 95% CrI (dotted lines). Open circles = observed presence (1) or absence (0) jittered in the vertical direction.

Often presence/absence data are obtained by visiting plots several times during a distinct period, for example, a breeding period, and then it is reported whether a species has been seen or not. If it has been seen and if there is no misidentification in the data, it is present, however, if it has not been seen we are usually not sure whether we have not detected it or whether it is absent. In the case of repeated visits to the same plot, it is possible to estimate the detection probability using occupancy models (MacKenzie et al., 2002) or point count models (Royle, 2004). In Section 14.3 we show how to fit an occupancy model using Stan.

Finally, logistic regression can be used in the sense of a discriminant function analysis that aims to find predictors that discriminate members of two groups (Anderson, 1974). However, if one wants to use the fitted value from such an analysis to assign group membership of a new subject, one has to take the prevalence of the two groups in the data into account. See Further Reading at the end of this chapter for methods such as cross-validation and area under the curve (AUC) used to discriminate two groups based on logistic regression.

8.4 POISSON MODEL

8.4.1 Background

The Poisson distribution is a discrete probability distribution that naturally describes the distribution of count data. If we know how many times something happened, but we do not know how many times it did not happen (in contrast to the binomial model, where we know the number of trials), such

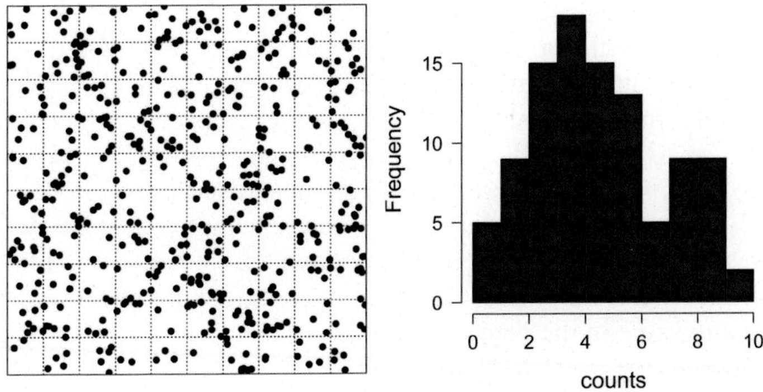

FIGURE 8-9 A natural process that produces Poisson distributed data is the number of raindrops falling (at random) into equally sized cells of a grid. *Left*: spatial distribution of raindrops, *right*: corresponding distribution of the number of raindrops per cell.

counts usually follow a Poisson distribution. Count data are positive integers ranging from 0 to ∞. A natural process that produces Poisson distributed data, for example, is the number of raindrops falling into cells of equal sizes (Figure 8-9).

The expected value of a Poisson distribution (e.g., the mean number of raindrops per cell) is denoted by λ and the probability of counting exactly x (e.g., raindrops in a cell) is given by

$$\mathrm{p}(x|\lambda) = \frac{\lambda^x}{x!}e^{-\lambda}$$

Instead of this formula we normally write the short version "$\sim$*Pois*" or use the R function `dpois`. One property of the Poisson distribution is that its variance equals the expected value: $\mathrm{var}(x) = \lambda$. This property can be used to quickly assess whether the spatial distribution of observations, for example, nest locations, is clustered, random, or equally spaced. Animal locations could be clustered due to coloniality, social, or other attraction. More equally spaced location may be due to territoriality.

Let x be the number of observations per grid cell; if $\mathrm{var}(x)/\mathrm{mean}(x) >> 1$ the observations are clustered, whereas if $\mathrm{var}(x)/\mathrm{mean}(x) << 1$, the observations are more equally spaced than expected by chance. Clustering will lead to overdispersion in the counts whereas more equally spaced locations will lead to underdispersion.

A Poisson distribution is positive-skewed (long tail to the right) if λ is small and it approximates a normal distribution for large λ. The Poisson distribution constitutes the stochastic part of a Poisson model. The deterministic part describes how λ is related to predictors. λ can only take on positive values. Therefore, we need a link function that transforms λ into the scale of the linear predictor (or, alternatively, an inverse link function that transforms the value from the linear predictor to nonnegative values). The most often used link function is the natural logarithm (log-link function).

This link function transforms all λ-values between 0 and 1 to the interval $-\infty$ to 0, and all λ-values higher than 1 are projected into the interval 0 to $+\infty$. Sometimes, the identity link function is used instead of the log-link function, particularly when the predictor variable only contains positive values and the effect of the predictor is additive rather than multiplicative, that is, when a change in the predictor produces an addition of a specific value in the outcome rather than a multiplication by a specific value. Further, the cauchit function can also be used as a link function for Poisson models.

8.4.2 Fitting a Poisson-Model in R

As an example, we fit a Poisson model with log-link function to a simulated data set containing the number of (virtual) aphids on a square centimeter (y) and a numeric predictor variable representing, for example, an aridity index (x, Figure 8-10). Real ecological data without overdispersion or zero-inflation and with no random structure are rather rare, thus we illustrate this model, which is the basis for more complex models, with simulated data. The model is:

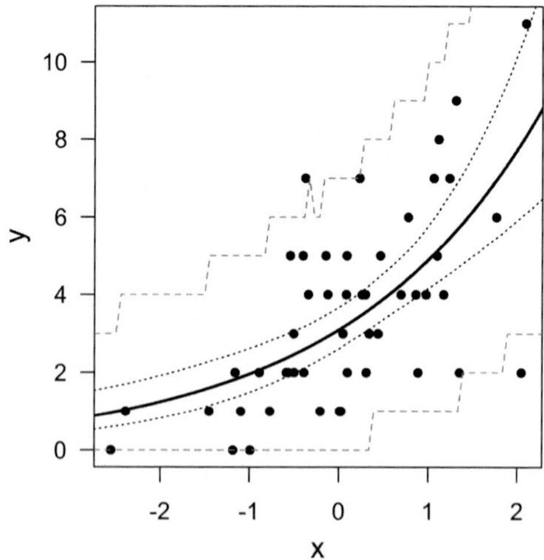

FIGURE 8-10 Simulated data (dots) with a Poisson regression line (solid), lower and upper bound of the 95% credible interval (dotted), and 95% interval of the posterior predictive distribution (dashed gray).

$$y_i \sim Pois(\lambda_i)$$

$$\log(\lambda_i) = \mathbf{X}_i \boldsymbol{\beta}$$

We use, similar to the R function `log`, the notation "log" for the natural logarithm. We fit the model in R using the function `glm` and use the argument "family" to specify that we assume a Poisson distribution as the error distribution. The log-link is used as the default link function. Then we add the regression line to the plot using the function `curve`. Further down we will add the credible and the prediction intervals to the plot.

```
plot(x,y, pch=16, las=1, cex.lab=1.4, cex.axis=1.2)
mod <- glm(y~x, family="poisson")
curve(exp(coef(mod)[1] + coef(mod)[2]*x), add=TRUE, lwd=2)
```

8.4.3 Assessing Model Assumptions

Because the residual variance in the Poisson model is defined by λ (the fitted value), it is not estimated as a separate parameter from the data. Therefore, we always have to check whether overdispersion is present. Ecological data are often overdispersed because not all influencing factors can be measured and included in the model. We explained in Section 8.2.3 what to do if overdispersion is present. As with the binomial model, in a Poisson model overdispersion is present when the residual deviance is larger than the residual degrees of freedom. This is because if we add one independent observation to the data, the deviance increases, on average, by one if the variance equals λ. If the variance is larger, the contribution of each observation to the deviance is, on average, larger than one.

We can check this in the model output:

```
mod
Call:  glm(formula = y ~ x, family = "poisson")
Coefficients:
(Intercept)            x
     1.1329       0.4574

Degrees of Freedom: 49 Total (i.e. Null); 48 Residual
Null Deviance:    85.35
Residual Deviance: 52.12  AIC: 198.9
```

The residual deviance is 52 compared to 48 degrees of freedom. This is perfect (of course, because the model is fit to simulated data). If we are not sure, we could include an observation-level random factor to see how large the extra variance in the data is (see Section 9.2). If there is substantial overdispersion, we could fit a quasi-Poisson model that includes a dispersion parameter. However, as explained previously, we prefer to include an observation-level random factor.

A further alternative to a quasi-Poisson model is a negative-binomial model (function `glm.nb` from the library MASS). In such a model, the variance is estimated from the data. However, the function `sim` does not work for negative-binomial models. The standard residual plots are obtained in the usual way:

```
par(mfrow=c(2,2))
plot(mod)
```

Of course, again, they look perfect because we used simulated data (plots not shown). In a Poisson model, as for the binomial model, it is easier to detect lack of model fit using posterior predictive model checking. For example, data could be simulated from the model and the proportion of 0 values in the simulated data could be compared to the proportion of 0 values in the observations to assess whether zero-inflation is present or not. Or, the variances could be compared between the simulated and observed data. We will introduce predictive model checking soon (Chapter 10).

8.4.4 Drawing Conclusions

For drawing inferences, we can use the function `sim` again:

```
n.sim <- 2000
bsim <- sim(mod, n.sim=n.sim)
apply(bsim@coef, 2, quantile, prob=c(0.025, 0.5, 0.975))
        (Intercept)           x
2.5%      0.9596814   0.2982989
50%       1.1340065   0.4535416
97.5%     1.3033275   0.6034479
```

The 95% credible interval of $\widehat{\beta}_1$ is 0.3 – 0.6. Given that an effect of 0.2 or larger on the aridity scale would be considered biologically relevant, we can be quite confident that aridity has a relevant effect on aphid abundance given our data and our model. With the simulations from the posterior distributions of the model parameters (stored in the object "bsim") we can obtain samples of the posterior distributions of fitted values for each of 100 x-values along the x-axis. Then we can draw the 95% credible interval of the regression line in Figure 8-10.

```
newdat <- data.frame(x=seq(-3, 2.5, length=100))
Xmat <- model.matrix(~x, data=newdat)
b <- coef(mod)
newdat$fit <- exp(Xmat %*% b)
fitmat <- matrix(ncol=n.sim, nrow=nrow(newdat))
for(i in 1:n.sim) fitmat[,i] <- exp(Xmat %*% bsim@coef[i,])
newdat$lwr <- apply(fitmat, 1, quantile, prob=0.025)
newdat$upr <- apply(fitmat, 1, quantile, prob=0.975)
lines(newdat$x, newdat$fit, lwd=2)
lines(newdat$x, newdat$lwr, lty=3)
lines(newdat$x, newdat$upr, lty=3)
```

To describe the posterior predictive distribution (for a specific x-value) we use the simulated fitted values. This means that for each of the 100 values along the x-axis that we have defined, we have 2000 fitted values (in the object fitmat), and for each of these we draw a random value from a Poisson distribution with λ equal to this fitted value.

```
predmat <- matrix(ncol=n.sim, nrow=nrow(newdat))
for(i in 1:n.sim) predmat[,i] <- rpois(nrow(newdat), fitmat[,i])
newdat$pred.lwr <- apply(predmat, 1, quantile, prob=0.025)
newdat$pred.upr <- apply(predmat, 1, quantile, prob=0.975)
lines(newdat$x, newdat$pred.lwr, lty=2, col=grey(0.5))
lines(newdat$x, newdat$pred.upr, lty=2, col=grey(0.5))
```

8.4.5 Modeling Rates and Densities: Poisson Model with an Offset

Many count data are measured in relation to a reference, such as an area or a time period or a population. For example, when we count animals on plots of different sizes, the most important predictor variable will likely be the size of the plot. Or, in other words, the absolute counts do not make much sense when they are not corrected for plot size: the relevant measure is animal density. Similarly, when we count how many times a specific behavior occurs and we follow the focal animals during time periods of different lengths, then the interest is in the rate of occurrence rather than in the absolute number counted. One way to analyze rates and densities is to divide the counts by the reference value and assume that this rate (or a transformation thereof) is normally distributed. However, it is usually hard to obtain normally distributed residuals using rates or densities as dependent variables.

A more natural approach to describe rates and densities is to use a Poisson model that takes the reference into account within the model. This is called an offset. To do so, λ is multiplied by the reference T (e.g., time interval, area, population). Therefore, $\log(T)$ has to be added to the linear predictor. Adding $\log(T)$ to the linear predictor is like adding a new predictor variable (the log of T) to the model with its model parameter (the slope) fixed to 1. The term "offset" says that we add a predictor but do not estimate its effect because it is fixed to 1.

$$y_i \sim Pois(\lambda_i T_i)$$

$$\log(\boldsymbol{\lambda}\mathbf{T}) = \log(\boldsymbol{\lambda}) + \log(\mathbf{T}) = \mathbf{X}\boldsymbol{\beta} + \log(\mathbf{T})$$

In R, we can use the argument "offset" within the `glm` function to specify an offset. We illustrate this using a breeding bird census on wildflower fields in Switzerland in 2007 conducted by Zollinger et al. (2013). We focus on the common whitethroat *Silvia communis*, a bird of field margins and fallow lands that has become rare in the intensively used agricultural landscape. Wildflower fields are an ecological compensation measure to provide food and nesting grounds for species such as the common whitethroat. Such fields are sown and then left unmanaged for several years except for the control of potentially problematic species (e.g., some thistle species, *Carduus* spp.). The plant composition and the vegetation structure in the field gradually changes over the years, hence the interest in this study was to determine the optimal age of a wildflower field for use by the common whitethroat.

We use the number of breeding pairs (bp) as the outcome variable and field size as an offset, which means that we model breeding pair density. We include the age of the field (age) as a linear and quadratic term because we expect there to be an optimal age of the field (i.e., a curvilinear relationship between the breeding pair density and age). We also include field size as a covariate (in addition to using it as the offset) because the size of the field may have an effect on the density; for example, small fields may have a higher density if the whitethroat can also use surrounding areas but uses the field to breed. Size (in hectares) was z-transformed before the model fit.

```
    data(wildflowerfields)
    dat <- wildflowerfields
mod <- glm(bp ~ age + I(age^2) + size.ha.z, offset=log(size.ha),
    data=dat, family=poisson)

mod
Call:  glm(formula = bp ~ age + I(age^2) + size.ha.z, family =
poisson, data = dat, offset = log(size.ha))

Coefficients:
(Intercept)        age      I(age^2)    size.ha.z
    -4.2294     1.5241       -0.1408      -0.5397
Degrees of Freedom: 40 Total (i.e. Null); 37 Residual
Null Deviance:    48.5
Residual Deviance: 27.75  AIC: 70.2
```

For the residual analysis and for drawing conclusions, we can proceed in the same way we did in the Poisson model (Sections 8.4.3, 8.4.4). From the model output we see that the residual deviance is smaller than the corresponding degrees of freedom, thus we have some degree of underdispersion. But the degree of underdispersion is not very extreme so we accept that the credible intervals will be a bit larger than "necessary" and proceed in this case.

After residual analyses, we can produce an effect plot of the estimated whitethroat density against the age of the wildflower field (Figure 8-11). And we see that the expected whitethroat density is largest on wildflower fields of age 4 to 7 years.

8.4.6 Frequentist Results

To draw inference in a frequentist way, as for the binomial model, the function `anova` and `drop1` can be used to do sequential and marginal likelihood ratio

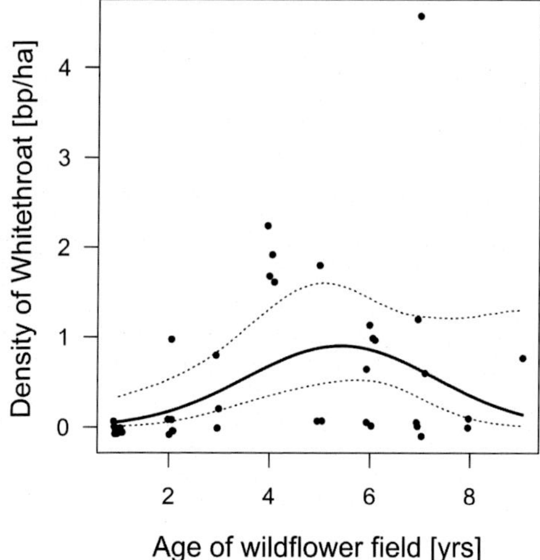

FIGURE 8-11 Whitethroat densities are highest in wildflower fields that are around 4 to 6 years old. Dots are the raw data, the bold line gives the fitted values (with the 95% credible interval given with dotted lines) for wildflower fields of different ages (years). The fitted values are given for average field sizes of 1.4 ha.

tests (Section 4.2.4). The function `predict` is also used as in the binomial model. Using `predict(mod)` produces fitted values on the scale of the link function (in our case, on the log-scale), while `predict(mod, type="response")` returns the values on the original scale (i.e., applying the inverse link function which, in this case, is `exp`).

FURTHER READING

Generalized linear models were introduced by McCullagh and Nelder (1989). Also, Agresti (2007) introduces the theoretical background of generalized linear models.

We highly recommend Gelman and Hill (2007) as a practical introduction to generalized linear models with R using Bayesian methods. Plenty of introductions to generalized linear models in R using frequentist methods are around, for example Wood (2006), Zuur et al. (2009), or Crawley (2007). Zuur et al. (2009) include a negative binomial example.

Rizzo (2008), Fielding and Bell (1997), and Pearce and Ferrier (2000) describe how to use cross-validation or area under the curve (AUC) measurements when using logistic regression as a discriminant function.

Introductions to occupancy and point count models using R and/or WinBUGS are given in Royle and Dorazio (2008) and Kéry and Schaub (2012). The R package unmarked (Fiske and Chandler, 2011) provides functions to fit occupancy and point count models using ML-methods.

Chapter 9

Generalized Linear Mixed Models

Chapter Outline

9.1 BINOMIAL MIXED MODEL

9.1.1 Background

To illustrate the binomial mixed model we have adapted a data set used by Grüebler et al. (2010) on barn swallow *Hirundo rustica* nestling survival (we have selected a nonrandom sample to be able to fit a simple model; hence, the results do not add unbiased knowledge about the swallow biology!). For 63 swallow broods, we know the clutch size and the number of the nestlings that fledged. The broods came from 51 farms, thus some of them had more than one brood. There are three predictors measured at the level of the farm: colony size (the number of swallow broods on that farm), cow (whether there are cows on the farm or not), and dung heap (the number of dung heaps, piles of cow dung, within 500 m of the farm).

The interest was to measure how swallows profit from insects that are attracted by livestock on the farm and by dung heaps. Broods from the same farm are not independent of each other. Also, the predictor variables were measured at the level of the farm, thus they are the same for all broods from a

http://dx.doi.org/10.1016/B978-0-12-801370-0.00009-5.

farm. We have to account for that when building the model by including farm as a random factor. The outcome variable consists of two values for each observation, as seen with the binomial model without random factors (Section 8.2.2): number of successes (fledge) and number of failures (chicks that died = clutch size minus number that fledged).

The random factor "farm" adds, to the intercept in the linear predictor, a farm-specific deviation b_g. These deviations are modeled as normally distributed with mean 0 and standard deviation σ_g.

$$
\begin{aligned}
y_i &\sim Binom(p_i, n_i) \\
\text{logit}(p_i) &= \beta_0 + b_{g[i]} + \beta_1 \text{ colonysize}_i + \beta_2 \text{ I}(\text{cow}_i = 1) + \beta_3 \text{ dungheap}_i \\
b_g &\sim Norm(0, \sigma_g)
\end{aligned}
$$

9.1.2 Fitting a Binomial Mixed Model in R

To fit the model in R we use the function `glmer`, which uses the Laplace approximation. The Laplace approximation is an analytic method to solve integrals, which is often used in Bayesian statistics to obtain the posterior distribution of parameters. We z-transform the covariates (subtraction of the mean and division by the standard deviation so that the transformed variable has a mean of 0 and a standard deviation of 1), which often facilitates convergence of the model. The notation for the random factor with only a random intercept is `(1|farm.f)`.

```
data(swallowfarms); dat <- swallowfarms
dat$colsize.z <- scale(dat$colsize)
dat$dung.z     <- scale(dat$dung)
dat$die        <- dat$clutch - dat$fledge
dat$farm.f     <- factor(dat$farm)
mod <- glmer(cbind(fledge, die) ~ colsize.z + cow + dung.z +
           (1|farm.f), data=dat, family=binomial)
mod
Generalized linear mixed model fit by maximum likelihood ['glmerMod']
Family: binomial ( logit )
Formula: cbind(fledge, die) ~ colsize.z + cow + dung.z + (1 | farm.f)
  Data: dat
     AIC       BIC    logLik  deviance  df.resid
282.5240  293.2397 -136.2620  272.5240        58
Random effects:
Groups Name        Std.Dev.
farm.f  (Intercept) 0.4536
Number of obs: 63, groups: farm.f, 51
Fixed Effects:
(Intercept)   colsize.z        cow      dung.z
   -0.09533     0.05087    0.39370    -0.14236
```

9.1.3 Assessing Model Assumptions

As always, we first look at the model fit before drawing inferences.

```
par(mfrow=c(2,2)) # divide the graphic window in 4 subregions

qqnorm(resid(mod)) # qq-plot of residuals
qqline(resid(mod))

qqnorm(ranef(mod)$farm.f[,1])   # qq-plot of the random effects
qqline(ranef(mod)$farm.f[,1])

plot(fitted(mod), resid(mod))  # residuals vs fitted values
abline(h=0)

dat$fitted <- fitted(mod)     # fitted vs observed values
plot(dat$fitted, jitter(dat$fledge/dat$clutch,0.05))
abline(0,1)
```

The residual plots look fine for a binomial mixed model (Figure 9-1). The two bottom plots do show a distinct pattern, but that is alright with mixed models: we tend to have negative residuals with small fitted values and positive residuals with large fitted values. That is an effect of the shrinkage, because fitted values for farms with high survival compared to the others are shrunk toward the overall mean, especially from farms with few data points. Similarly, farms with small observed survival will have large negative residuals.

In addition to the residual plots shown in Figure 9-1 we also plotted the residuals against each predictor variable and could not detect indications of serious violations of the assumptions.

We have experienced, sometimes, that the function `glmer` produces wrong results without giving any warning or error. Such a failure may be diagnosed by checking the mean of each random effect. These means (only one in our case) should be close to 0:

```
mean(ranef(mod)$farm.f[,1])
-0.001690303
```

That seems acceptable. Because the farm effects are added to the overall mean to obtain the farm-specific fitted values, the estimate for the overall mean (on the logit-scale) will be 0.0017 too high. This translates to an overestimation of the overall mean survival of about 0.09% at the intercept, which is negligible in our study.

```
# mean survival estimate by model
t.should <- plogis(fixef(mod)["(Intercept)"])
```

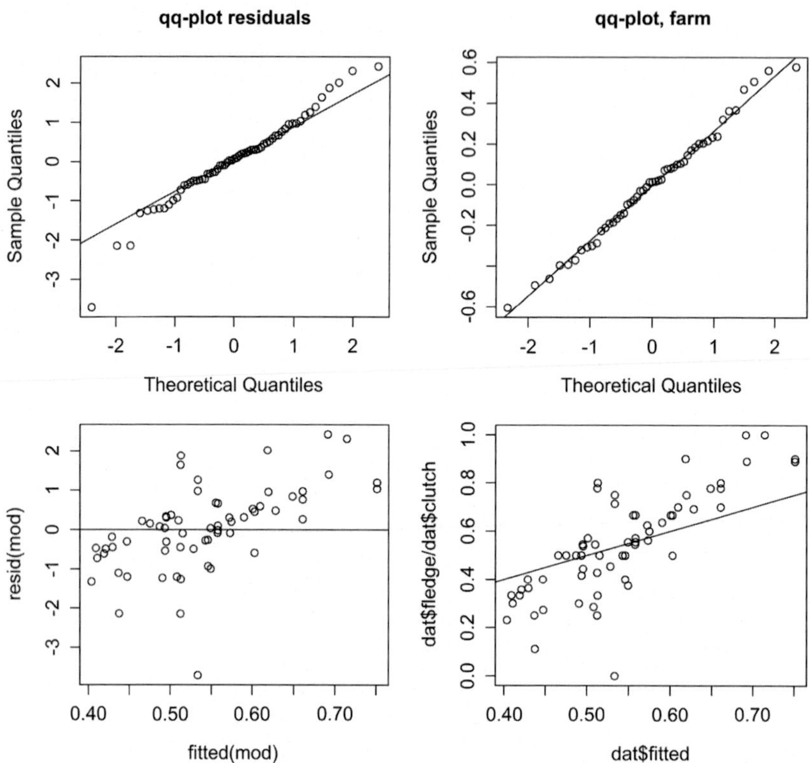

FIGURE 9-1 Diagnostic plots for the swallow nestling survival model. *Upper left*: QQ plot of the residuals; *upper right*: QQ plot of the random effects; *lower left*: residuals versus fitted values; *lower right*: observed versus fitted values.

```
# mean survival corrected for fitting failure
t.is    <- plogis(fixef(mod)["(Intercept)"] - 0.001690303)
(t.should - t.is) / t.should
0.0008853673
```

Note: a mean of the estimates of a random factor of, for example, 0.1 would translate to quite a substantial error of 5% (given the logit link function).

This problem of a nonzero mean random effect seems to appear more often when the model structure does not fit well to the data. For example, it appeared when we fitted a model to simulated data but added a random factor that was not related to the data. Thus, if the mean of the estimates of a random factor is far from 0, we should try to find a more realistic model.

To check for overdispersion, we can use a function that we downloaded from the R helplist (`dispersion_glmer` in blmeco; see its help page for the reference).

```
dispersion_glmer(mod)
[1] 1.192931
```

We get a value of 1.19, which suggests some, but tolerable, overdispersion (values over 1.4 would suggest more serious overdispersion). However, a value not indicating overdispersion does not guarantee that the model fits well. Sometimes, a combination of zero-inflation and overdispersion can lead to scale parameters close to 1, even though there is a clear lack of fit. Posterior predictive model checking can reveal such structures (Chapter 10).

9.1.4 Drawing Conclusions

The function `sim` draws random samples from the posterior distribution of the model parameters. We can use these simulations to obtain 95% CrI for the model parameters. For some versions of the package arm (e.g., version 1.6–10), the fixef slot of the object generated by sim (we usually name it bsim) does not contain the parameter names. In this case, we can add column names manually, for example, as shown in the third line that follows.

```
nsim <- 2000
bsim <- sim(mod, n.sim=nsim)
colnames (bsim@fixef) <- names(fixef(mod)) # for arm 1.6-10
fixef(mod)
(Intercept)   colsize.z         cow      dung.z
-0.09532525  0.05087098  0.39370024 -0.14236419
apply(bsim@fixef, 2, quantile, prob=c(0.025,0.975))
       (Intercept)   colsize.z          cow       dung.z
2.5%   -0.4665512  -0.1818510  -0.03813838  -0.34584466
97.5%   0.2819780   0.2757242   0.82054376   0.05972346
```

The credible intervals indicate large uncertainty for all parameters given their effect sizes. We can draw effect plots to better interpret the result

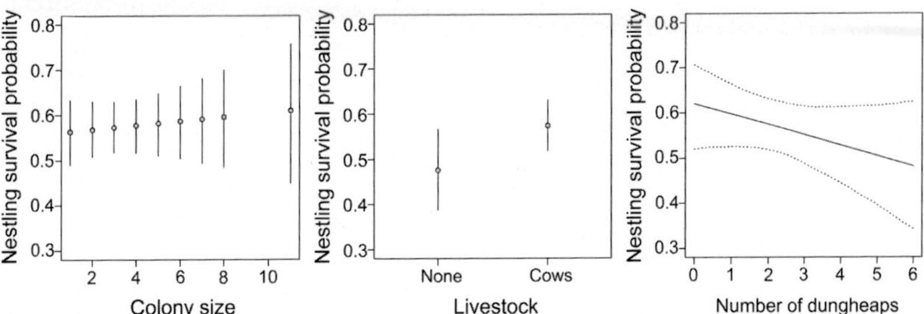

FIGURE 9-2 Nestling survival probability in relation to colony size, livestock presence, and the number of dung heaps around the farm. Given are the fitted values with 95% credible intervals. The effect of colony size is shown conditional on livestock being present and the number of dung heaps being 2. For showing the effect of livestock, colony size was set to 3.3, and the number of dung heaps to 2, and the effect of dung heaps is shown with colony size set to 3.3 and livestock present.

(Figure 9-2). Remember that such an effect plot shows the effect of one predictor while holding constant the other predictors on specific values (e.g., their means). Therefore, the effect plot shows the effect, after controlling for the effects of all other predictors. This is relevant when the predictors are correlated (i.e., in unbalanced or collinear data). In such cases, a plot of the raw data against the predictor often looks more or less different than the effect plot, because raw data are not corrected for all the other influencing factors. A look at the pairs plot that shows the correlation between all predictors can be helpful (see Section 4.2.7).

With our example, we observe the expected effect of livestock on the farm: with cows, the nestling survival was about 57%, but only 48% on farms with no livestock. Nestling survival seems to decrease by around 10% from 0 to 6 dung heaps within 500 m of the nest. (Note that in the original, unmanipulated data this effect was positive.) The effect of colony size is difficult to judge due to the large degree of uncertainty.

9.2 POISSON MIXED MODEL

9.2.1 Background

The differences between the normal linear mixed model presented in Section 7.2 and the Poisson mixed model is that the distribution of the observations around the expected value λ is now a Poisson distribution instead of a normal distribution, and that the linear predictor is (usually) related to the logarithm of the expected value rather than to the expected value itself.

$$
\begin{aligned}
y_i &\sim Pois(\lambda_i) \\
\log(\lambda_i) &= \beta_0 + b_{g[i]} + \beta_1 \text{ predictor } A_i + \beta_2 \text{ predictor } B_i + \ldots \\
b_g &\sim Norm\left(0, \sigma_b^2\right)
\end{aligned}
$$

We added a random group effect to the intercept, thus a group-specific intercept β_0 is modeled. Of course, any other parameter in the fixed effects part of the model can be modeled depending on group, if the data allow (or require).

9.2.2 Fitting a Poisson Mixed Model in R

To illustrate the Poisson mixed model, we go back to the whitethroat data introduced in Section 8.4.5. There, we selected one census year of the study and focused on one species, the common whitethroat. Now, we use all years and we model the number of species found in each wildflower field, including, among others, stonechat *Saxicola rubicola*, yellowhammer *Emberiza citrinella*, and skylark *Alauda arvensis*.

Here, we do not use field size as an offset to account for unequal field sizes because species number is not directly linked to the area (unlike the density of breeding pairs as used in Section 8.4.5). Rather, the increase of species number with field size is expected to gradually level off as the field size increases, a connection known as the species-area relationship. We do have to take field size into account, but we include it as a linear and quadratic term, and not as an offset; depending on the data, it might be advisable to add more polynomial terms to model the species-area curve or use a nonlinear model.

To familiarize us with the data set again, we look at it using the function `str`:

```
data(wildflowerfields)
dat <- wildflowerfields
str(dat)
'data.frame':  136 obs.  of  8 variables:
$ field        : int 1 1...          # ID of the wildflower field
$ year         : int 2006 2007... # year of bird count
$ age          : int 8 9...          # age of the wildflower field in years
$ bp           : int 1 1...          # number of whitethroat breeding pairs
$ X            : int 526025 526025... # X-coordinate of the field
$ Y            : int 166425 166425... # Y-coordinate of the field
$ size         : int 148 148...         # size of the wildflower field [are]
$ Nspec        : int 7 6 2 4 5...        # number of species in the field
```

The aim is to estimate the optimal age of the wildflower field regarding the number of species that use the field. We provide more exploratory data analyses in Section 9.2.5—for example, regarding correlations between explanatory variables. For now, we model the species number using age and size as fixed factors and year and field as random factors. Both age and size are expected to have a nonlinear relationship with the number of species, thus we include polynomials of each. However, using the simple polynomials, up to the third degree of these variables did not work well: the model fitting algorithm did not converge.

This reminds us of an intricacy often encountered, especially with generalized linear (mixed) models: collinear predictors can make model convergence difficult, and age.z^2 and age.z^3 (age.z = centered and scaled age) are heavily correlated with each other, and the same holds for size.z^2 and size.z^3. Using orthogonal polynomials (see Section 4.2.9) solves this problem. For the effect plot we want to draw afterwards, we need to store the poly object, and we add the orthogonal polynomials as new variables to the data frame:

```
t.poly.age <- poly(dat$age,3)
dat$age.l  <- t.poly.age[,1]
dat$age.q  <- t.poly.age[,2]
dat$age.c  <- t.poly.age[,3]
```

Similarly, orthogonal linear and quadratic effects were calculated for size (not shown here), and the following model can be fit using the function glmer with the family argument set to poisson:

```
mod <- glmer(Nspec ~ age.l + age.q + age.c + size.l + size.q +
           (1|year.f) + (1|field.f), data=dat, family=poisson)
```

```
mod
Generalized linear mixed model fit by maximum likelihood (Laplace
Approximation) ['glmerMod']
 Family: poisson ( log )
Formula: Nspec ~ age.l + age.q + age.c + size.l + size.q +
(1 | year.f) + (1|field.f)
   Data: dat
     AIC       BIC    logLik  deviance  df.resid
485.9526  509.2538 -234.9763  469.9526       128
Random effects:
 Groups Name         Std.Dev.
 field.f   (Intercept) 0.2849
 year.f    (Intercept) 0.1335
Number of obs: 136, groups: field, 67; year, 8
Fixed Effects:
(Intercept)     age.l      age.q     age.c    size.l     size.q
     0.7435    4.0966    -4.7101    2.8929    2.6328    -0.8874
```

9.2.3 Assessing Model Assumptions

Plotting the data together with the model is one important tool to assess model fit; however, it is not sufficient. We do the usual residual analysis; the plots, not shown here, do not indicate serious problems.

```
qqnorm(resid(mod)) # qq plot of residuals
qqline(resid(mod))

qqnorm(ranef(mod)$year.f[,1])             # qq plot of random effects
qqline(ranef(mod)$year.f[,1])
qqnorm(ranef(mod)$field.f[,1])
qqline(ranef(mod)$field.f[,1])

scatter.smooth(fitted(mod), resid(mod)) # fitted versus residuals

plot(fitted(mod),dat$Nspec)               # data versus fitted values
```

To check whether overdispersion is present we can again use the function dispersion_glmer (see Section 9.1.3); it yields a value of 0.99—that is, there is no indication of overdispersion. An alternative way to look at the issue is to fit a model with an observation level random factor added. We can then have a look at the variance of this factor and judge from that whether we think overdispersion could be important; in our case, this seems not to be the case:

```
dat$obsid <- factor(1:nrow(dat))
modod <- glmer(Nspec ~ age.l + age.q + age.c + size.l + size.q +
               (1|year.f) + (1|field.f) + (1|obsid), data=dat,
               family=poisson)
modod
[...]
Random effects:
 Groups Name        Std.Dev.
 obsid   (Intercept) 0.0000
 field.f (Intercept) 0.2849
 year.f  (Intercept) 0.1335
[...]
```

The additional variance is estimated to be 0. There is apparently negligible overdispersion.

9.2.4 Drawing Conclusions

The uncertainty measurements for the parameter estimates are obtained from their posterior distributions simulated by `sim`.

```
nsim <- 2000
bsim <- sim(mod, n.sim=nsim)
apply(bsim@fixef, 2, quantile, prob=c(0.025,0.5,0.975))
      (Intercept)   age.l     age.q    age.c   size.l     size.q
2.5%      0.55300  2.3891   -6.3780   1.2966   1.0409   -2.35111
50%       0.74840  4.0820   -4.7063   2.8711   2.6342   -0.91090
97.5%     0.94287  5.9022   -2.9683   4.4356   4.1916    0.61581
```

The fitted values and their uncertainties are also derived from the simulated values of the posterior distributions of the model parameters. Take care to use the right inverse link function to transform the linear predictor to the scale of the response. Here, we have to use `exp`, since we used the log-link. Also, any transformations that had been done to the predictors must be done in the same way if we want to predict on the original scale of a predictor. Because we have used orthogonal polynomials, we do the same transformation to the age-values for which we want to get fitted values, using the function `predict` (Section 4.2.9). The effect plot for age shows that fields around five years old are associated with the highest number of species (Figure 9-3).

```
newdat <- data.frame(
   year = mean(dat$year),
   age = seq(min(dat$age), max(dat$age), length.out=100),
   size = mean(dat$size))
```

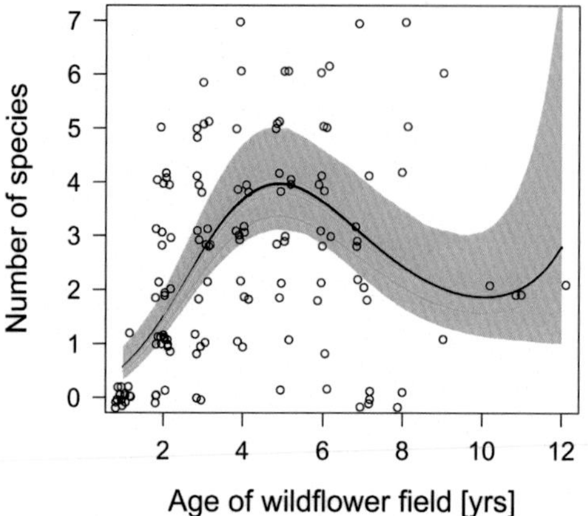

FIGURE 9-3 Species number in relation to the age of wildflower fields with average regression line over all fields and years (bold line) including the 95% credible interval (gray), and year-specific regression lines (orange lines). Circles are the raw data.

```
newdat$age.l  <- predict(t.poly.age, newdat$age)[,1]
newdat$age.q  <- predict(t.poly.age, newdat$age)[,2]
newdat$age.c  <- predict(t.poly.age, newdat$age)[,3]
newdat$size.l <- predict(t.poly.size, newdat$size)[,1]
newdat$size.q <- predict(t.poly.size, newdat$size)[,2]

Xmat <- model.matrix(~ age.l + age.q + age.c + size.l + size.q,
                     data=newdat)
newdat$fit <- exp(Xmat %*% fixef(mod)) # exp = inverse link function
```

Now, we calculate for each fitted value in newdat 2000 values that are random draws from their posterior distributions. The 2.5% and 97.5% quantiles of these values are used as lower and upper limits of the 95% credible interval.

```
fitmat <- matrix(nrow=nrow(newdat), ncol=nsim)
for(i in 1:nsim) fitmat[,i] <- exp(Xmat %*% bsim@fixef[i,])
newdat$lwr <- apply(fitmat, 1, quantile, prob=0.025)
newdat$upr <- apply(fitmat, 1, quantile, prob=0.975)
```

Now, we can use newdat$age, newdat$fit, newdat$lwr, and lwedat$upr to produce an effect plot on the original scale of age:

```
plot(Nspec ~ age, data=dat)
polygon(c(newdat$age, rev(newdat$age)),
    c(newdat$lwr, rev(newdat$upr)),
    col=grey(0.7), border=NA)     # 95% CrI given as a shadow
```

```
lines(newdat$age, newdat$fit, lwd=2) # population mean
# add separate lines for each year:
for(i in 1:nlevels(dat$year.f)) {
  t.year <- levels(dat$year.f)[i]
  x <- seq(min(dat$age[dat$year.f==t.year]),
          max(dat$age[dat$year.f==t.year]), length=100)
  x.l <- predict(t.poly.age, x)[,1] # analogous transformation
  x.q <- predict(t.poly.age, x)[,2] # to get orthogonal linear,
  x.c <- predict(t.poly.age, x)[,3] # quadratic and cubic terms
  y <- exp(fixef(mod)[1] + ranef(mod)$year.f[i,1] +
          fixef(mod)[2]*x.l + fixef(mod)[3]*x.q + fixef(mod)[4]*x.c)
  lines(x,y, col="orange",lwd=1.2)
  }
```

9.2.5 Modeling Bird Densities by a Poisson Mixed Model Including an Offset

To present the Poisson mixed model with an offset, we again use the whitethroat data introduced in Section 8.4.5 and used in Sections 9.2.2 through 9.2.4, too. Now, we again focus on the common whitethroat, but unlike in Section 8.4.5 we use all years, which means that many of the wildflower fields were monitored repeatedly. To account for these repeated measurements, the wildflower field ID is used as a random factor.

The variables in the data file we use here are field (ID of the wildflower field), year (census year), age (age of the wildflower field in years), bp (number of whitethroat breeding pairs), X and Y (the coordinates of the field), and size (the size of the field in areas, i.e., 10×10 m).

First, we do some exploratory analyses to get an overview of the structure of the data set. Specifically, we are interested in how balanced the data are and whether the predictor variables year and age are correlated, or whether there are other structures we have to take into account when modeling whitethroat density.

```
data(wildflowerfields); dat <- wildflowerfields
table(table(dat$field)) # how many cases with n observations per field?
 1  2 3 4 5
26 27 6 2 6
```

From 26 of the 67 wildflower fields, only one bird census exists. Twenty-seven fields were monitored in 2 years, and 14 were monitored for 3 to 5 years. To see which wildflower field was monitored in which year we can type the next line:

```
table(dat$field, dat$year) # output not shown
```

The highest numbers of wildflower fields were monitored during the years 2006 and 2007.

```
table(dat$year)
2004  2005  2006 2007  2008  2009  2010  2011
   1     1    41   41    11    16    13    12
```

The main variable of interest in this study is the age of the wildflower field; remember that the question was at which age wildflower fields are optimal for birds. We, therefore, check whether age is confounded with other variables, such as year. We use boxplots, and we see that the median ages per year do not show a trend, despite the fact that the three oldest wildflower fields were all monitored during the last three years of the study. We also do not find a significant correlation between size and age of the wildflower fields (Figure 9-4). Thus, the data seem to be suited for analysis of age effects on whitethroat densities.

```
boxplot(age ~ year, dat, ylab="Age of wildflower field")
scatter.smooth(dat$size, dat$age)
```

Now, we can construct our model to analyze whitethroat densities in the wildflower fields in relation to age, year, and size of the wildflower field. The number of whitethroat territories y_i is directly related to the size of the wildflower field. We, therefore, include size as an offset in the model. We further include a linear trend of year as well as a random year effect. The linear trend accounts for systematic changes in whitethroat densities over the years whereas the random year effect accounts for random between-year variance, for example, due to different weather situations.

Similar to the analysis done with just one year (Section 8.4.5) we include polynomials of "age" up to the third degree because we expect an optimal age of the wildflower field for whitethroat density. Also, we again would like to see whether the size of a wildflower field affects whitethroat density, thus we include a linear trend of size (i.e., size is the offset but also a predictor).

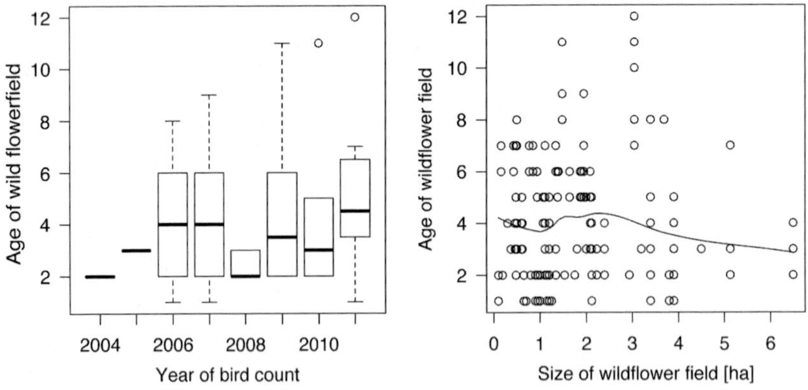

FIGURE 9-4 Age of the wildflower fields in relation to year and size of the field.

Finally, we include the field ID as a random factor because some fields have been monitored in more than one year, so that these data are not independent. The model formula looks like this:

$$
\begin{aligned}
y_i &\sim Pois(\lambda_i \text{size}_i) \\
\log(\lambda_i \text{size}_i) &= \beta_0 + \beta_1 \text{year}_i + \beta_2 \text{age}_i + \beta_3 \text{age}_i^2 + \beta_4 \text{age}_i^3 \\
&\quad + \beta_5 \text{size}_i + b_{\text{ID}[i]} + d_{\text{year}[i]} + \log(\text{size}_i) \\
b_{\text{ID}} &\sim \text{Norm}(0, \sigma_b) \\
d_{\text{year}} &\sim \text{Norm}(0, \sigma_d)
\end{aligned}
$$

Before fitting such a complicated model, it is recommended to transform the offset variable to a sensible scale, to standardize covariates, and, possibly, to use orthogonal polynomials (Section 4.2.9). Otherwise, the fitting algorithm often may not converge and estimates could become unreliable due to correlations between the model parameters or due to some values of the linear predictor taking on values beyond computer accuracy. For example, if we measure the size of the fields in square meters and use this variable as the offset, the fitted number of territories will be close to zero; similarly, if we include a polynomial of a covariate with large values, the polynomials may be too large for the computer.

A sensible scale for measuring whitethroat density is the number of breeding pairs per hectare. Therefore, we transform the size variable originally measured in ares (10 × 10 m) to hectares (100 × 100 m) to be used as the offset. For the linear size effect, we use the standardized size variable size.z. We also standardize year and use orthogonal polynomials of age.

```
dat$size.ha <- dat$size/100
dat$size.z <- as.numeric(scale(dat$size))
dat$year.z <- as.numeric(scale(dat$year))
t.poly.age <- poly(dat$age,3)
dat$age.l <- t.poly.age[,1]
dat$age.q <- t.poly.age[,2]
dat$age.c <- t.poly.age[,3]
dat$field.f <- factor(dat$field)
dat$year.f <- factor(dat$year)
```

Now, we are ready to fit the model:

```
mod <- glmer(bp ~ year.z + age.l + age.q + age.c + size.z +
            (1|field.f) + (1|year.f) + offset(log(size.ha)),
            family=poisson, data=dat)
mod
Generalized linear mixed model fit by maximum likelihood (Laplace
Approximation) ['glmerMod']
Family: poisson ( log )
```

```
Formula: bp ~ year.z + age.l + age.q + age.c + size.z + (1 | field.f) +
  (1 | year.f) + offset(log(size.ha))
   Data: dat
      AIC       BIC     logLik  deviance  df.resid
 308.7927  332.0940  -146.3964  292.7927       128
Random effects:
 Groups Name        Std.Dev.
 field.f  (Intercept) 0.425
 year.f   (Intercept) 0.000
Number of obs: 136, groups: field.f, 67; year.f, 8
Fixed Effects:
(Intercept) year.z   age.l    age.q    age.c    size.z
   -0.8529  0.2656  4.2290  -5.7332  3.0469  -0.2655
```

We see that the between-year variance in whitethroat density is negligible. Before drawing conclusions we look at the residuals.

```
par(mfrow=c(2,2))
scatter.smooth(fitted(mod), resid(mod))
scatter.smooth(dat$year.z, resid(mod))
scatter.smooth(dat$age.z, resid(mod))
scatter.smooth(dat$size.z, resid(mod))
```

From the plots in Figure 9-5 we cannot see that we have missed important structures in the data. Also, the QQ plots of the residuals and the two random factors do not look very bad (not shown). To check whether the data are overdispersed, we can include an observation level random factor:

```
dat$obsid <- factor(1:nrow(dat))
modod <- glmer(bp ~ year.z + age.l + age.q + age.c + size.z +
   (1|field.f) + (1|year.f) + (1|obsid) + offset(log(size.ha)),
   family=poisson, data=dat)
modod
Random effects:
Groups Name        Std.Dev.
obsid  (Intercept) 0.000
field.f (Intercept) 0.425
year.f (Intercept) 0.000
Number of obs: 136, groups: obsid, 136; field.f, 67; year.f, 8
```

We see that the extra variance between the observations is essentially zero. Thus, we do not need the observation level random factor.

Finally, we check whether spatial correlation is an issue in these data.

```
library(gstat); library(sp)
spdata <- data.frame(resid=resid(mod), x=dat$X, y=dat$Y)
coordinates(spdata) <- c("x", "y")
bubble(spdata, "resid", col=c("blue", "orange"), main="Residuals")
```

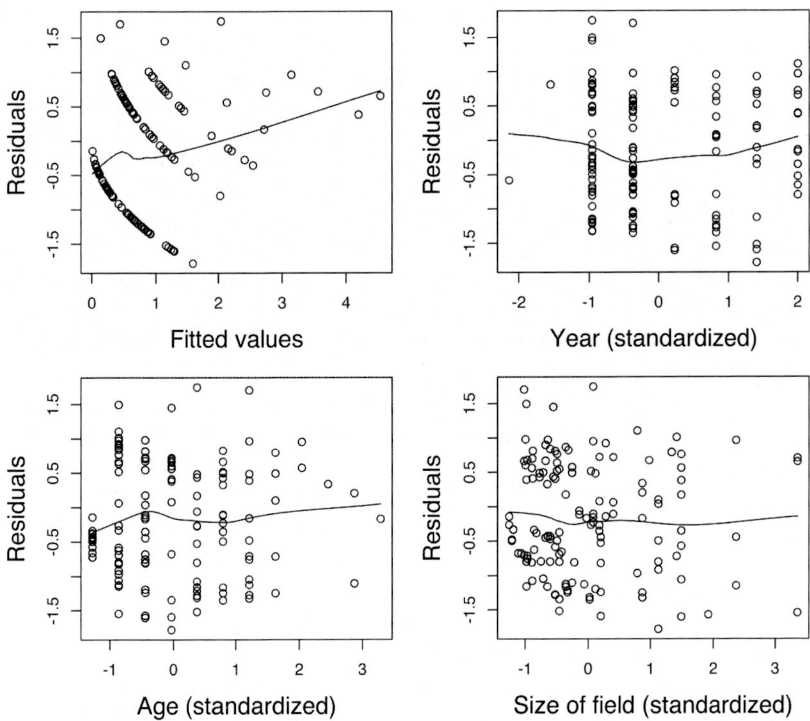

FIGURE 9-5 Residual plots of the whitethroat model: Residuals versus (1) fitted values, (2) standardized year, (3) standardized age, and (4) standardized size.

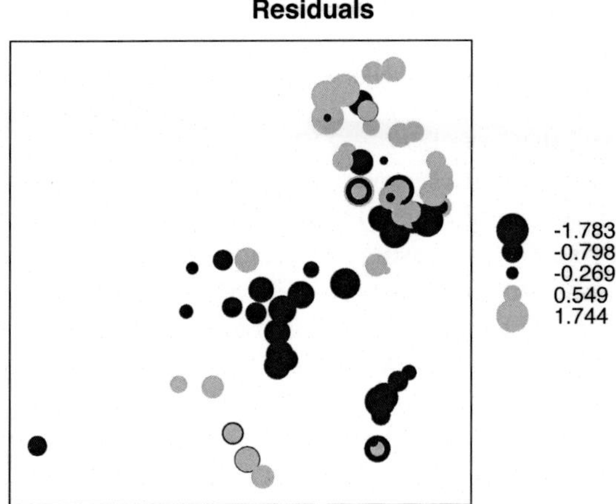

FIGURE 9-6 Spatial distribution of the residuals in the whitethroat data.

Indeed, we see that the residuals are spatially correlated (Figure 9-6). The positive residuals cluster in two patches (upper right corner and bottom center) whereas in the center of the study area negative residuals are overrepresented. Such a pattern means that the observations are not independent of each other. As a consequence, the uncertainty of the parameter estimates is underestimated (due to pseudoreplication). However, in the semivariogram of the residuals the spatial correlation is only weakly discernible. Therefore, we proceed here assuming no spatial correlation. But, in the real world, we would probably proceed in the following way: First, try to find a factor that explains the pattern such as a habitat or landscape characteristics and, second, include subregions within the study area as a random factor. If this does not help, use a method presented in Chapter 13.

For the moment, we trust the semivariogram suggesting only very weak spatial correlation and start drawing conclusions. Let's first have a look at the 95% credible intervals of the parameters.

```
nsim <- 2000
bsim <- sim(mod, n.sim=nsim)
apply(bsim@fixef, 2, quantile, prob=c(0.025, 0.5, 0.975))
      (Intercept)    year.z   age.l     age.q     age.c     size.z
2.5%    -1.11672  0.060595  1.1712  -8.5458  0.30477  -0.49780
50%     -0.84925  0.261393  4.1953  -5.7838  3.04104  -0.26055
97.5%   -0.59134  0.485135  7.2399  -2.8823  5.70323  -0.02071
```

We see a slight increase of the number of territories over the years. Because we have used a log-link function, the effects are not additive. The exponential of the coefficient (exp(0.263) = 1.304) is the multiplicative change in the outcome variable. Thus, when year.z increased by 1 (which corresponds to the standard deviation of the original variable year, `sd(dat$year)` = 1.7 years), whitethroat density increased by 30%. We also see a slight negative relationship between whitethroat density and the size of the wildflower field. With each increase of the field size by `sd(dat$area)` = 139 are = 1.4 ha, density decreases to exp(−0.261) = 0.77 = 77%.

The nonlinear age effect is more difficult to describe just from looking at the estimated model parameters. Therefore, we plot the effect of age. We calculate fitted whitethroat densities (expected number of breeding pairs per ha) for several different values of age while holding the other two predictors constant. To do so, we prepare a new data frame containing all age values for which we would like to predict. We further add the variable "year.z" and set it to 0. Because year.z is a standardized variable, the value 0 corresponds to the mean of the data. As we would like to predict whitethroat density for 1 ha, we insert for the variable "size.z" the value that corresponds to 1 ha, which is 100 minus the mean of the original variable "size" divided by its standard

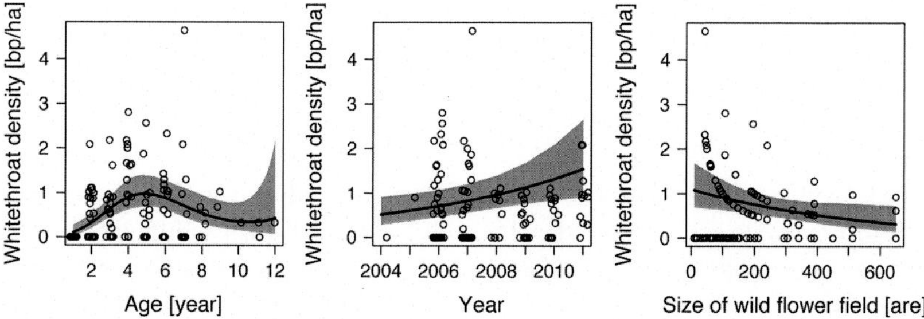

FIGURE 9-7 Whitethroat densities on sown wildflower fields with respect to age of the field, year, and size of the field. Circles = observed densities [bp/ha], solid line = fitted values with 95% credible interval (gray area).

deviation. Because we have used orthogonal age polynomials, we have to transform age in the new data frame as we have done it with the original variable, using `predict` (compare to Section 4.2.9).

As we have done before, we construct the model matrix "Xmat" and obtain the predicted values for the new predictor variables in newdat. Then we do this for all nsim simulated model parameters to obtain a 95% credible interval of the fitted values. At last, we plot the data together with the fitted values. We do this in turn for showing the effects of age, year, and size of the wildflower field (Figure 9-7). The R code that follows produces, as an example, the effect of age.

```
newdat <- data.frame(age=seq(1,12, by=0.1), size=100)
newdat$year.z <- 0 # corresponds to the mean of dat$year
newdat$size.z <- (newdat$size-mean(dat$size))/sd(dat$size)
newdat$age.l <- predict(t.poly.age,newdat$age)[,1]
newdat$age.q <- predict(t.poly.age,newdat$age)[,2]
newdat$age.c <- predict(t.poly.age,newdat$age)[,3]

Xmat <- model.matrix(~ year.z + age.l + age.q + age.c + size.z ,
                     data=newdat)
b <- fixef(mod)
newdat$fit <- exp(Xmat %*% b) # exp = inverse link function
fitmat <- matrix(ncol=nsim, nrow=nrow(newdat))
for(i in 1:nsim) fitmat[,i] <- exp(Xmat %*% bsim@fixef[i,])

newdat$lwr <- apply(fitmat, 1, quantile, prob=0.025)
newdat$upr <- apply(fitmat, 1, quantile, prob=0.975)

plot(dat$age, dat$bp/dat$size.ha, xlab="Age [year]",
     ylab="Whitethroat density [bp/ha]", las=1, cex.lab=1.4,
     cex.axis=1.2, type="n")
```

```
# draw 95% CrI as a shade:
polygon(c(newdat$age, rev(newdat$age)), c(newdat$lwr,
        rev(newdat$upr)),  border=NA, col=grey(0.5))
lines(newdat$age, newdat$fit, lwd=2)
points(jitter(dat$age), dat$bp/dat$size.ha)
```

The left panel in Figure 9-7 shows what the data tell us about the relationship between whitethroat density and the age of sown wildflower fields. Our original question was, at what age of the wildflower fields do we find the highest density of whitethroat? We can answer this question more directly by giving the age at which whitethroat density is maximized as a point estimate with 95% credible interval.

In Figure 9-7 we see that the maximum is between 4 and 5 years. To get the exact point of the maximum, we need to take the first derivative from the cubic function (for that, we refit the model using normal polynomials of the z-transformed age rather than orthogonal polynomials), set it to 0 and solve it. We do this for the 2000 simulated sets of model parameters (defining 2000 different cubic functions) to obtain 2000 simulated values from the posterior distribution of the optimal age. In such cases, where the calculation of the derived parameter involves several steps, we find it convenient to write an R function that does the whole calculation.

We will also find similar preprogrammed functions on the internet when we, for example, search within the R webpage (type "r-project.org: maximum of polynomial function" into Google). However, if it is not too difficult, we prefer to write our own functions, because then we know exactly what we have done. In the following function, we get the x- and y-values of the maximum of an exponentiated cubic curve. The exponentiated cubic curve corresponds to the regression line for age in our Poisson model.

```
maxofcubicfun <- function(Intercept, x){
  b <- x[1]
  c <- x[2]
  d <- x[3]
  D <- 4*(c^2 - 3*b*d)
  if(D<0){
    xmax <- ymax <- NA
    }
  if(D>0){
    xzero <- (-2*c + c(-1,1)*sqrt(D))/(6*d)
    yzero <- exp(Intercept + b*xzero + c*xzero^2 + d*xzero^3)
    index <- yzero==max(yzero)
    xmax <- xzero[index]
    ymax <- yzero[index]
    }
  return(c(xmax,ymax))
  }
```

We can apply this function to the cubic function in Figure 9-7.

```
maxofcubicfun(b[1],b[3:5])
[1] 0.3367687 0.8312031
```

The first number is the *x*-coordinate of the maximum and the second number is the corresponding maximal mean number of breeding pairs per ha. Of course, we have to back-transform the first number because we have *z*-transformed age before the model fit:

```
maxofcubicfun(b[1],b[3:5])[1]*sd(dat$age) + mean(dat$age)
[1] 4.892504
```

Now, we do this for all the 2000 sets of model coefficients in the bsimobject (we have to rerun `sim` on the model with the normal polynomials for age and call it "bsim2").

```
postoptage.z <- numeric(nsim)
for(i in 1:nsim) postoptage.z[i] <-
  maxofcubicfun(fixef(bsim2)[i,1],fixef(bsim2)[i,3:5])[1]
postoptage <- postoptage.z*sd(dat$age) + mean(dat$age)
```

And then we extract the 95% CrI. We also calculate the proportion of regression lines that did not have a maximum at all, which was only 0.15%, thus, we can be quite sure that there is an optimal age somewhere between 4 and 7 years.

```
quantile(postoptage, prob=c(0.025,0.975), na.rm=TRUE)
    2.5%     97.5%
4.262726 6.220102
sum(is.na(postoptage))/nsim  # proportion with no maximum
[1] 0.0015
```

FURTHER READING

In a review on generalized linear mixed models (GLMM) in ecology, Bolker et al. (2008) recommended using Bayesian methods to calculate uncertainty estimates.

The R package MCMCglmm (Hadfield, 2010) provides functions to fitting fairly complex GLMMs, such as models for data with correlation structure caused by genetic relationships (pedigree or phylogeny). GLMMs are the basis of many more complicated ecological models. Therefore, we find very good introductions in books like the spatial capture-recapture book by Royle et al. 2014.

Chapter 10

Posterior Predictive Model Checking and Proportion of Explained Variance

Chapter Outline

10.1 POSTERIOR PREDICTIVE MODEL CHECKING

Only if the model describes the data-generating process sufficiently accurately can we draw relevant conclusions from the model. It is therefore essential to assess model fit: our goal is to describe how well the model fits the data with respect to different aspects of the model. In this book, we present three ways to assess how well a model reproduces the data-generating process: (1) residual analysis (Chapter 6), (2) posterior predictive model checking (this chapter) and (3) prior sensitivity analysis (Chapter 15).

Posterior predictive model checking is the comparison of replicated data generated under the model with the observed data. The aim of posterior predictive model checking is similar to the aim of a residual analysis, that is, to look at what data structures the model does not explain. However, the possibilities of residual analyses are limited, particularly in the case of nonnormal error distributions. For example, in a logistic regression, positive residuals are always associated with $y = 1$ and negative residuals with $y = 0$. As a consequence, temporal and spatial patterns in the residuals will always look similar to these patterns in the observations and it is difficult to judge whether the model captures these processes adequately. In such cases, simulating data from the posterior predictive distribution of a model and comparing these data with the observations (i.e., predictive model checking) gives a clearer insight into the performance of a model.

We follow the notation of Gelman et al. (2014) in that we use "replicated data", y^{rep} for a set of n new observations drawn from the posterior

Bayesian Data Analysis in Ecology Using Linear Models with R, BUGS, and Stan
http://dx.doi.org/10.1016/B978-0-12-801370-0.00010-1.

predictive distribution for the specific predictor variables x of the n observations in our data set. When we simulate new observations for new values of the predictor variables, for example, to show the prediction interval in an effect plot, we use y^{new}.

The first step in posterior predictive model checking is to simulate a replicated data set for each set of simulated values of the joint posterior distribution of the model parameters. Thus, we produce, for example, 2000 replicated data sets. These replicated data sets are then compared graphically, or more formally by test statistics, with the observed data. The Bayesian p-value offers a way for formalized testing. It is defined as the probability that the replicated data from the model are more extreme than the observed data, as measured by a test statistic. In case of a perfect fit, we expect that the test statistic from the observed data is well in the middle of the ones from the replicated data. In other words, around 50% of the test statistics from the replicated data are higher than the one from the observed data, resulting in a Bayesian p-value close to 0.5. Bayesian p-values close to 0 or close to 1, on the contrary, indicate that the aspect of the model measured by the specific test statistic is not well represented by the model. Note that the Bayesian p-value has a different meaning than the frequentist p-value (for the latter, see Section 3.3).

Test statistics have to be chosen such that they describe important data structures that are not directly measured as a model parameter. Because model parameters are chosen so that they fit the data well, it is not surprising to find p-values close to 0.5 when using model parameters as test statistics. For example, extreme values or quantiles of y are often better suited than the mean as test statistics, because they are less redundant with the model parameter that is fitted to the data. Similarly, the number of switches from 0 to 1 in binary data is better suited to check for autocorrelation than the proportion of 1s among all the data. Other test statistics could be a measure for asymmetry, such as the relative difference between the 10 and 90% quantiles, or the proportion of zero values in a Poisson model.

We like predictive model checking because it allows us to look at different, specific aspects of the model. It helps us to judge which conclusions from the model are reliable and to identify the limitation of a model. Predictive model checking also helps to understand the process that has generated the data.

We use the analysis of the whitethroat breeding density in wildflower fields of different ages for illustration (example from Chapter 9). The aim of this analysis was to identify an optimal age of wildflower fields that serves as good habitat for the whitethroat. In Chapter 9, we used a Poisson mixed model with an offset to describe the relationship between whitethroat density and the age of wildflower fields. Here we omit the year as a random variable from the model, because it was estimated to be close to 0 and therefore has little influence on the overall model fit. This omission has a negligible effect on the conclusions but reduces programming effort needed for the posterior predictive model checking. The model we use here is:

```
mod <- glmer (bp~ year.z + age.l + age.q + age.c + size.z +
  (1|field) + offset(log(size.ha)), family=poisson, data=dat)
```

The R code for the variable transformations is given in Section 9.2.5. We simulate 2000 sets of parameter values from their joint posterior distribution.

```
nsim <- 2000
bsim <- sim(mod, n.sim=nsim)
```

Then, we calculate a fitted value for each observation in the data set and for each simulated set of model parameters. These 2000 fitted values for each observation are stored in the object fitmat. In the last step, we draw for each observation in the data set and for each of the 2000 fitted values, a random number y_i^{rep} from a Poisson distribution using the function `rpois`.

```
fitmat <- matrix(ncol=nsim, nrow=nrow(dat))
yrep <- matrix(ncol=nsim, nrow=nrow(dat))
Xmat <- model.matrix(mod)
for(i in 1:nsim) {
  fitmat[,i] <- exp(Xmat %*% bsim@fixef[i,] +
    bsim@ranef$field[i,match(dat$field, levels(dat$field)),1])*
    dat$size.ha
  yrep[,i] <- rpois(nrow(dat), lambda=fitmat[,i])
}
```

As is visible from the R code, we add the random effect estimates for each field to the linear predictor, because we simulate new observations for the specific fields we have sampled in our data set (conditional on the fields in the data set). However, it would also be possible to simulate data for new fields to check whether the model captures the between-field variance reasonably. To do so, random values from a normal distribution with a mean of zero and the between-field variance would be added instead of the estimated field effects for the fields in the data. Thus, the term `+ bsim@ranef$field[i,match(dat$field, levels(dat$field)),1]` would be replaced by `+ rnorm(nrow(dat), 0, sd(bsim@ranef$field[i,,1]))`. Further, note that if the model contains random slopes, the random effects have to be added to the coefficients (`bsim@fixef[i,]`) prior to matrix multiplication, and then, the matrix multiplication has to be done for each level of the random effect separately. Finally, note that the fitted value is multiplied by the size of the fields to take the offset into account.

The matrix yrep contains 2000 columns, each representing one replicated data set from the model. We can compare some of these replicated data sets with the observed data using histograms (Figure 10-1). It seems as if the replicated data sets are more skewed than the observations.

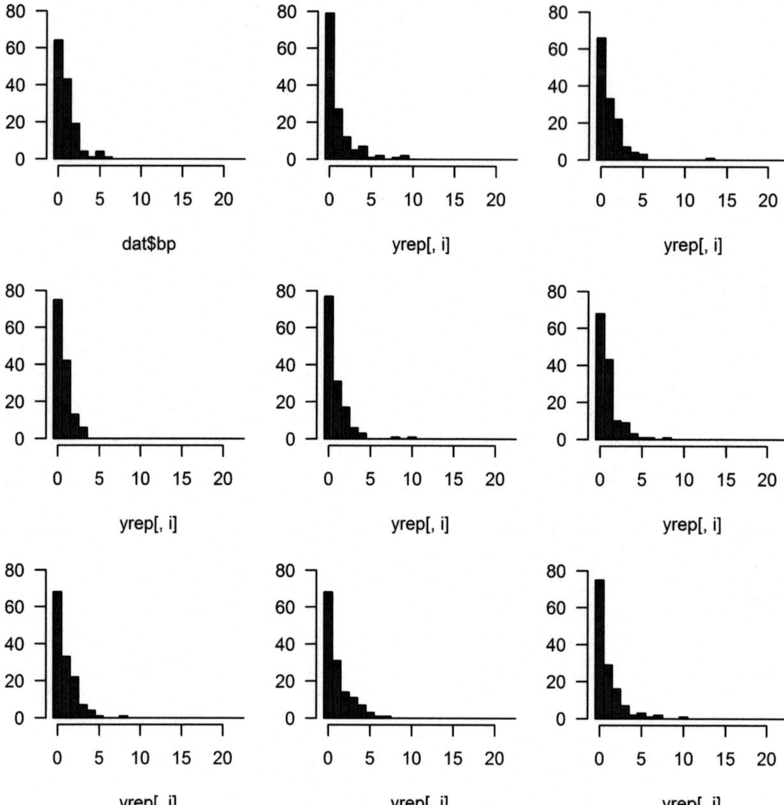

FIGURE 10-1 Histograms of the observed data (dat$bp) and 8 of the 2000 replicated data sets. The arguments breaks and ylim have been used in the function hist to produce the same scale of the *x*- and *y*-axis in all plots. This makes comparison among the plots easier.

The proportion of zero counts could be a sensitive test statistic for this data set. First, we define a function "propzero" that extracts the proportion of zero counts from a vector of count data. Then we apply this function to the observed data and to each of the 2000 replicated data sets. At last, we extract the 1 and 99% quantile of the proportion of zero values of the replicated data.

```
propzeros <- function(x) sum(x==0)/length(x)
propzeros(dat$bp)      # prop. zero values in observed data
[1] 0.4705882
pzerossim <- apply(yrep, 2, propzeros)     # prop. zero values in yrep
quantile(pzerossim, prob=c(0.01, 0.99))
       1%       99%
0.4044118 0.6250000
```

The observed data set contains 47% zero counts. This proportion was between 40 and 63% in the replicated data set. The Bayesian *p*-value is 0.85.

```
sum(pzerossim>=propzeros(dat$bp))/nsim
[1] 0.8495
```

What about the upper tail of the data? Let's look at the 90% quantile.

```
quantile(dat$bp, prob=0.9)     # for observed data
90%
  2
q90sim <- apply(yrep, 2, quantile, prob=0.9)     # for simulated data
table(q90sim)
q90sim
  2      2.5      3    3.5      4
701      218   1053     12     16
```

The 90% quantile of the observations is 2. Out of the 2000 replicated data sets, 701 had the same 90% quantile and in 1299 data sets this quantile was larger. However, because of the discreteness of the data and the low variance in the 90% quantiles of the replicated data sets, this test statistic may be not extremely sensitive. For example, the Bayesian *p*-value is strongly dependent on whether we use the "larger than" or "larger" sign.

```
mean(q90sim>=2)
[1]  1
mean(q90sim>2)
[1]  0.6495
```

Instead of looking at the 90% quantile of the observations *y*, we could look at the 90% quantile of the residuals to assess whether the variance in the upper tail is adequately captured by the model. To do so, we calculate residuals for the observed data for all the 2000 models described by the 2000 simulated sets of model parameters, and we calculate the standardized residuals for the replicated data sets. Then, we extract the 90% quantile for each set of residuals and compare them.

```
obsresid <- matrix(nrow=nrow(dat), ncol=nsim)
for(i in 1:nsim) obsresid[,i] <- (dat$bp-fitmat[,i])/sqrt(fitmat[,i])
yrepresid <- (yrep-fitmat)/sqrt(fitmat)
q90obsresid <- apply(obsresid, 2, quantile, prob=0.9)
q90yrepresid <- apply(yrepresid, 2, quantile, prob=0.9)
```

The plot of the observed 90% quantiles of the residuals versus the 90% quantiles of the residuals for the replicated data shows that the 90% quantiles of the residuals for the replicated data might be slightly higher on average than the ones for the observed data (Figure 10-2). But the Bayesian *p*-value is not too far from 0.5.

```
mean(q90yrepresid>=q90obsresid)
[1] 0.7305
```

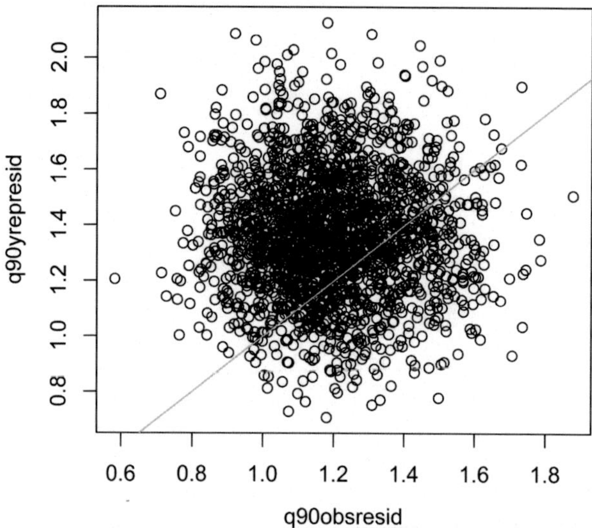

FIGURE 10-2 The 90% quantiles of the residuals for the replicated data versus the 90% quantiles of the residuals for the observed data. The orange line is the $y = x$ line.

We also can look at the spatial distribution of the data and the replicated data. The variables X and Y are the coordinates of the wildflower fields. We can use them to draw transparent gray dots sized according to the number of breeding pairs

```
par(mfrow=c(3,3), mar=c(1,1,1,1))
plot(dat$X, dat$Y, pch=16, cex=dat$bp+0.2, col=rgb(0,0,0,0.5),
  axes=FALSE)
box()
r <- sample(1:nsim, 1)    # draw 8 replicated data sets at random
for(i in r:(r+7)){
  plot(dat$X, dat$Y, pch=16, cex=yrep[,i]+0.2,
          col=rgb(0,0,0,0.5), axes=FALSE)
  box()
}
```

The spatial distribution of the replicated data sets seems to be similar to the observed one at first look. With a second look, we may detect two patterns: (1) higher values can occur in the replicated data than in the observed data, and (2) in the middle of the study area the model may predict slightly larger numbers than observed (Figure 10-3). The first pattern is familiar, because we have seen it already by looking at the 90% quantiles of the data and at the residuals. The second pattern may motivate us to find the reason for the poor fit if the main interest is whitethroat density estimates. Are there important elements in the landscape that influence whitethroat densities and that we have not yet taken into account in the model? However, our main interest is finding

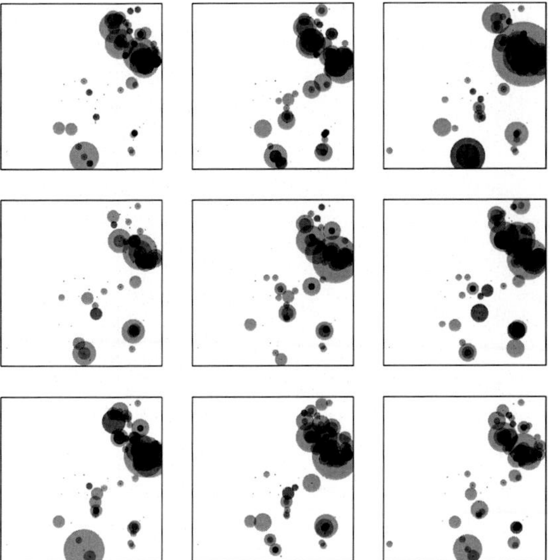

FIGURE 10-3 Spatial distribution of the whitethroat breeding pair counts and of 8 randomly chosen replicated data sets with data simulated based on the model. The smallest dot corresponds to a count of 0, the largest to a count of 20 breeding pairs. The panel in the upper left corner shows the data, the other panels are replicated data from the model.

the optimal age of wildflower fields for the whitethroat. Therefore, we look at the mean age of the 10% of the fields with the highest breeding densities.

To do so, we first define a function that extracts the mean field age of the 10% largest whitethroat density values, and then we apply this function to the observed data and to the 2000 replicated data sets.

```
magehighest <- function(x) {
  q90 <- quantile(x/dat$size.ha, prob=0.90)
  index <- (x/dat$size.ha)>=q90
  mage <- mean(dat$age[index])
  return(mage)
}

magehighest(dat$bp)
[1] 4.4
magesim <- apply(yrep, 2, magehighest)
quantile(magesim, prob=c(0.01, 0.5,0.99))
      1%      50%       99%
3.666667 4.733333 5.857143
```

The mean age of the 10% of the fields with the highest whitethroat densities is 4.4 years in the observed data set. In the replicated data set it is between 3.7 and 5.9 years. The Bayesian *p*-value is 0.80. Thus, in around 80% of the replicated data sets the mean age of the 10% fields with the highest whitethroat densities was higher than the observed one (Figure 10-4).

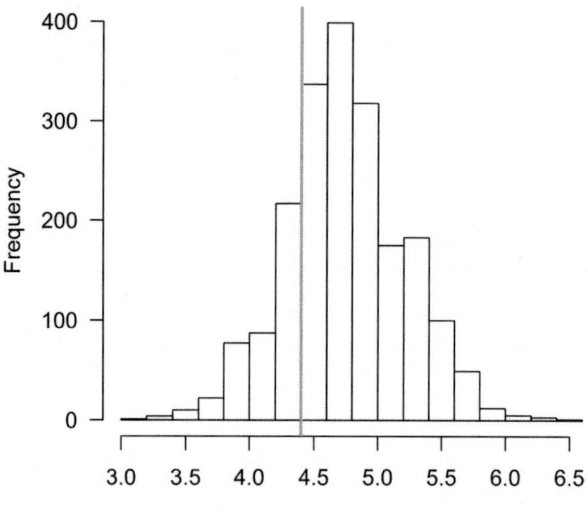

FIGURE 10-4 Histogram of the average age of the 10% wildflower fields with the highest breeding densities in the replicated data sets. The orange line indicates the average age for the 10% fields with the highest observed whitethroat densities.

```
mean(magesim>=magehighest(dat$bp))
[1] 0.8005
hist(magesim)
abline(v=magehighest(dat$bp), col="orange", lwd=2)
```

We can summarize the fit of different aspects of the model in a table (Table 10-1). We conclude that the model fits in most important aspects reasonably well but prediction intervals may have upper limits that are too high.

TABLE 10-1 Results of Posterior Predictive Model Checking for Four Different Test Statistics for the Whitethroat Data

Test statistics	Observed data	Replicated data (1–99% quantiles)	Bayesian *p*-value
Proportion of zero	47%	40–63	0.85
90% quantile of y	2	2–3.5	0.65–1
90% quantile of the residuals	1.19	0.86–1.92	0.73
Mean age of 10% fields with highest densities	4.4	3.7–4.7	0.80

10.2 MEASURES OF EXPLAINED VARIANCE

Why do we discuss the measure of explained variance, R^2, in the same chapter as posterior predictive model checking? From a Bayesian perspective, these two things do not have much in common. However, the book is structured more like a classical introduction to applied data analyses than an introduction to Bayesian data analyses. And, classically, people will look for R^2 in the "goodness of fit" section, because it measures the proportion of the variance that can be explained by the model. Therefore, the small section on R^2 is located in the same chapter as posterior predictive model checking. But, as you will see in the following, we use the R^2 value more as an alternative measure of effect size than as an overall goodness of fit measure.

Classically, the R^2 value, also called coefficient of determination or squared multiple correlation, is used as a measure of explained variance. In the case of a one-level normal linear model (a normal linear model without random effects), the R^2 value is the proportion of the sum of squared differences from the overall mean that is explained by the model, or 1 minus the proportion of unexplained (residual) sum of squares ($\text{SSR} = \sum_{i=1}^{n}(y_i - \widehat{y})^2$):

$$R^2 = 1 - \frac{SSR}{SST}$$

where SST is the sum of the squared differences from the overall mean (total sum of squares, Section 4.2.2).

If the residual variance is close to the overall variance, R^2 is close to 0. Then, the model does not explain a lot of variance. The smaller the residual variance, the closer R^2 is to 1. This version of R^2 approximates 1 when the number of model parameters approximates sample size even if none of the predictor variables correlates with the outcome. It is exactly 1 when the number of model parameters equals sample size, because n measurements can be exactly described by n parameters. The adjusted R^2,

$$R^2 = \frac{\text{var}(y) - \widehat{\sigma}^2}{\text{var}(y)}$$

takes sample size n and the number of model parameters k into account. Therefore, the adjusted R^2 is recommended as a measurement of the proportion of explained variance (e.g., Draper & Smith, 1998; Glantz & Slinker, 2001). Recall that $\widehat{\sigma}^2$ is the sum of the squared residuals divided by $n - k$.

In the case of mixed models or generalized linear (mixed) models, the definition of a measure similar to R^2 or the adjusted R^2 is not straightforward. There are many different versions of R^2 values (e.g., Cameron & Windmeijer, 1997; Menard, 2000; Nakagawa & Schielzeth, 2013). All these R^2 values summarize the explained variance with one or two numbers. However, for hierarchical models (such as mixed models) such numbers are more complicated to interpret, because these models have more than one level. In addition, in these models different proportions of variance can be explained for different levels of the model (Gelman & Pardoe, 2006). For example, a numeric predictor x can show a high correlation with the group means of y but it does not necessarily have to be correlated with y within the groups. On the other hand, x could show no correlation with the group means but within the groups, x is highly correlated with y. Having measures of explained variance for different levels of the data can assist with biological interpretation.

To illustrate this, we use a random slope mixed model that has predictors on the data level and on the group level. Anthes et al. (2014) investigated sperm depletion across consecutive copulations in the sea slug *Chelidonura sandrana*. The outcome variable was the square root of the number of sperm a donor (this sea slug species is a hermaphrodite, we therefore speak of a sperm donor rather than a male) transferred to the receiver per copulation. Each sperm donor was observed over four consecutive copulations. Donor-level predictor variables were the (centered) weight (weight.c) and a relative body size measurement (relSize.c). Data-level predictors were the number (matN.c) and the duration (dur.c) of copulations. The main interest of the study was sperm depletion measured as the effect of the copulation number on the number of transferred sperm and how this depletion differed between the donors, for example, of different weights. Therefore, the interaction of weight and mating number and donor-specific random slopes for copulation number were included in the model. The outcome variable was square-root transformed to achieve normally distributed residuals.

```
data(spermdepletion)
dat <- spermdepletion
mod <- lmer(sqrt(totalsperm) ~ weight.c + matN.c + relSize.c + dur.c +
  weight.c:matN.c + (matN.c|focal), data=dat)
mod
Linear mixed model fit by REML ['lmerMod']
Formula: sqrt(totalsperm) ~ weight.c * matN.c + relSize.c + dur.c +
(matN.c | focal)
   Data: dat
REML criterion at convergence: 2130.963
Random effects:
Groups    Name         Std.Dev. Corr
focal     (Intercept) 13.203
          matN.c        3.473   -0.89
Residual              11.015
Number of obs: 264, groups: focal, 66
Fixed Effects:
    (Intercept)      weight.c       matN.c      relSize.c       dur.c
        41.2498        6.5589      -6.9716        28.9171      0.7427
weight.c:matN.c
        -2.2914
```

Sperm counts substantially declined across four successive copulations indicating sperm depletion. Sperm depletion occurred at a disproportionately faster rate in heavier slugs. The estimated effects are, for average sized slugs, −7.0 (95% CrI: −8.5 to −5.4), and the effect of the average weight of the slugs on the depletion rate is −2.3 (−3.3 to −1.3). The units of these effects are square roots of the number of sperms per mating, and square roots of the number of sperms per mating and mg, respectively. Only specialists of mating behavior in sea slugs will be able to judge whether these effects are biologically relevant or not. For other people, it may be helpful to relate these effects to the overall variance in the specific level of the data, that is, to give estimates for the proportion of variances explained at the different levels of the data.

The sperm depletion model has three variance components, and thus the proportion of explained variances exist on three different levels: (1) the data level (what proportion of the total variance in the data is explained by the model?), (2) in the intercept at the focal individual level, and (3) in the slope at the focal individual level (what proportion of between-individual variance in the intercept and the slope is explained by the model?).

First, we calculate the proportion of variance that the model can explain at the data level (the level of the measurements). This is the variance that all predictors (fixed and random effects together) can explain of the total variance in the data. This R^2 is 1 minus the residual variance divided by the variance in the data: $1 - \frac{var(y-\widehat{y})}{var(y)}$. In a Bayesian framework, we average this measurement over the posterior distribution of the fitted values, $\widehat{y}$. To do so, we first calculate the 2000 fitted values for each observation from the 2000 simulated values from the posterior distribution of the model parameters. Note that we have to add the random intercepts (first parameter) and random slopes for mateN.c (third parameter) for each focal individual to the fixed effects before calculating the fitted values (matrix multiplication). This calculation has to be done in a loop for every observation (index j) separately, because the fixed effects differ between the observations. In the last step, we calculate the residuals for every fitted value as the difference between the observed square roots of the sperm counts and the fitted values.

```
nsim <- 2000
bsim <- sim(mod, n.sim=nsim)
Xmat <- model.matrix(mod)
y.hat <- matrix(nrow=nrow(dat), ncol=nsim)
resid.y <- matrix(nrow=nrow(dat), ncol=nsim)
for(i in 1:nsim){
  bsimi <- matrix(fixef(bsim)[i,], nrow=nrow(dat),
                  ncol=length(fixef(mod)),byrow=TRUE)
  bsimi[,1] <- bsimi[,1] +
     ranef(bsim)$focal[i,match(dat$focal, levels(dat$focal)),1]
```

```
  bsimi[,3] <- bsimi[,3] +
     ranef(bsim)$focal[i,match(dat$focal, levels(dat$focal)),2]
  for(j in 1:nrow(dat)){
    y.hat[j,i] <- Xmat[j,]%*%bsimi[j,]
    resid.y[j,i] <- sqrt(dat$totalsperm[j])-y.hat[j,i]
  }
}
```

The object resid.y now contains 2000 sets of residuals from which we extract 2000 residual variances (using the function `apply`). These 2000 residual variances describe the posterior distribution of the residual variance. We use the mean of these values to obtain R^2.

```
1-mean(apply(resid.y, 2, var))/var(sqrt(dat$totalsperm))
[1] 0.7483652
```

The model explains around 75% of the total variance in the square root of the sperm counts. However, this includes the proportion of variance that can be explained by the variance among the focal individuals. This corresponds to the "conditional" R^2 in Nakagawa and Schielzeth (2013). Not all of this explained variance can be allocated to fixed effects.

Therefore, we calculate a second R^2 that measures the proportion of between-focal individual variance in the intercepts that can be explained by the fixed effects. To do so, we compare the between-individual variance that cannot be explained by the fixed effects (the estimated between-individual variance in the intercept, `apply(ranef(bsim)$focal[,,1], 1, var)`) with the variance in the individual-specific means. The latter is the sum of the fitted values on the focal level (without taking into account the individual-specific random effects) and the individual-specific random effects. It is obtained in R by first defining a new data frame that contains one row for each focal individual (datfocal). This new data frame contains for each focal individual, the individual-specific measurements (weight.c and relSize.c) and the means of the repeated measurements within the focal individuals for matN.c and dur.c.

```
datfocal <- aggregate(dat$weight.c, list(focal=dat$focal), mean)
names(datfocal)[names(datfocal)=="x"] <- "weight.c"
datfocal$matN.c <- aggregate(dat$matN.c, list(focal=dat$focal),
                             mean)$x
datfocal$relSize.c<-aggregate(dat$relSize.c,list(focal=dat$focal),
                             mean)$x
datfocal$dur.c<-aggregate(dat$dur.c, list(focal=dat$focal), mean)$x
```

We prepare the matrix a.hat to store the 2000 values that describe the posterior distribution for each fitted value on the focal level. Then, we fill up this matrix columnwise. Note that here, we do not add the random effects, because these fitted values should show how we can explain the focal means solely based on the fixed effects.

```
a.hat <- matrix(nrow=nrow(datfocal), ncol=nsim)
Xmat <- model.matrix(~ weight.c * matN.c + relSize.c + dur.c,
                      data=datfocal)
for(i in 1:nsim){
    a.hat[,i] <- Xmat %*% fixef(bsim)[i,]
  }
```

Then we compare the between-individual variance that cannot be explained by the fixed effects (i.e., the estimated between-individual variance) with the total variance in the means per focal individual (i.e., the sum of the fitted value and the random focal effect). Note that we have to transpose (R function `t`, use the rows as columns) the a.hat matrix because the rows in the a.hat matrix correspond to the focal individuals and the columns to the simulation, whereas the random effects in the ranef slot of bsim are assorted in the opposite way (rows = simulations and columns = focal individuals).

```
1-mean(apply(ranef(bsim)$focal[,,1], 1, var))/
  mean(apply(t(a.hat)+ranef(bsim)$focal[,,1], 1, var))
[1] 0.3853014
```

Thus, around 39% of the among-focal individual variance in the square root of sperm number can be explained by the individual-specific predictors weight and relative size and the data-level measurement dur.c. The variable matN.c does not contribute to the proportion of explained variance here, because there is no between-individual variance in this variable.

In contrast, the data-level measurement duration.c has variance both within a focal individual and between the individuals. As a consequence, this variable explains variance on both levels (the data level and the focal individual level). Sometimes, such variables can even have different effects on the different levels. For example, individuals that mate for a long time could be the ones that have fewer sperm but the longer the same individual mates, the more sperm it transfers. Thus, we have a negative relationship between duration and number of sperm on the between-individual level but a positive one within individual. In such cases, the model we used is not useful to describe the effect of duration.c because it estimates a weighted average (because of partial pooling) of the two effects, which may not make sense biologically. Then, we might consider estimating the between- and within-individual effects separately, for example, as described by van de Pol and Wright (2009).

From these results, we have seen that the order of the copulation in the sequence of copulations (matN.c) is negatively related to the square root of the number of sperm that are transferred. This indicates sperm depletion. We also saw that the strength of this depletion depends on the weights of the slugs and that it varies between the focal individuals (random slope). What proportion of variance in the slopes (= effects of matN.c) can be explained by weight? We obtain this R^2 by comparing the between-focal individual variance in the slope (the variance that we cannot explain by the fixed effects) with the variance in the slopes per focal individual that we estimated based on their weights. To do so, we first prepare the matrix b.hat that we fill up with the estimated slopes per focal individual from the weight measurements. We calculate these slopes for all 2000 simulated sets of model parameters to describe the uncertainty in these slope estimates.

```
b.hat <- matrix(nrow=nrow(datfocal), ncol=nsim)
for(i in 1:nsim){
  b.hat[,i] <- fixef(bsim)[i,"matN.c"] +
    datfocal$weight.c*fixef(bsim)[i,"weight.c:matN.c"]
}
```

Then, we obtain the R^2 value as previously described.

```
1-mean(apply(ranef(bsim)$focal[,,2], 1, var))/
  mean(apply(t(b.hat)+ranef(bsim)$focal[,,2], 1, var))
[1] 0.5424302
```

Thus, around 54% of the between-individual variance in the effect of copulation number can be explained by the weight of the slugs.

R^2 values help with interpretation of the results at hand but they are hardly comparable between different studies, because they depend on the temporal and spatial scale of a study and on the overall variance in the outcome variable. Therefore, we do not see the R^2 value as a general measure of "goodness of fit" but rather as a quantity that measures how estimated effects relate to the total variance in the data and thus helps in understanding and interpreting the results. However, R^2 values should not replace effect size estimates along with their uncertainties and information about the overall variance in the data.

FURTHER READING

One of the first authors who compared simulated data under the model with the observed data to assess model fit was Guttman (1967). This idea has been used several times in both Bayesian and frequentist frameworks, see, for example, references in Gelman et al. (2014).

A practical and theoretical introduction to posterior predictive model checking is in Chapter 6 in Gelman et al. (2014). Chambert et al. (2014) show how predictive model checking can be used in ecological modeling to gain important insights into the system and to understand how well the model performs for the system. Example applications in ecology are Link and Sauer (2002), Schofield et al. (2013), and Korner-Nievergelt et al. (2014).

Gelman and Pardoe (2006) and Gelman and Hill (2007) explain how level-specific R^2 values can be calculated using BUGS. Nakagawa and Schielzeth (2013) propose, in addition to the conditional R^2 value, a marginal R^2 value as a quantity of biological interest for GLMMs.

van de Pol and Wright (2009) describe a simple method to disentangle within-group from between-group effects in mixed models.

Chapter 11

Model Selection and Multimodel Inference

Chapter Outline

11.1 WHEN AND WHY WE SELECT MODELS AND WHY THIS IS DIFFICULT

Model selection and multimodel inference are delicate topics! During the data analysis process we sometimes come to a point where we have more than one model that adequately describes the data (Chapters 6 and 10), and that are potentially interpretable in a sensible way. The more complex a model is, the better it fits the data and residual plots (Chapter 6) and predictive model checking (Chapter 10) look even better. But, at what point do we want to stop

http://dx.doi.org/10.1016/B978-0-12-801370-0.00011-3.

adding complexity? There is no unique answer to this question, except that the choice of a model is central to science, and that this choice may be based on mathematical criteria and/or on expert knowledge (e.g., Gelman & Rubin, 1995, 1999; Anderson, 2008; Claeskens & Hjort, 2008; Link & Barker, 2010; Gelman et al., 2014; and many more). Biologists should build biologically meaningful models based on their experience of the subject. Consequently, thinking about the processes that have generated the data is a central aspect of model selection.

Why is model selection difficult? With increasing complexity of the model, the bias (i.e., the systematic difference between a parameter estimate and its true values) decreases, whereas the uncertainty of the parameter estimates increases. To reduce bias, we would like to increase the complexity of the model (e.g., by increasing the number of predictors and building more realistic models) but with models that are too complex (compared to sample size) the uncertainty in the parameter estimates becomes so large that the model will not be useful. Hence, if the process studied is complex, a large sample size is needed. However, sample sizes are usually low compared to the complexity of the processes that have generated them. Then, less complex models may capture the information in the data, at least for some of the processes, more clearly than a model too complex for the data at hand.

Where the optimum in the tradeoff between bias and uncertainty lies depends on the purpose and the context of the study. In some studies, it is extremely important to keep bias small, whereas in other studies high precision is more useful. The first type of models are sometimes called "confirmatory" and the latter "predictive" (Shmueli, 2010). Mathematical techniques for selecting a model or for basing the inference on several models simultaneously (multimodel inference) are particularly helpful for constructing predictive models. For example, estimation of the total population of wallcreepers *Tichodroma muraria* in Switzerland may be done based on a habitat model fitted to population count data on a sample of 1 km^2 plots (e.g., from a monitoring program).

The purpose of the model is to predict wallcreeper abundance for the nonsampled plots. The sum of the predicted abundances over all plots gives an estimate for the total population of wallcreepers in Switzerland. The aim of the model in this example is to produce precise predictions. Therefore, we aim at selecting, among the available habitat variables, the ones that produce the most reliable predictions of wallcreeper abundance. In this case, the predictive performance of a model is more important than the estimation of unbiased effect sizes, hence, we aim at reducing the uncertainty of the estimates, rather than including all possible relevant predictors. Many of the widely available habitat variables will not have a direct relationship with the biology of the wallcreeper and their estimated effect sizes will be difficult to interpret anyway. Therefore, it is less of a problem when the estimated effects are biased as long as the predictions are reliable.

In contrast, in a study that aims at better understanding a system, for example, when we are interested in the effect of body size on the number of sperm a donor sea slug can produce (see Chapter 10, Section 10.2), it is important to measure the effect of body size on sperm numbers with minimal bias. Because bias decreases the more realistic the model is, we may prefer to use more complex models in such cases. Then, we would use predictive model checking and biological reasoning about the process that generated the data, rather than a mathematical model selection method. Using the latter method, final models often contain structures (e.g. fixed effects) that are only characteristic for the specific data set but have no general validity (even if out-of-data predictive performance has been used, the model is still fitted to a finite number of observations). Theoretically, we should construct the model before looking at the data. However, to do so, we need to know the system that we study in advance, which may be difficult since we often collect data to learn about the system in the first place.

Bias enters in scientific studies at many places ranging from what questions we ask, to the tools that are available for data collection, to the way we present the results. The data we collect are the ones we think are important (or the ones that can be logistically and physically collected) but not the ones that may be relevant in the system. A nonbiased study would be one that does not start with a question, where all the data are collected and then mathematical criteria are used to sort out the relationships. Of course, it is impossible to collect data for all factors relevant to the system. Therefore, we formulate specific questions and carefully design our study, at the cost of introducing bias due to the way we think about the system. Selecting models after having collected the data introduces another bias. At this stage, the bias is due to specificity of the data (overfit). But sometimes it is useful to reduce model complexity by a formal model selection or multimodel inference method. Table 11-1 lists some situations that evoke a model selection or multimodel inference problem and how we may address the problem.

The out-of-data predictive performance of a model is certainly very important in studies such as the preceding wallcreeper example. But also in basic research studies, such as the slug example, measurements of the predictive performance can help in understanding a model. A model that is too simple will lead us to make poor predictions in many cases because it does not account for all important processes, whereas a too complex model (compared to the sample size) will make poor predictions because parameter estimates are uncertain. Unfortunately (or fortunately?), there does not yet exist a single method for measuring predictive performance that is generally considered the gold standard. That makes model selection difficult. Here, we try to summarize and present some methods that we found useful for specific problems. A plenitude of articles and books discuss model selection and multimodel inference much more exhaustively than we do here (Section 11.5).

TABLE 11-1 Situations (after residual analysis and predictive model checking) That Evoke Model Selection or Multimodel Inference with Useful Methods

Situations where a mathematical criteria can help ranking the models	Useful methods
Omit interactions from a model for better interpretability	Decide whether the interaction is biologically relevant, compare effect sizes, or use WAIC, or cross-validation
There is a random factor (e.g., an observation-level random factor to account for overdispersion) that should be omitted to simplify the model (reduce computing time, increase convergence performance)	Decide whether the estimated variance parameter is relevant. Was there a good reason for including the variance parameter in the model? If yes, do not delete it from the model. If no, try WAIC.
The residuals are spatially correlated and no theory about the shape of the correlation function exists: choose a correlation function (e.g., linear, exponential, spherical)	WAIC may help, but as it has difficulties in structured models, such as spatial models (Gelman et al. 2014), AIC may be a pragmatic alternative. Use predictive model-checking to see which model best captures the data's spatial distribution.
Different models represent different biological hypotheses, which mutually exclude each other. It should be measured which one has the highest support given the data	Bayes factor and posterior model probabilities are nice tools for that. However, results are more easily interpreted if prior information is available. In case of no prior information, WAIC may be a better alternative. If the interest is in derived parameters (e.g., fitted values, predictions, population estimates), model averaging may be considered.
Describe a nonlinear effect using polynomials: How many polynomials should be included?	Thorough residual analysis, WAIC, or cross-validation or, using a generalized additive model or a nonlinear model instead of the polynomials may be considered.
Many explanatory variables and only the important ones should be included to reduce uncertainty	Do not remove variables of interest (aim of the study) from the model. For nuisance variables, cross-validation, WAIC, LASSO, ridge regression, or inclusion probabilities may be useful. Choose method depending on the purpose of the study (Figure 11-2).
There are more explanatory variables than observations	LASSO and ridge regression may be helpful.

Note: This list is neither intended to be exhaustive nor dogmatic.

In many cases, models have specific purposes. Based on results from a statistical model, decisions are made and actions are started. These decisions and actions are associated with benefits and risks. The benefits and risks define the "utility function". Useful models are associated with high values of the utility function. Such methods are beyond the scope of this book, but we highly recommend focusing on the purpose of the model when comparing (or selecting) models. A primer to decision analysis can be found in Yokomizo et al. (2014).

11.2 METHODS FOR MODEL SELECTION AND MODEL COMPARISONS

11.2.1 Cross-Validation

The principle of cross-validation is to fit the model using only a part of the data (training data) and to make predictions based on the data that are held out. The differences between the observed and predicted values are measured. Frequentists use the mean squared differences (MSE) or the root of the MSE (RMSE) as a measure of predictive fit (cross-validation score). Bayesians sum the logarithm of the predictive density values of the holdout data points using the model fitted to the training data (see following). To do a cross-validation, one observation at a time can be held out (leave-one-out cross-validation) or several observations can be held out simultaneously (k-fold cross-validation, where k is the number of groups held out in turn). How many observations should be held out in turn depends on the data structure. The holdout data should be independent of the training data (e.g., leave out an entire level of a random factor; the data held out should not be spatially or temporally correlated with the training data).

In the case of unstructured (independent) data, each single observation is left out in turn. For hierarchical or grouped data (mixed models) the cross-validation procedure has to be programmed in R by hand, because the way the data are partitioned into training and holdout sets is unique to each data set. For independent observations (LM or GLM), the function `loo.cv` calculates the Bayesian leave-one-out cross-validation estimate of predictive fit (described in Gelman et al., 2014). It leaves out one observation y_i at a time, and fits the model to the remaining data y_{-i}. Then, it integrates the predictive density of y_i over the posterior distribution of the model parameters. The log of this value is summed over all observations (i.e., leaving out each observation in turn) to return the leave-one-out cross-validation estimate of predictive fit. The larger this value is, the better the prediction fits the left-out observations. Because all n models are based on $n - 1$ observations only, the predictive fit is underestimated, especially when sample size is small. This bias can be corrected by setting the argument "bias.corr" to TRUE.

We illustrate Bayesian leave-one-out cross-validation using the "pondfrog1" data. This data set contains the size of frog populations in different ponds with different characteristics. It includes simulated data. Therefore, the "true" model is known and model comparison using different methods may be illustrative. Let's say that we need to estimate the total frog population in a large area with around 5000 small ponds. It is not feasible to count the number of frogs in 5000 ponds. Therefore, we would like to predict the number of frogs for the 5000 ponds from a sample of 130 ponds, for which the frog populations have been counted, and pH, water depth, and average temperature have been measured. The aim is to use these variables as predictors for the number of frogs for the remaining 4870 ponds. Measuring water depth, pH, and temperature is expensive. Therefore, we would like to include only those predictors that are really necessary.

First, we fit eight candidate models (because these are simulated data, we know that model 1 is the "true" one).

```
data(pondfrog1)
mod1 <- glm(frog ~ ph + waterdepth + temp, data=pondfrog1,
            family=poisson)
mod2 <- glm(frog ~ waterdepth + temp, data=pondfrog1,
            family=poisson)
mod3 <- glm(frog ~ ph + temp, data=pondfrog1, family=poisson)
mod4 <- glm(frog ~ ph + waterdepth , data=pondfrog1,
            family=poisson)
mod5 <- glm(frog ~ ph , data=pondfrog1, family=poisson)
mod6 <- glm(frog ~ waterdepth , data=pondfrog1, family=poisson)
mod7 <- glm(frog ~ temp, data=pondfrog1, family=poisson)
mod8 <- glm(frog ~ 1 , data=pondfrog1, family=poisson)
```

The leave-one-out cross-validation predictive performance of the different models can be compared.

```
loo.cv(mod1, bias.corr=TRUE, nsim=1000)
$LOO.CV
[1] -402.6925

$bias.corrected.LOO.CV
[1] -402.6491

$minus2times_lppd
[1] 805.2983

$est.peff
[1] 4.678689
```

The output gives the leave-one-out cross-validation estimate of predictive fit, the bias-corrected leave-one-out cross-validation, which is recommended for small sample sizes, and the minus two times leave-one-out cross-validation estimate, which corresponds to a cross-validated deviance. We also get an estimate for the number of effective parameters (est.peff), which is the difference between the non-cross-validated predictive fit and the cross-validated predictive fit (LOO.CV). It measures "overfit", that is, how much the model fits better to the data used for model fitting compared to new data. The LOO.CV values for the other models were −404.7, −1194.6, −2799.3, −3254.6, −2782.3, −1188.7, −3252, thus the first two models seem to have the highest predictive fit.

Cross-validation in a Bayesian framework is extremely computing time intensive, because n models have to be fitted, and for each model the posterior distribution of the parameter estimates needs to be simulated. Frequentist cross-validation requires less computing time, because no posterior distributions have to be simulated. And, for normal linear models, a leave-one-out cross-validation can be computed using matrix algebra without having to refit any model (e.g., see Wood, 2006). This may be a fast alternative.

11.2.2 Information Criteria: Akaike Information Criterion and Widely Applicable Information Criterion

The Akaike information criterion (AIC) and the widely applicable information criterion (WAIC) are asymptotically equivalent to cross-validation (Stone, 1977; Gelman et al., 2014). AIC is minus two times the log likelihood (the "frequentist" likelihood, see Chapter 5) plus two times the number of model parameters (Akaike, 1974):

$$AIC = -2\log\left(L\left(\widehat{\theta^{ML}}\middle|y\right)\right) + 2K = -2\log\left(p\left(y\middle|\widehat{\theta^{ML}}\right)\right) + 2K$$

where $\widehat{\theta^{ML}}$ is the vector of maximum likelihood estimates of the model parameters and K is the number of parameters.

The first quantity measures the model fit. Because of the factor "−2", the smaller the value, the better the model fits the data. The second quantity penalizes for overfit: for each parameter, a value of two is added. Generally, the smaller the AIC, the "better" is the predictive performance of the model. Philosophically, AIC is an estimate of the expected relative distance between the fitted model and the unknown true mechanism that actually generated the observed data (Burnham & Anderson, 2002).

When sample size is small, a higher penalty term is needed and a corrected AIC value is more reliable:

$$AIC_c = AIC + 2\frac{K(K+1)}{n-K-1}$$

As AIC_c approximates AIC for large samples, it is recommended to us AIC_c in all cases. For simplicity, we use the name "AIC" for all variants of it.

The AIC can be used to rank models similar to the cross-validation estimate of predictive fit. Different models can be weighted according to their AIC values and AIC weights are used to obtain an averaged model, or averaged derived parameters. Because AIC is based on point estimates of the model parameters rather than their posterior distribution, the WAIC is preferred by Bayesians. Further, AIC does not work in cases with informative prior distributions and for hierarchical models (AIC-variants have been proposed for hierarchical models; Burnham & Anderson, 2002; Vaida & Blanchard, 2005). AIC weights correspond to Bayesian posterior model probabilities given specific prior model probabilities (Anderson, 2008).

The WAIC is a Bayesian version of the AIC. It behaves asymptotically similar to the Bayesian cross-validation (Watanabe, 2010). It also works for hierarchical models and when informative priors are used. Its formula looks similar to the one for the AIC, but instead of the log-likelihood for the ML-point estimates of the model parameters, WAIC uses the logarithm of the pointwise predictive density (the likelihood function averaged over the posterior distribution of the model parameters and its logarithm summed over all observations, see Chapter 5). The penalization term for overfit is an estimate of the number of effective parameters.

$$WAIC = -2\log\left(p\left(y|\theta_{post}\right)\right) + 2p_{WAIC}$$

where θ_{post} is the joint posterior distribution of the model parameters and p_{WAIC} an estimate of the number of effective parameters. p_{WAIC} can be calculated in two different ways, therefore, two different WAIC values exist (Gelman et al., 2014). Both are implemented in the function `WAIC` but the WAIC2 more strongly resembles the cross-validation and is, therefore, recommended.

```
WAIC(mod1, nsim=1000)
$lppd
[1] -397.9756

$pwaic1
[1] 4.274796

$pwaic2
[1] 4.658856

$WAIC1
[1] 804.5007

$WAIC2
[1] 805.2689
```

The lppd is the log pointwise posterior predictive density (a measure of fit; the better the model fits the data the larger it is). pwaic1 and pwaic2 are the two estimates for the effective number of parameters, a measure for the increase in model fit that is solely due to the number of parameters, and that is used to estimate WAIC1 and WAIC2. The WAIC2 values for the other models in our frog example were 809.3, 2386.4, 5599.3, 6514.0, 5563.2, 2376.9, 6515.6. Thus the conclusions are similar as with the cross-validation.

A technical warning: To compare models, the models must have been fitted using the same data set. R omits observations with missing values when fitting a linear model! Thus, if a predictor variable contains missing values, R fits the model to data excluding the missing cases. We only see this in a small note in the summary output, which is easily overlooked. When excluding the predictor variable with missing values from the model, R fits the model to a larger data set. As a result, the two models cannot be compared because they have been fitted to different data. Therefore, we have to make sure that all models we compare have been fitted to the same data set. Such a line of code may prevent unjustified comparisons:

```
dat2 <- dat[complete.cases(dat),]
```

where “dat” contains the outcome variable and all predictors (but no other variables with NAs, or too many cases might be deleted).

We are not aware of anyone who has used WAIC weights in a way similar to AIC weights. This may be because WAIC is quite new or because, in a Bayesian framework, posterior model probabilities are available (Section 11.2.4). These probabilities quantify, based on probability theory, the probability that a model is the true (or, better, “the most realistic”) model given a set of models, the data, and the prior model probabilities.

11.2.3 Other Information Criteria

The Bayesian information criterion (BIC) was introduced by Schwarz (1978) and discussed by Hoeting et al. (1999). Its purpose is not to measure the predictive performance of a model, but to obtain posterior model probabilities. The results obtained by BIC approximate results obtained by Bayes factors given specific conditions (e.g., see Link & Barker, 2010 and Section 11.3). It is no longer considered to be a Bayesian method (Gelman & Rubin, 1995). It differs from the AIC only in the penalization term that takes sample size into account.

$$BIC = -2\log\left(p\left(y\middle|\widehat{\theta^{ML}}\right)\right) + \log(n)K$$

Even though it is similar to AIC, its purpose is different. The assumptions made by BIC, particularly concerning the prior model probabilities, are so specific that in many cases it may not do what we expect. Thus, constructing Bayes factors using specific prior model probabilities may be preferred (Section 11.3).

The deviance information criterion DIC (Spiegelhalter et al., 2002) corresponds to the AIC for models fitted in a Bayesian framework. However, unlike the WAIC, it conditions on the point estimates of the model parameters $\overline{\theta}$ rather than averages over their posterior distribution. It is

$$DIC = -2\log\left(p\left(y\middle|\overline{\theta}\right)\right) + 2p_D$$

where p_D is the effective number of parameters. The first term is the deviance calculated for the mean of the posterior distribution. The effective number of parameters is the difference between the posterior mean deviance and the deviance of the mean parameters. The smaller the DIC, the higher the predictive performance of the model. For some hierarchical models DIC does not seem to behave reliably (e.g., Millar, 2009).

11.2.4 Bayes Factors and Posterior Model Probabilities

The Bayes factor corresponds to the likelihood ratio, but instead of evaluating the likelihood function for the ML-estimates, the likelihood function is integrated over the posterior distribution of the model parameters. The Bayes factor is useful when choosing one model over another when these models represent discrete, mutually exclusive hypotheses. This is a very rare situation. Furthermore, based on Bayes factors, posterior model probabilities for each model in a set of models can be calculated. Such posterior model probabilities are useful because they reflect the support in the data for each model.

Unfortunately, Bayes factors and the posterior model probabilities are very sensitive to the specification of prior model probabilities, and even to the specification of prior distributions for the model parameters. Therefore, we recommend applying Bayes factors only with informative priors. Applying Bayes factors requires some experience with prior distributions; an introduction to Bayes factors is given in, for example, Link and Barker (2010) or Gelman et al. (2014). In Section 11.3, we introduce the R package BMS, which allows calculation of posterior model probabilities. There, we also show that the specification of the prior distributions for the model coefficients and the models is essential.

11.2.5 Model-Based Methods to Obtain Posterior Model Probabilities and Inclusion Probabilities

Model-based selection methods include the process of model selection in the model fitting process. Many techniques exist to do model-based model selection. O'Hara and Sillanpää (2009) and Hooten and Hobbs (2014) review a number of these techniques. In the next section, we introduce the least absolute shrinkage and selection operator (LASSO) and ridge regression. Here, we briefly introduce reversible jump Markov chain Monte Carlo simulation (RJMCMC), an indicator variable selection to obtain inclusion probabilities and posterior model probabilities.

RJMCMC is a simulation technique that allows sampling from the posterior distribution of the model parameters of different models simultaneously. The simulation algorithm jumps between the different models in such a way that the number of "visits" it spends at the different models is proportional to their posterior model probabilities. Model averaging (Section 11.3) can be done based on such simulations. The posterior inclusion probability is the posterior probability that a specific variable is included in the model given the prior distributions for the parameters, the set of models, and the data. These probabilities correspond to the proportion of RJMCMC iterations from the models with the specific variable. The software WinBUGS provides RJMCMC for linear regression type models (see e.g., Gimenez et al., 2009). For other types of models, RJMCMC algorithms have to be programmed in R or another programming language (e.g., see King and Brooks, 2002; Gimenez et al., 2009).

Another method to obtain predictor inclusion probabilities is to multiply each parameter in the model by a latent (unobserved) indicator variable. This latent indicator variable is assumed to be Bernoulli distributed. The parameter of the Bernoulli distribution corresponds to the inclusion probability and can be estimated in addition to the parameter values. This method has been called "indicator variable selection". Because such models contain a lot of parameters, informative prior information for the model parameters helps to obtain a reliable model fit (e.g., Carlin and Chib, 1995; BUGS code is presented in the supplementary material of O'Hara and Sillanpää, 2009 or Royle, 2009).

11.2.6 "Least Absolute Shrinkage and Selection Operator" (LASSO) and Ridge Regression

In this section, we introduce two model-based methods that can be used to reduce the model dimension (i.e., the number of parameters): the LASSO (least absolute shrinkage and selection operator; Tibshirani, 1996) and ridge regression. The principle of these methods is to constrain the model coefficients within the model, so that some of them either shrink toward zero or become zero. This reduces the model dimensions because explanatory variables whose coefficients become zero are "omitted" from the model. An advantage of these methods compared to information criteria such as WAIC is that the data analyst can choose the degree of dimension reduction.

To illustrate, we apply LASSO to a multiple regression with the logarithm of the number of frogs as the outcome variable and pH, temperature, and water depth as numeric predictors (data "pondfrog1"):

$$y_i \sim Norm(\mu_i, \sigma_i)$$

$$\mu_i = \beta_0 + \beta_1 \mathrm{pH}_i + \beta_2 \mathrm{waterdepth}_i + \beta_3 \mathrm{temperature}_i$$

Amphibian specialists would leave all three variables in the model because they know that these variables are important limiting factors for amphibians. However, to illustrate LASSO, let's imagine for the moment that we had no idea about which of these variables are important (and we could have over 100 such variables!). To filter out the important variables, we add the constraint that the sum of the absolute values for the slope parameters does not exceed a specific value. The mathematical notation for this constraint is:

$$\sum_{i=1}^{3} |\beta_i| \leq t$$

As a result, depending on t, some of the coefficients collapse to zero. Note that we do not add any constraint for the intercept. Also, it is important that all variables, including the outcome variable, are z-transformed so that the scale of all variables is the same.

The R package lasso2 provides functions to fit LASSO in a frequentist framework. The Bayesian LASSO is constructed by specifying a Laplace distribution as a prior distribution for the model coefficients β_i instead of having independent prior distributions for each β_i (e.g., see Hooten & Hobbs 2014). Bayesian LASSO models can be fitted in BUGS or Stan (O'Hara and Sillanpää 2009). Because it is so easy to use, and because it illustrates nicely the principle of LASSO, here, we use the frequentist version of LASSO, implemented in lasso2.

```
library(lasso2)
data(pondfrog1)
```

We z-transform all variables. We also log-transform the outcome variable to make the residuals look normally distributed.

```
pondfrog1$ph.z <- as.numeric(scale(pondfrog1$ph))
pondfrog1$waterdepth.z <- as.numeric(scale(pondfrog1$waterdepth))
pondfrog1$temp.z <- as.numeric(scale(pondfrog1$temp))
pondfrog1$frog.z <- as.numeric(scale(log(pondfrog1$frog+1)))
```

The function `l1ce` is used to fit a LASSO model. The degree of constraint is specified in the argument "bound". Here, we use a sequence of different values for the constraint ranging from 0.02 to 1 to visualize the effect of the constraint later; "bound" is a relative value and has to be between 0 and 1; 1 means no constraint, which is equivalent to the model

```
lm(frog.z ~ ph.z + waterdepth.z + temp.z, data=pondfrog1).
mod <- l1ce(frog.z ~ ph.z + waterdepth.z
        + temp.z, data=pondfrog1, bound=c(1:50)/50)
```

The output of this model contains the parameter estimates for each of the 50 different values of the constraint. We can visualize the estimated coefficients dependent on the constraint (Figure 11-1).

```
plotlasso <- plot(mod)
matplot(plotlasso$bound[,"rel.bound"], plotlasso$mat[,-1], type="l",
     lwd=2, xlim=c(0, 1.4), ylim=c(-0.55,0.95), xlab="Constraint",
     ylab="Coefficient", col=c(1, "orange", "blue"), las=1, xaxt="n")
axis(1,seq(0,1,by=0.2))
text(cbind(1.03, coef(mod[50])[-1]), labels(mod), adj=0)
```

For a high constraint (left end of the *x*-axis in Figure 11-1), only temperature is in the model. Water depth enters at a medium constraint, and pH only enters with a very weak or almost no constraint, which means that pH may be much less important for the prediction of frog numbers compared to temperature and water depth. Thus, besides serving as a model reduction tool, LASSO can also help in understanding the model and the roles of the different predictors. If the model contains factors, group lasso needs to be used; see packages grpreg and grplasso.

Ridge regression is similar to LASSO. The difference is the form of the constraint. Instead of constraining the sum of the absolute values of the model coefficients, the sum of the squared model coefficients is constraint:

$$\sum_{i=1}^{3} \beta_i^2 \leq t$$

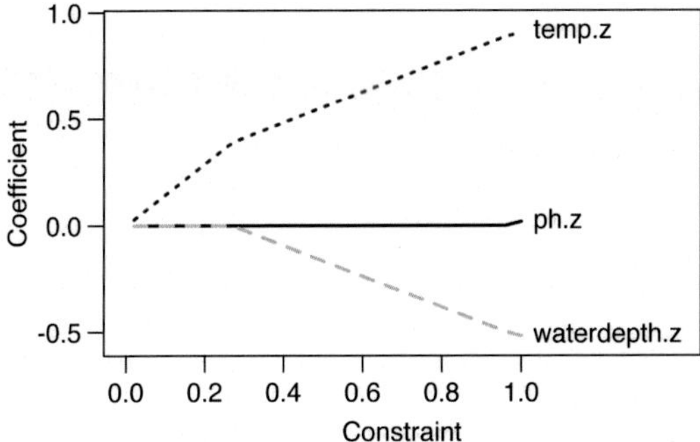

FIGURE 11-1 Estimates of the model coefficients for the three variables plotted against the LASSO constraint for the frog example data. A smaller *x*-value corresponds to a stronger constraint.

In ridge regression, model coefficients do not become exactly zero but they are shrunk toward zero. Thus, the coefficients of "unimportant" predictors will be closer to zero than when they are estimated without the constraint (i.e., simply using `lm`).

The Bayesian equivalent to ridge regression is to specify a prior distribution for the β_i, for example, $\beta_i \sim Norm(0, \tau_i)$. The priors for τ_i then specify how strongly the parameter estimates are shrunk toward zero. Different types of prior distributions for the coefficients and for τ_i have been used; see, for example, Brown et al. (2002), MacLehose et al. (2007), or Armagan and Zaretzki (2010).

LASSO and ridge regression can handle situations with more predictor variables than observations. Such data sets are common in genetics, for example, when the expression of >10,000 genes is measured in a sample of n individuals with n typically much smaller than 10,000, and the aim is to pinpoint the genes whose expressions correlate with specific phenotypes of the individuals. Another advantage of LASSO and ridge regression is that the degree of the constraint can be chosen in relation to what the model is used for, that is, in a decision theoretic framework (e.g., Brown et al., 2002; Wan, 2002).

11.3 MULTIMODEL INFERENCE

Multimodel inference is the process of drawing inference while taking model selection uncertainty into account. The principle is to average the model parameters over all (or at least several) possible models weighted by model probabilities. Often, people have used AIC weights (Burnham & Anderson, 2002). Here, we show how to do model averaging in R based on posterior model probabilities using the R package BMS (Zeugner, 2011). But the R code given in the following also works when the posterior model probabilities are replaced with AIC weights if these are preferred.

Bayesian multimodel inference provides much flexibility in how to weigh different aspects of the model via the priors. The priors for the model parameters and the model probabilities strongly influence posterior model probabilities often in not so easily understandable ways. For example, when we use flat priors for all parameters, we would expect that larger models would get higher posterior model probabilities. However, this can interfere with the prior for the model probabilities, when we specify these to favor smaller models and vice versa. Here, it is not possible to give a comprehensive overview of all the possible priors and their usage. We only give a simple example to communicate the principles of Bayesian multimodel inference. To start using Bayesian multimodel inference, we recommend Hoeting et al. (1999) as well as the tutorials and the online material for the BMS package (Zeugner, 2011).

Model averaging yields averaged parameter estimates, but it is also possible to obtain averaged model predictions, or averaged estimates for any other derived parameter, with credible intervals. In the following, we show

how to obtain averaged model predictions with credible intervals according to the methods described in Hoeting et al. (1999) using the package BMS and the function `sim`. We again use the pond frog example (as in Section 11.2.6).

```
library(BMS)
```

First, create the data matrix with the dependent variable in the first column and all other columns containing numeric values of the predictors; factors, if present, need to be transformed to dummy variables.

```
pondfrog1$y <- log(pondfrog1$frog+1)
X <- pondfrog1[, c("y", "ph", "waterdepth", "temp")]
```

Before we start doing multimodel inference using the function `bms`, we need to specify priors for the model coefficients and for the model probabilities. The priors for the model probabilities are given in the argument "mprior". "uniform" means uniform prior probabilities for all models M_i, $p(M_i) \propto 1$ (a horizontal line at 1, which is an improper prior, see Chapter 15). For the intercept and the residual variance, the function `bms` also uses improper priors, that is, $\propto 1$ for the intercept and $\propto 1/\sigma$ for the residual standard deviation.

For the model coefficients, the function `bms` uses the so-called Zellner's g prior, which is defined by the parameter g. It describes a multivariate normal distribution for the coefficients, with expected values of zero for all coefficients and a variance-covariance structure defined by the variance in the data and the parameter g. The larger g is, the less certain the researcher is that the coefficients are zero. A Zellner's g prior results in t-distributions for the posterior distributions for the coefficients with means equal to $g/(1-g)\widehat{\beta^{LS}}$, where $\widehat{\beta^{LS}}$ is the ordinary least-squares estimate that we would obtain using `lm`. The default "UIP" (unit information prior) sets $g = n$ (sample size). Using $g = n$ and uniform priors for the model probabilities leads to asymptotically equal results as when using BIC.

The function `bms` fits all possible models and calculates posterior inclusion probabilities (PIP) of each variable as well as the posterior means and standard deviations of the model-averaged parameters.

```
bmsmod <- bms(X, mprior="uniform", g="UIP")
                 PIP    Post Mean     Post SD  Cond.Pos.Sign  Idx
waterdepth 1.00000 -0.3023819 0.0107712              0    2
temp       1.00000  0.1916262 0.0038883              1    3
ph         0.12901  0.0062116 0.0233886              1    1
```

The variables in the output are ordered according to their PIP. The most important variables come first. The second column gives the posterior means of the parameters averaged over the models. It corresponds to a weighted mean of the estimates from the models with the model probabilities as weights. By default, the posterior mean is calculated over all models using the value 0 for models excluding the variable (unconditional posterior mean).

By setting the argument condi.coef = TRUE, only the models including a variable are used to calculate the posterior mean. The column Post SD is the posterior standard deviation of the parameter estimates. The next column gives the posterior probability that the parameter value is positive conditional on the inclusion of this parameter. Thus, we can infer from the output that water depth and temperature are important predictors (inclusion probabilities close to 1), and that water depth has a negative and temperature a positive effect on the size of the frog population in a pond.

We extract posterior model probabilities (PMP) using `topmodels.bma`:

```
topmodels.bma(bmsmod)[,1:5] # we extract only the 5 best models
                     3          7         1          5          2
ph             0.00000    1.00000    0.0000     1.0000     0.0000
waterdepth     1.00000    1.00000    0.0000     0.0000     1.0000
temp           1.00000    1.00000    1.0000     1.0000     0.0000
PMP (Exact)    0.87098    0.12901    0.0000     0.0000     0.0000
PMP (MCMC)     0.87098    0.12901    0.0000     0.0000     0.0000
```

The header gives the name of the models (how these names are constructed is a mystery to us). The first three lines specify the models: 1 means that the variable is included in the model, 0 means the variable is not included in the model. The last two lines give the model probabilities obtained by two different methods (see Zeugner, 2011 for details).

A model-averaged parameter estimate for waterdepth, for example, is a weighted average of the parameter estimates for the effect of water depth in the first and second model (with weights 0.87 and 0.13; the weights for the other models are essentially 0, therefore, we do not consider them further). The first model contains water depth and temperature as predictors; the second model contains all three variables. Thus, in the first model, the parameter measures the effect of water depth while holding temperature constant, whereas the same parameter in the second model measures the effect of water depth while holding temperature and pH constant. These two different measurements are only expected to have the same value if water depth and pH are uncorrelated. However, in most studies (at least in ecology) the different variables of a model are not uncorrelated, and the interpretation of an estimated effect depends on what other variables (or interactions and polynomials) are in the model. Therefore, the interpretation of a value that is a weighted mean ("Post Mean" in the output above) of qualitatively different estimates becomes obscure.

However, if the primary interest of the analysis is to estimate derived quantities, such as fitted values or population size (e.g., as in the wallcreeper in Switzerland example, compare Section 11.1), rather than model coefficients, then model-averaged fitted values or population size estimates may be useful because such averaged estimates take the model selection uncertainty into account.

When we use simulated values from posterior distributions of a derived parameter, it is easy to average it over different models: we just draw random samples from the posterior distributions of the derived parameter from the different models, with sampling effort per posterior distribution proportional to the model probabilities (PMP). Then, we simply pool these random samples to get a sample from the posterior distribution of the model-averaged derived parameter. Here is an example of how to average the predicted number of frogs for a new pond where the frogs have not been counted. This new pond has a water depth of 4.2 m, average temperature of 7.5°C, and a pH of 7.2.

First, we extract all models using the function `topmodels.bma` and save these in the object "allmods". Second, we save the model probabilities as "modweights".

```
allmods <- topmodels.bma(bmsmod)
modweights <- allmods["PMP (Exact)",] # extract model weights
```

Third, we prepare a new data frame that contains the measurements of the predictor variables for the new pond.

```
newdat <- data.frame(ph=7.2, waterdepth=4.2, temp=7.5)
```

Fourth, we calculate the number of simulated values in the averaged posterior distribution that should come from each model. In total, we would like to have 2000 simulations, thus the number of values per model equals the model probability times 2000.

```
nsim <- 2000
nsimpm <- round(nsim*modweights)
```

Fifth, we prepare the matrices to store the random values from the posterior distribution of the model-averaged predicted value, and to store the random values from the posterior predictive distribution to obtain a model-averaged prediction interval.

```
predmat <- matrix(nrow=nrow(newdat), ncol=sum(nsimpm))
predictivemat <- matrix(nrow=nrow(newdat), ncol=sum(nsimpm))
```

Now, we fit each model in turn and simulate from the posterior distribution of the prediction and from the posterior predictive distribution. This needs some coding. To fit each model in turn, we construct the model formula according to the columns in the matrix "allmods" that indicate which terms are in the model. Subsequently, we use the function `sim` to draw a number (proportional to the PMP) of random values from the joint posterior distribution of the model parameters. For each draw of the model parameters, we store the predicted value in the matrix "predmat", and one random value from the predictive distribution in the matrix "predictivemat".

```
for(i in 1:length(nsimpm)){
  nsimm <- nsimpm[i]
  if(nsimm==0) next
  termsinmodi <- rownames(allmods)[c(allmods[1:
    (nrow(allmods)-2),i],0,0)==1]
  if(length(termsinmodi)<1) termsinmodi <- "1"
  fmla <- as.formula(paste("y ~ ", paste(termsinmodi, collapse="+")))
  modi <- lm(fmla, data=pondfrog1)
  bsim <- sim(modi, n.sim=nsimpm[i])
  Xmat <- model.matrix(fmla[c(1,3)], data=newdat)
  for(r in 1:nsimpm[i]){
    predmat[,c(0, cumsum(nsimpm))[i]+r] <- Xmat %*% bsim@coef[r,]
    predictivemat[,c(0, cumsum(nsimpm))[i]+r] <-
      rnorm(1, predmat[,c(0, cumsum(nsimpm))[i]+r], bsim@sigma[r])
    }
  }
```

At last, we summarize the posterior distribution of the model-averaged prediction by giving the mean, 2.5%, and 97.5% quantiles (i.e., the credible interval), and we add the posterior 95% interval of the predictive distribution. We back-transform these values to the original scale (the transformation was log(frog+1)).

```
newdat$pred  <- exp(apply(predmat, 1, mean))-1
newdat$lower <- exp(apply(predmat, 1, quantile, prob=0.025))-1
newdat$upper <- exp(apply(predmat, 1, quantile, prob=0.975))-1
newdat$pred.lower <- exp(apply(predictivemat, 1, quantile,
  prob=0.025))-1
newdat$pred.upper <- exp(apply(predictivemat, 1, quantile,
  prob=0.975))-1
newdat
```

```
 ph  waterdepth  temp    pred   lower   upper  pred.lower  pred.upper
7.2         4.2   7.5  7.4709  7.0319  7.9374      4.8308     11.1983
```

This result means that in the new pond we expect between 5 and 11 frogs with probability 0.95. The average number of frogs in ponds with water depth, pH, and temperature values equal to the new pond is 7.5 frogs with a 95% credible interval of 7.0 to 7.9. This prediction takes into account that we are not sure, for example, whether we should use pH as a predictor variable or not. However, if we know from other studies that pH is a very important predictor for frog populations, but we have no idea how important water depth and temperature are, we could use this information to do the model averaging by specifying informative prior distributions.

To do so, we set the "mprior" argument to "pip", which means that we want to specify the prior model probabilities via the prior inclusion probabilities of the predictors. The actual values we want to use as prior inclusion probabilities are given in the argument "mprior.size". The order corresponds to the order of the variable in the data. pH is our first variable, therefore the first element is 0.9 to declare that we are 90% sure that pH should be used as a predictor. If we absolutely want to keep pH in the model, we could fix the PIP for pH to 1 setting the argument "fixed.reg" to "pH". But let's assume here that we are only 90% sure whether we would like to have pH in the model.

```
bmsmod <-bms(X, mprior="pip", g="UIP", mprior.size=c(0.9,0.5,0.5))
```

	PIP	Post Mean	Post SD	Cond.Pos.Sign	Idx
waterdepth	1.0000000	−0.30274610	0.010769335	0	2
temp	1.0000000	0.19170784	0.003883739	1	3
ph	0.5713913	0.02751055	0.042857717	1	1

Now, the inclusion probability of pH increased from 0.13 to 0.57. This shows how important prior model probabilities are when calculating posterior model probabilities. This is a good way to use existing knowledge in the analyses. However, if we do not have prior knowledge, setting the priors is a delicate process. In the preceeding example, we saw that the prior model probability did not have a noticeable influence on the prediction of the number of frogs for the new pond: The estimated mean using the informative priors was 7.4 (calculations not shown; the corresponding value for uniform priors was 7.5 before) and the 95% CrI and the prediction interval were the same as with uniform priors. This sensitivity analysis may increase the confidence we have in our predictions. But we still feel that real prior information would improve this example. Such prior information should be drawn from independent studies or experts and sufficently documented. Once priors are constructed, type `vignette("bms")` into the R console to find help on how to implement specific priors.

Another remark we would like to add: It has been questioned whether it makes sense to assign the hypothesis H: $\beta = 0$ a positive value (e.g., Gelman et al., 2014), because the probability that a continuous parameter is exactly 0 is always 0. This is at odds with the notion of posterior inclusion probabilities lower than 1. For this reason, it may be preferable to use a model that shrinks the coefficients toward 0 (rather than excluding predictors), such as in ridge regression.

11.4 WHICH METHOD TO CHOOSE AND WHICH STRATEGY TO FOLLOW

As outlined in the first section of this chapter, before starting model selection and multimodel inference a few questions should be asked: What is the type of the study? Maybe the model is defined by the study design and a formal model selection method is not needed or may even be harmful because model selection increases bias. Is the sample size large compared to the number of model parameters? If yes, a formal model selection process may also not be necessary because precision is sufficiently high to draw conclusions from the model. If sample size is low, formal model selection may be important to extract the relevant information in the data. What is the aim of a study? For a basic research study that aims to understand the system, we may be more reluctant to apply a formal model selection method, because unbiased estimates of the effects are important. In contrast, if the study has an applied purpose, formal model selection, particularly in combination with decision theoretic approaches, can strongly increase the usefulness of a model.

The lower part of Figure 11-2 gives an overview of some of the commonly used methods for doing multimodel inference. The blue lines connect related concepts and methods. The upper part of the figure show different characteristics of studies. The orange lines connect characteristics of studies with methods that are suggested given the specific characteristics of the study. The figure is intended to provide some guidance in deciding what method to choose, but it will not give unambiguous answers! For example, most studies cannot be classified as wholly experimental or observational, since most studies will contain some variables that were manipulated (treatments, variables of primary interest) as well as nuisance variables (variables of secondary interests). In such cases, the different types of variables may be treated differently with respect to model selection. We usually recommend keeping variables of primary interest in the model (independent of their effect sizes), whereas model reduction/optimization/averaging may be done with the nuisance variables.

A parsimonious model does not necessarily perform best in all situations. On one hand, if important factors have to be ignored in a model just because of parsimony, the parameter estimates can be biased. Therefore, it may be better to include more parameters and accept that the uncertainty is increased (Gelman & Rubin, 1995; Forstmeier & Schielzeth, 2011). On the other hand, if we would like to extract the information inherent in the data (i.e., identify factors that are important based on the information in the data), we may go for the most parsimonious model and admit that we cannot say anything about the factors deleted from the model, and that estimated effect sizes of the remaining factors may have some bias, that is, they may be less reliable (Burnham & Anderson, 2002; Anderson, 2008).

Once it is decided to search for a parsimonious model, that is, to do model selection, the strategy to follow needs consideration. None of the stepwise methods (backward, forward, both, or blockwise inclusions/exclusions) can be recommended. They may be fast and easy to apply, but stepwise procedures, by definition, mean that important models may not be included in the analysis and effect sizes will generally be overestimated (e.g., Whittingham et al., 2006). Therefore, it is generally better to start with a set of meaningful models. It is important that all models in the set fit the data reasonably well (residual analysis or predictive model checking needs to be done for each model, or at least for the key models in the analysis). Then, these models can be ranked or averaged.

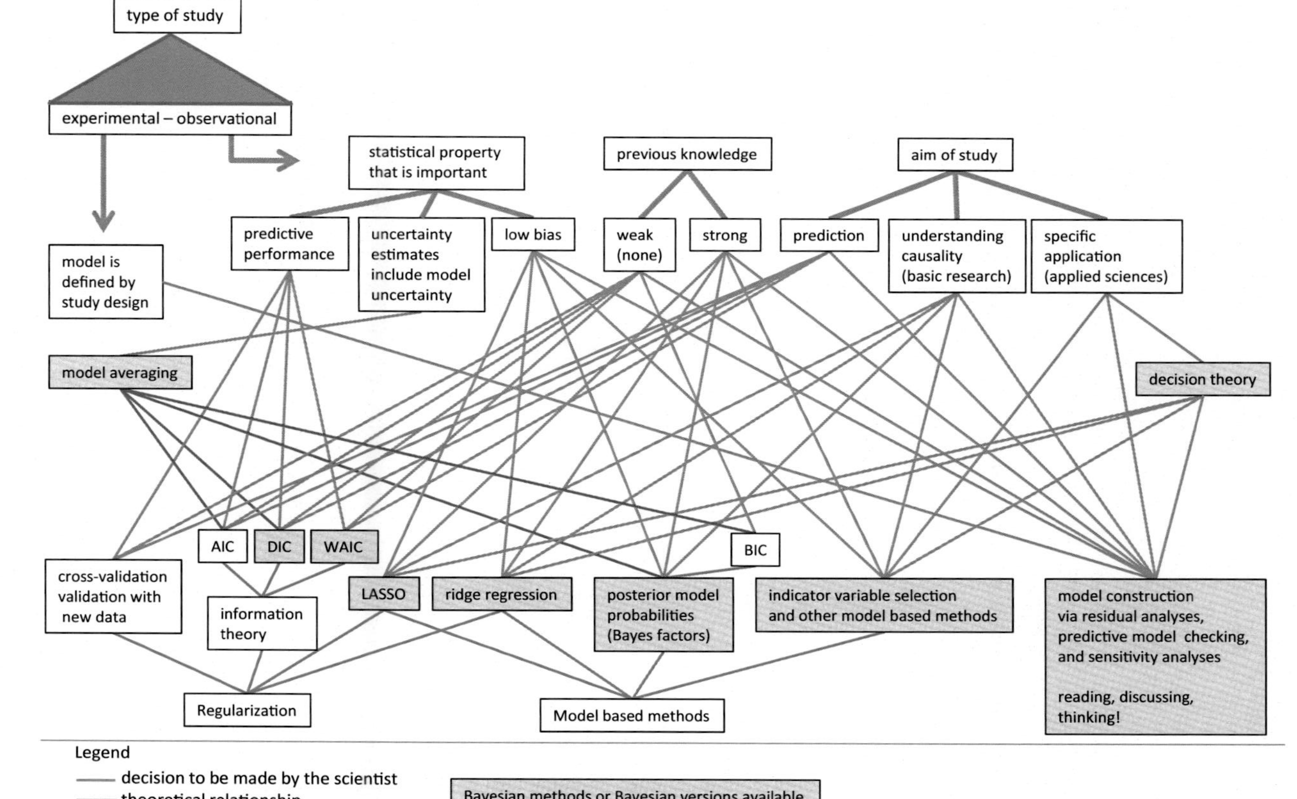

FIGURE 11-2 Schematic overview of commonly used model selection and multimodel inference methods, and how they are related. The boxes on the top categorize the types of studies the techniques may be applied to.

FURTHER READING

Kadane and Lazar (2004) discuss various different versions of the Bayes factor and many more possibilities for Bayesian model selection. We can highly recommend the up-to-date review of the commonly used model selection methods in Bayesian statistics by Hooten and Hobbs (2014). They nicely explain the common framework of the information criteria, cross-validation, LASSO, and ridge regression, which is called regularization.

Wasserman (2000) reviews AIC, BIC, and Bayes factor in a Bayesian framework. He also reviews model averaging. Gelman and Rubin (1995) describe that BIC is not a Bayesian method, explain why Bayes factors are only defined when prior distributions are proper, and how Bayes factors are sensitive to prior distributions.

More details on multimodel inference in a Bayesian framework are provided in Link and Barker (2010). Hoeting et al. (1999) give a detailed description on how to do Bayesian model averaging. More information on the computation of LASSO and ridge regression is found at *http://staffhome.ecm.uwa.edu.au/~00043886/software/lasso.html.*

The classical reference for information theory is Burnham and Anderson (2002). Anderson (2008) is a more digested version for practitioners, particularly ecologists. Buckland et al. (1997) present model averaging as a tool to draw inference while taking model selection uncertainty into account. Claeskens and Hjort (2008) give a mathematical introduction to a variety of different information criteria and model selection techniques. Johnson and Omland (2004) review model selection in ecology and evolution and present advantages of model averaging.

Claeskens and Hjort (2003) developed a focused information criteria, FIC, that aims at optimizing model selection in relation to one parameter of interest.

Grueber et al. (2011a) show how to do model averaging based on AIC weights for linear models including generalized linear mixed models. They also give R code in a worked example using the R package MuMIn (Bartoń, 2011). But note that the way Grueber et al. (2011a) obtain the confidence intervals for the averaged fitted values is wrong, see corrigendum Grueber et al. (2011b). The package AICcmodavg (Mazerolle, 2014) provides R functions to do multimodel inference on a variety of different model types. Different strategies to select models among a large number of models are discussed by Ripley (2004).

A generalization of the ordinary cross-validation is used to find the smoothing parameters in additive models (Wood, 2006).

Arnold (2009) shows that the AIC can decrease when unimportant (random) variables are added to a model ("pretending variable problem"). Whittingham et al. (2006) shows that stepwise backward model selection leads to overestimated effect sizes.

Chapter 12

Markov Chain Monte Carlo Simulation

Chapter Outline

12.1 BACKGROUND

Markov chain Monte Carlo (MCMC) simulation techniques were developed in the mid-1950s by physicists (Metropolis et al., 1953). Later, statisticians discovered MCMC (Hastings, 1970; Geman & Geman, 1984; Tanner & Wong, 1987; Gelfand et al., 1990; Gelfand & Smith, 1990). MCMC methods make it possible to obtain posterior distributions for parameters and latent variables (unobserved variables) of complex models. In parallel, personal computer capacities increased in the 1990s and user-friendly software such as the different programs based on the programming language BUGS (Spiegelhalter et al., 2003) came out. These developments boosted the use of Bayesian data analyses, particularly in genetics and ecology.

To what detail does a biologist need to understand the mathematics behind MCMC? The application of MCMC using software such as WinBUGS, OpenBUGS, JAGS, or Stan does not require a detailed knowledge of the mathematical algorithms that are used in the background. Similarly, it is not required to know the mathematical details of an iteratively reweighted least-squares method when applying the function `glm` to fit a GLM. However, we need to know why an algorithm might fail, how to judge whether the results produced by the software are reliable, and how to extract conclusions from an MCMC output. Note that in the following we will use "BUGS" as a synonym for the different BUGS-based programs such as WinBUGS, OpenBUGS, or JAGS. We will present all our examples with OpenBUGS, but most of them will also work with WinBUGS and JAGS (see Section 12.2). Additionally, we

http://dx.doi.org/10.1016/B978-0-12-801370-0.00012-5.

will introduce the program Stan. The program Stan is based on its own language with a strong focus on computational efficiency (see Section 12.3).

Of course, it is never a bad thing to learn the mathematics behind MCMC algorithms. But this is not what this book aims for. We rely on software that has been written by a team of mathematicians and computer specialists, who, for example, understand why computers can solve logit(1e−17) = logit(1×10^{-17}) but fail with logit(1 − 1e−17). In other words, we are convinced that people with backgrounds in computer sciences and mathematics can program MCMC algorithms much better and more efficiently than we (background in biology and applied statistics) could do. If you are keen to program MCMC algorithms yourself, you find some references in the Further Reading section at the end of this chapter. The R package NIMBLE (r-nimble.org) allows programming with BUGS models and even provides the opportunity to use other algorithms than the default MCMCs to efficiently solve more complicated (hierarchical) models.

A chain of values x_t, with $t = 1, \ldots, T$ (T is the total number of iterations), is a Markov chain if x_t depends (only) on x_{t-1}. The dependency of x_t on x_{t-1} is determined by a model that includes a stochastic component. Markov chains have some further mathematical properties that we do not discuss here; interested readers can refer to the Further Reading section. There are many different MCMC algorithms used in Bayesian data analyses, among which the most famous ones are Gibbs sampling and the Metropolis–Hastings algorithm. MCMC algorithms are designed so that the Markov chains converge into the joint posterior distribution of the model parameters given the model, the data, and prior distributions. For models with more than one parameters, the Gibbs sampler draws a value for parameter β_k from the conditional posterior distribution $p(\beta_k | \beta_1, \beta_2, \ldots, \beta_{k-1}, \beta_{k+1}, \ldots, \beta_K, \mathbf{y})$. One iteration, therefore, includes as many random draws as there are parameters in the model; in other words, the chain for each parameter is updated by using the last value sampled for each of the other parameters, which is referred to as "full conditional sampling".

The Gibbs sampler does not work well for some types of models. In those cases, the Metropolis–Hasting (M–H) algorithm may do better. The principle of the M–H algorithm is to first sample from the "proposal distribution" that is a crude approximation of the posterior distribution. Second, the sampled values are accepted or rejected based on an acceptance probability that leads to the accepted values being a sample from the posterior distribution (Metropolis et al., 1953; Hastings, 1970). The Hamiltonian Monte Carlo (HMC) is a more sophisticated and more efficient MCMC algorithm (Duane et al., 1987; Neal, 1994). It is a combination of MCMC and deterministic simulation methods. It first uses the derivative of the log posterior density as a gradient to find the region of the posterior distribution with high mass. And then it moves, or better, "jumps", around the posterior distribution (see Gelman et al., 2014). The no-U-turn sampler (NUTS) that is implemented in Stan is a specific type of Hamiltonian Monte Carlo (Hoffman & Gelman, 2012; Betancourt, 2013).

To start the MCMC simulations, initial values have to be given for every model parameter and for every latent variable. These initial values should not be too far away from a typical set of parameters (where posterior density is high) because the MCMC algorithms need a lot of iterations before convergence if the initial values are far in the tail of the posterior distribution. HMC is usually better at handling widely spread initial values. If different sets of initial values are used for the different chains, convergence can be better assessed (Kass et al., 1998).

To assess whether the Markov chains have converged, we plot the traces. A trace plot is a plot where the value of a draw is plotted against the iteration number of that draw. It is not possible to prove convergence but there are signs that clearly indicate nonconvergence. Nonconvergence can be seen by comparing several Markov chains in a single plot. Therefore, the golden rule is to always run more than one chain that started with different initial values. When the different chains end up in different distributions, we have a clear sign of nonconvergence. An example of a trace plot with two chains starting with different initial values is shown in Figure 12-1. After convergence, the different chains overlaid within the same plot should not be distinguishable as different chains.

In addition to trace plots, several statistics have been developed to detect nonconvergence in MCMCs. Widely used statistics is the $\widehat{R}$ that is based on the comparison of the variance across chains, or fractions of chains, with the

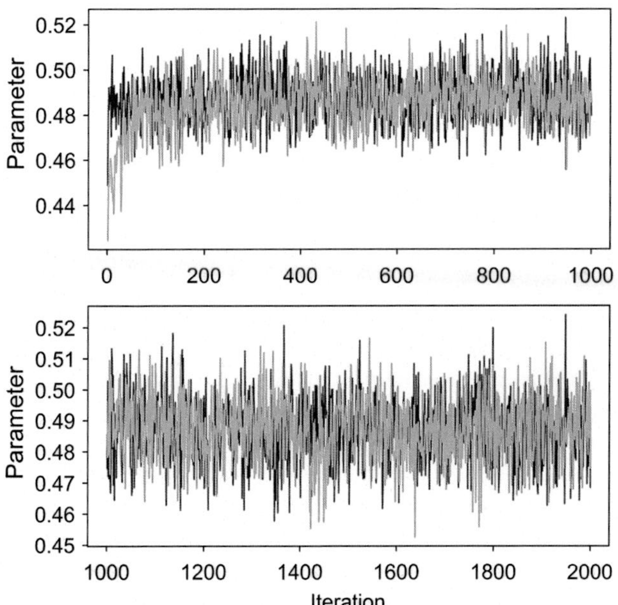

FIGURE 12-1 Trace plots (value vs. iteration) for the first 1000 iterations (upper panel), and for iterations 1000 to 2000 (lower panel). Here, the chains look converged after around 100 iterations.

variance within chains, or within-chain fractions (Brooks & Gelman, 1998; Brooks, 1998; Gelman et al., 2014). The $\widehat{R}$ described by Gelman et al. (2014) and implemented in Stan is called the "potential scale reduction factor" because it is an estimate for the factor by which the scale (square root of the variance) of the current distribution (values from all chains pooled) might be reduced if the number of iterations approaches infinity. Different authors and different software calculate $\widehat{R}$ in slightly different ways, but all $\widehat{R}$ have in common the fact that values much larger than 1 indicate nonconvergence. Where we define a threshold depends on the problem at hand. An often used threshold is 1.02 (Brooks & Gelman, 1998). However, if the Monte Carlo error is acceptable and the effective sample size reasonable, $\widehat{R}$ values lower than 1.1 may be acceptable (Gelman et al., 2014).

The starting period, that is, the iterations before convergence, are called burn-in (in BUGS) or warm-up (in Stan). The simulations done during the warm-up period in Stan are unlike the ones during the burn-in period in BUGS, in that they are not a Markov chain, hence the different names. Simulations from the burn-in and warm-up are discarded before inference is drawn from the sampled values. The length of the burn-in strongly differs between models. Complex models, especially those including random effects, usually have much longer burn-ins than simple models. The simplest way to decide on the burn-in length is to have a look at the trace plots that include several chains starting with different initial values, and discard the period when the chains have not mixed.

Markov chains, particularly Gibbs sampling and M–H algorithms, often have a strong autocorrelation. That means the value at iteration t is closer to the value at iteration $t - 1$ than it is, for example, to the value at $t - 100$. This is particularly the case when parameters of the model are correlated, that is, when the conditional posterior distribution of one parameter looks very different between different values of the other parameters. In such cases, the value generated at time t will be relatively similar to the one at time $t - 1$ because the combination of values of the other parameters only allow to sample from a part of the marginal posterior distribution (i.e., the conditional posterior distribution). As a result, a higher number of iterations are required to obtain a reliable sample of the marginal posterior distribution. When using HMC, the autocorrelation usually is much lower than when Gibbs or M–H algorithms are used.

Autocorrelation in MCMCs is not a problem per se, one just needs to make sure that the number of effective samples (see following) is large enough to get reliable information about the marginal posterior distributions. However, computer space can become a problem when a high number of values for a large number of parameters have to be stored. If this problem is encountered, the chains can be thinned, that is, only every k-th iteration is saved. Of course, when doing this, some information is thrown away—see, for example, Link and Eaton (2012). Thinning increases the relative number of effective samples compared to the total number of values stored. Thus, by thinning, the information per draw that is stored is increased but the information the MCMC simulation is acquiring per time unit is decreased.

The effective sample size is a measure of the number of independent samples from the posterior distribution. It is calculated by the number of iterations divided by a measurement for the correlation between the stored draws. For the latter, within and across-chain information is used; see, for example, Carlin and Louis (2000), Ntzoufras (2009), or Gelman et al. (2014).

The Monte Carlo error is the standard deviation divided by the square root of the effective sample size. It measures the uncertainty in the parameter estimates that is due to simulation. In general, Monte Carlo error is proportional to the inverse of the number of iterations drawn. Thus, it can (must!) be reduced to a negligibly small value by increasing the number of iterations.

How fast the chains cover the whole marginal posterior distribution is called "mixing". Chains that mix well are those that produce homogenous samples from the marginal posterior distribution after only a few iterations, whereas chains that mix badly need a high number of iterations until a reliable sample of the marginal posterior distribution is available. Bad mixing may be caused by high autocorrelation.

Once we are convinced that the Markov chains have converged and that the Monte Carlo error is negligibly small, we can draw inferences from the chains. The values of the chains are a sample of simulated values from the joint posterior distribution of the model parameters. Thus, we have exactly the same thing as we had obtained earlier using the function `sim`. The means can be extracted as parameter estimates and quantiles can be used to describe uncertainty (e.g., 95% credible interval). Or, the posterior distributions can be visualized in histograms. The whole sample of simulated values can be used to propagate the uncertainty to derived parameters such as fitted values or to show isolated effects of single variables. Further, we can use them to predict future observations and to do predictive model checking for assessing the fit of the model.

12.2 MCMC USING BUGS

Several software programs belong to the "Bayesian inference using Gibbs sampling" (BUGS) family. These programs do Gibbs sampling, M–H algorithms, and slice sampling. WinBUGS is the best known among them (Lunn et al., 2009; Lunn et al., 2013). It can be freely downloaded and installed from http://www.mrc-bsu.cam.ac.uk/software/bugs. However, WinBUGS is currently not being developed further. Active BUGS versions are OpenBUGS (www.openbugs.net; Spiegelhalter et al., 2007) and JAGS (mcmc-jags.sourceforge.net). All three are similar in that the model is specified in the BUGS language and they can be called from R using specific functions of the packages R2WinBUGS (Sturtz et al., 2005), BRugs, R2OpenBUGS, or R2jags (Su & Yajima, 2012). In the following we present all of our examples with OpenBUGS.

The principle of these programs is similar. We provide the model, the data, and the initial values. The software then chooses, based on the model, an appropriate MCMC algorithm, tunes the algorithm according to the model and the data if needed, and starts sampling. We define how many iterations should be simulated and the software produces files that contain the generated values for each model parameter. The various software differ in technical details of the MCMC algorithms and how they are tuned. Thus, speed can be different depending on the model, and the error messages differ. Sometimes the same model that can be fitted using one algorithm fails with another. It can, therefore, be helpful to try out all three programs when developing and debugging a complex model.

BUGS can be used as standalone software or via R. Here, we first show how to use OpenBUGS as standalone software and then introduce the R code to run BUGS from within R. For debugging a model code it is sometimes useful to be familiar with running the model from within BUGS because BUGS will set the cursor at the right place of a syntax error, if there is one. When we run BUGS from R, BUGS opens and we can work within BUGS as if it was a standalone program.

WinBUGS is made to run on Windows. On a Mac, we use the software Parallels, which runs Windows on Mac. Because WinBUGS writes files to the folder where we put the program, it may not work if it is in the preinstalled program folder due to write permission problems in Windows. Therefore, it may be easiest to install the program in a personal documents folder. OpenBUGS and JAGS are fully compatible with Windows, Linux, and Mac.

12.2.1 Using BUGS from OpenBUGS

To fit a model using BUGS, we first have to prepare a text file that contains the model written in the BUGS language. The BUGS language is similar to the R language with some important differences. First of all, there are only a limited number of functions available. They are listed in the BUGS manual (Spiegelhalter et al., 2003; Lunn et al., 2013). Only limited possibilities exist to do vector or matrix algebra. Instead of vector algebra, loops are used. The variance in the normal distribution is given as precision (=1/variance). Note that the parameterization of the normal and also other distributions is done differently in different software. Therefore it is worth checking before coding a distribution. Finally, the "=" is not allowed as an assignment sign and "<-" must be used.

A linear regression in BUGS language looks as follows:

BUGS Code for a Normal Linear Regression (file: "normlinreg.txt")

```
model {
  ## priors
  sigmaRes ~ dt(0,1,1)I(0,)
  tauRes <- pow(sigmaRes, -2)
  beta0 ~ dnorm(0, 0.04)
  beta1 ~ dnorm(0, 0.04)

  ## likelihood
  for(i in 1:n){
    y[i] ~ dnorm(mu[i], tauRes)
    mu[i] <- beta0 + beta1*x[i]
  }
}
```

In the likelihood section, we describe the model for every observation separately, thus the loop occurs across n observations.

For each model parameter, a prior distribution has to be specified. Prior distributions are discussed in Chapter 15. Here, we use the folded t-distribution for the residual standard deviation (sigmaRes) that has around 90% of its mass below 6.3. This means that, before having looked at the data, we think that it is unlikely that the residual standard deviation is larger than 6.3. The `I(0,)` truncates the t-distribution at 0 so that it contains only values larger than or equal to 0. The precision (tauRes) is defined as the inverse variance, that is, 1 over the square of the standard deviation. We have to calculate the precision because, in BUGS, the normal distribution is defined by the mean (first parameter in `dnorm`) and the precision (second parameter in `dnorm`). For the intercept and the slope parameters, we use normal distributions with most of their mass between -10 and 10 because we think that plausible values are within that range (see Chapter 15).

The data have to be given as a list in the model file or in a separate text file. The list must contain only the data used in the model. If we provide data that are not used in the model, OpenBUGS will crash (but not JAGS).

```
list(n=50, y=c(18.7604, 10.6176, .....), x=c(26.1818, 10.2029,
....))
```

Finally, initial values have to be given as a list either within the same text file as the model and the data or as a separate file:

```
list(beta0=-0.1395, beta1=-0.6966, sigmaRes=0.3192)
```

Then, we are ready to start OpenBUGS. To fit a model, we have to go through the following steps in exactly the order given as follows.

---

1. Check the model: Open the text file from within OpenBUGS (e.g., "normlinreg.txt") with the model specification (in most cases "as text"). Set the cursor in the word "model" in the first line of the text file. In the menu "Model" choose "Specification…" and click "check model". If the model specification is correct, we see the note "model is syntactically correct" in the lower left corner of the OpenBUGS window.
2. Open the file with the data. Set the cursor in the word "list" at the beginning of the file and click "load data" in the model specification window. If OpenBUGS says "data loaded", proceed.
3. Choose the number of chains: Increase the number of chains to at least 2 (3 in our example).
4. Click the "compile" button. Proceed when OpenBUGS returns "model compiled".
5. Load initial values: Open the files with the initial values. Set the cursor in the word "list" of the first initial values and click "load inits". Repeat this with all the remaining initial values until all chains are initialized. When the initial values do not cover all parameters we will have to click "gen inits" to finishing the initiating ("initial values generated, model initialized"). When using "gen inits", OpenBUGS will generate initial values from the prior distribution.
6. Then we can start the MCMC simulations by clicking the "update" button (menu "Model" -> "update…"). Run, for example, 1000 updates. Wait until OpenBUGS has finished.
7. Specify which parameters should be monitored. For parameters that are monitored, the sampled values are saved and can later be used for drawing inference. For all the other parameters, the samples from the posterior distribution are discarded. Choose "Samples…" from the "Inference" menu. For each parameter that should be monitored, write the parameter name into the node window and click "set".
8. Further updates: Go back to the "Update…" window and increase the number of updates (e.g., to 10,000; press "update").
9. Inspect the MCMCs: In the sample monitor tool, specify a model parameter (or use "*" for all parameters) and choose from different options to look at the MCMCs, such as history plot, autocorrelation plot, Brooks—Gelman—Rubin statistics (Brooks & Gelman, 1998), or the summary statistics. If we would like to delete the first part of the chains (burn-in), we can specify the cutoff point in the "beg" window.

12.2.2 Using BUGS from R

Using BUGS from within R saves a lot of time because no clicking is needed. Load the package R2OpenBUGS, and specify the working directory. A short path should be used for the working directory. We use a folder that we prepared for temporary storing of BUGS files.

```
library(R2OpenBUGS)
bugsworkingdir <- "…/BUGS"     # replace dots with your BUGS working
                                  directory
```

We use the same ordinary linear regression from the previous chapter as an example. The model is saved as "normlinreg.txt" in the BUGS working directory.

The initial values can be specified as a function that creates a list with the initial values for each model parameter.

```
inits <- function() {
  list(beta0=runif(1, -2, 2),
       beta1=runif(1, -2, 2),
       sigmaRes=runif(1, 0.1,2))
}
```

This inits function will be applied for each Markov chain. Thus, when using functions that produce random numbers from a specified distribution (e.g., `runif`, `rnorm`, `rpois`) within the inits function, different starting values for every chain will be produced. We can test it by applying `inits()` several times. We then need to put the data into a list. As an example data set, we use the example fake data that were produced in Section 4.1.2.

```
datax <- list(n=length(y), y=y, x=x)
```

Now we are almost ready to run OpenBUGS. We just need the names of the parameters that we want to monitor in a character vector, and then we can fit the model in OpenBUGS using the function `bugs` of the package R2OpenBUGS:

```
parameters <- c("beta0", "beta1", "sigmaRes")
fit <- bugs(datax, inits, parameters, model.file="normlinreg.txt",
            n.thin=1, n.chains=2, n.burnin=5000, n.iter=10000,
            debug=FALSE, working.directory=bugsworkingdir)
```

Beside the list of data, the initial values function, the names of the parameters that should be monitored, and the name of the model file, we specify how many chains are run (n.chains), how many iterations are done in total (n.iter), how many iterations are assigned to the burn-in and, thus, will be deleted from the final sample (n.burnin). In the argument n.thin we specify the degree of thinning of the chains. A value of n.thin = 1 means that all iterations are saved; a value of n.thin = 2 means that every second iteration is used. If the debug argument is set to TRUE, OpenBUGS will stay open when an error occurs, so that we can search for the error from within OpenBUGS. If we have used the installer to install OpenBUGS, R finds the software OpenBUGS itself. In the case that R does not find OpenBUGS, we can give the path to the folder where OpenBUGS is installed using the argument OpenBUGS.pgm. In Windows, we may have to include the name OpenBUGS.exe in the path, for example, "c:/MyDocuments/OpenBUGS232/OpenBUGS.exe".

Once OpenBUGS has finished simulations, we close OpenBUGS (only necessary if debug = TRUE). In R, we can check whether we find indications of nonconvergence by plotting the trace plots and have a look at $\widehat{R}$.

```
par(mfrow=c(3,1))
historyplot(fit, "beta0")
historyplot(fit, "beta1")
historyplot(fit, "sigmaRes")
```

In the trace plots, there is no indication of nonconvergence (Figure 12-2) and all $\widehat{R}$ are lower than 1.02. To have a look at the parameter estimates we use:

```
print(fit$summary, 2)
```

```
            mean   sd   2.5%    25%    50%    75%  97.5% Rhat n.eff
beta0       2.82 2.63  -2.18   0.98   2.77   4.63   8.02 1.01  9800
beta1       0.60 0.12   0.36   0.52   0.61   0.69   0.84 1.01  1400
sigmaRes    4.57 0.47   3.75   4.24   4.53   4.86   5.64 1.00  2900
deviance  293.38 2.45 290.50 291.60 292.80 294.50 299.70 1.01   220
```

The sample of simulated values from the joint posterior distribution produced by OpenBUGS can be used to draw the regression line with a 95% credible interval using the following R code (Figure 12-3).

```
newdat <- data.frame(x=seq(10, 30, length=100))
Xmat <- model.matrix(~x, data=newdat)
fitmat <- matrix(ncol=fit$n.sim, nrow=nrow(newdat))
for(i in 1:fit$n.sim) fitmat[,i] <- Xmat %*% c(fit$sims.list$beta0[i],
                                              fit$sims.list$beta1[i])
newdat$lower <- apply(fitmat, 1, quantile, prob=0.025)
newdat$upper <- apply(fitmat, 1, quantile, prob=0.975)
plot(y~x, pch=16, las=1, cex.lab=1.4, cex.axis=1.2, type="n",
   main="")
polygon(c(newdat$x, rev(newdat$x)), c(newdat$lower,
         rev(newdat$upper)), col=grey(0.7), border=NA)
abline(c(fit$mean$beta0, fit$mean$beta1), lwd=2)
box()
points(x,y)
```

12.3 MCMC USING STAN

Stan is similar to BUGS: a program that draws random samples from the joint posterior distribution of the model parameters given a model, the data, prior

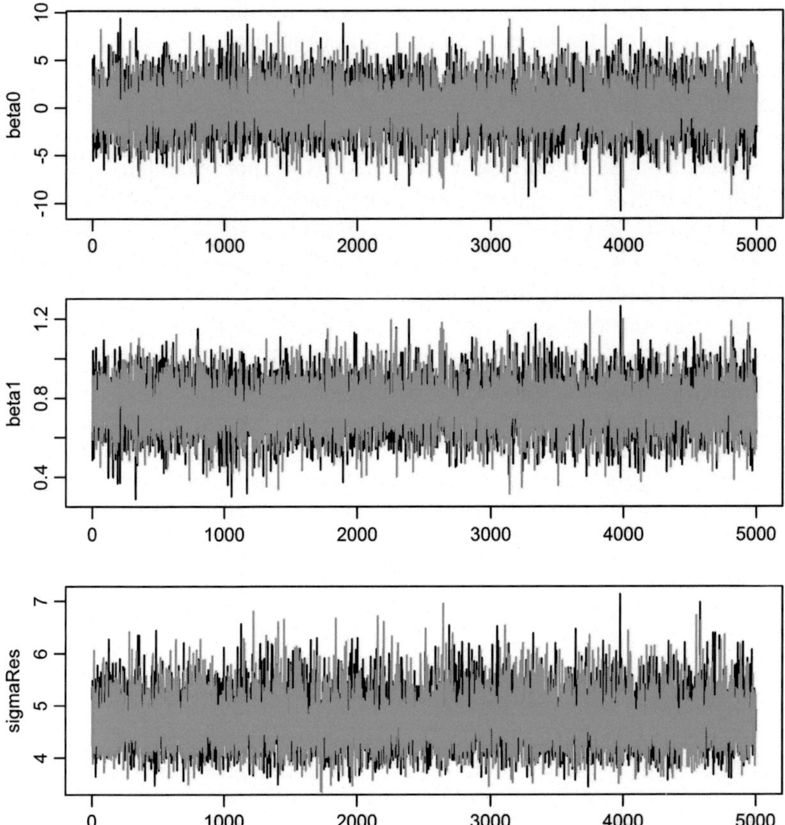

FIGURE 12-2 Trace plots of the Markov chains for the three model parameters. Two chains were run, depicted in different colors.

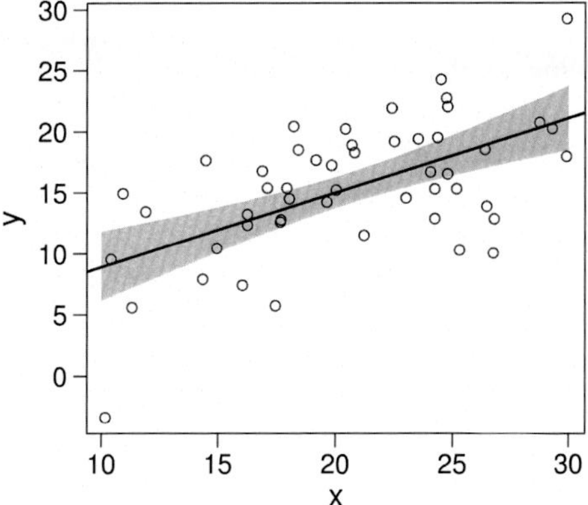

FIGURE 12-3 Regression line with 95% credible interval (shaded gray). The circles depict the data.

distributions, and initial values. To do so, it uses the "no-U-turn sampler," which is a type of Hamiltonian Monte Carlo simulation (Hoffman & Gelman, 2012; Betancourt, 2013), and optimization-based point estimation. These algorithms are more efficient than the ones implemented in BUGS programs and they can handle larger data sets. Stan works particularly well for hierarchical models (Betancourt & Girolami, 2013). For example, the mark-recapture model with random effects presented in Section 14.5 used less than 2 hours in Stan to obtain a similar effective sample size as WinBUGS produced after several days. This motivated us to learn Stan in addition to BUGS and to cover Stan in this book.

Stan runs on Windows, Mac, and Linux. It is freely available from http://mc-stan.org/. Stan can be used via the R interface rstan. Stan is automatically installed when the R package rstan is installed. For installing rstan, it is advised to follow closely the system-specific instructions. Stan requires that a C++ compiler is installed on the computer. Therefore, before downloading rstan we have to install a C++ compiler (e.g., Rtools). How to do this is explained very well on the Stan website. Besides the interface to R, interfaces to Python, PyStan, and to the shell, CmdStan, are available.

The use of Stan is similar to BUGS. The model has to be defined in a text file, the data prepared within R, and then Stan is called from R. However, the Stan language differs from BUGS and it is stricter, forcing the user to write unambiguous models. A comprehensive manual introduces the Stan language (Stan Development Team, 2014), downloadable from mc-stan.org.

A Stan code defining a model consists of different named blocks. These blocks are (from first to last): data, transformed data, parameters, transformed parameters, model, and generated quantities. The blocks must appear in this order. The model block is mandatory; all other blocks are optional. In the data block, the type, dimension, and name of every variable has to be declared. Optionally, the range of possible values can be specified. For example, `vector[N] y;` means that y is a vector (type real) of length N, and `int<lower=0> N;` means that N is an integer with nonnegative values (the bounds, here 0, are included). Note that the restriction to a possible range of values is not strictly necessary but this will help specifying the correct model and it will improve speed. We also see that each line needs to be closed by a column sign. In the parameters block, all model parameters have to be defined. The coefficients of the linear predictor constitute a vector of length 2, `vector[2] beta;`. Alternatively, `real beta[2];` could be used. The sigma parameter is a one-number parameter that has to be positive, therefore `real<lower=0> sigma;`.

The model block contains the model specification. Unlike BUGS, Stan functions can handle vectors. Therefore, we do not have to loop through all observations. Here, we use a Cauchy distribution as a prior distribution for sigma. This distribution can have negative values, but because we defined the lower limit of sigma to be 0 in the parameters block, the prior distribution actually used in the model is a truncated Cauchy distribution (truncated at zero). In Chapter 15 we explain how to choose prior distributions.

Stan Code for a Normal Linear Regression (file: "linreg.stan")

```
data {
  int<lower=0> N;
  vector[N] y;
  vector[N] x;
}
parameters {
  vector[2] beta;
  real<lower=0> sigma;
}
model {
  //priors
  beta ~ normal(0,5);
  sigma ~ cauchy(0,5);
  // likelihood
  y ~ normal(beta[1] + beta[2] * x, sigma);
}
```

Further characteristics of the Stan language that are good to know include: The variance parameter for the normal distribution is specified as the standard deviation (like in R but different from BUGS, where the precision is used). If no prior is specified, Stan uses a uniform prior over the range of possible values as specified in the parameter block. Variable names must not contain periods, for example, "x.z" would not be allowed, but "x_z" is allowed. To comment out a line, use double forward-slashes // ("#" in BUGS and R). The Stan website provides a large collection of model code that can be downloaded and adapted.

We fit the model to the data that has been produced in Chapter 4. There, the vectors x and y have been simulated and n is the sample size. Stan needs a vector containing the names of the data objects. In our case, x, y, and N are objects that exist in the R console.

```
library(rstan)
datax <- c("N", "x", "y")
```

The function `stan` starts Stan and returns an object containing MCMCs for every model parameter. We have to specify the name of the file that contains the model specification, the data, the number of chains, and the number of iterations per chain we would like to have. The first half of the iterations of each chain is declared as the warm-up. During the warm-up, Stan is not simulating a Markov chain, because in every step the algorithm is adapted. After the warm-up the algorithm is fixed and Stan simulates Markov chains.

```
fit <- stan(file = "linreg.stan", data=datax, chains=10,
          iter=1000)
```

And we look at the parameter estimates.

```
print(fit, c("beta", "sigma"))
Inference for Stan model: linreg.
10 chains, each with iter=1000; warmup=500; thin=1;
post-warmup draws per chain=500, total post-warmup draws=5000.

         mean se_mean  sd  2.5%  25%  50%  75% 97.5% n_eff Rhat
beta[1]  1.7     0.1  2.4  -2.9  0.2  1.7  3.3   6.5  1334    1
beta[2]  0.7     0.0  0.1   0.5  0.6  0.7  0.8   0.9  1200    1
sigma    4.8     0.0  0.5   4.0  4.5  4.8  5.2   6.0  1236    1

Samples were drawn using NUTS(diag_e) at Wed Apr 23 12:11:06 2014.
For each parameter, n_eff is a crude measure of effective sample size,
and Rhat is the potential scale reduction factor on split chains (at
convergence, Rhat=1).
```

The `print` function summarizes the MCMC output. It gives the mean and the standard deviation (sd) of the marginal posterior distribution of each parameter. "se_mean" is the standard deviation divided by the square root of the effective sample size, which is a Monte Carlo error. Then, some quantiles, the effective sample sizes, and $\hat{R}$ are listed (see Section 12.1). The last two statistics do not indicate nonconvergence. The trace plots can be plotted by the following R code (figures not shown); they also show no indication of nonconvergence.

```
traceplot(fit, "beta")
traceplot(fit, "sigma")
```

The function `extract` produces an object that contains all the simulated values from the posterior distributions in a list that can be used for deriving fitted values with credible intervals and predictive distributions. This object looks essentially the same as the object produced by the `sim` function (we used to call this object "bsim"), except that it has an additional element, the vector "lp_" with simulated values from the log posterior distribution multiplied by a constant.

```
modsims <- extract(fit)
str(modsims)
List of 3
 $ beta : num [1:5000, 1:2] -1.278 0.724 -0.958 2.431 1.183 ...
  ..- attr(*, "dimnames")=List of 2
  .. ..$ iterations: NULL
  .. ..$   : NULL
 $ sigma: num [1:5000(1d)] 4.18 5.29 4.86 5.4 4.39 ...
  ..- attr(*, "dimnames")=List of 1
  .. ..$ iterations: NULL
 $ lp__ : num [1:5000(1d)] -102 -101 -101 -104 -100 ...
  ..- attr(*, "dimnames")=List of 1
  .. ..$ iterations: NULL
```

Thus, the following code to obtain the fitted values with a 95% credible interval should look familiar.

```
newdat <- data.frame(x=seq(10, 30, length=100))
Xmat <- model.matrix(~x, data=newdat)
b <- apply(modsims$beta, 2, mean)
newdat$fit <- Xmat %*% b
nsim <- nrow(modsims$beta)
fitmat <- matrix(ncol=nsim, nrow=nrow(newdat))
for(i in 1:nsim){
  fitmat[,i] <- Xmat %*% modsims$beta[i,]
}
newdat$fitlwr <- apply(fitmat, 1, quantile, prob=0.025)
newdat$fitupr <- apply(fitmat, 1, quantile, prob=0.975)
```

When we plot these values it looks essentially the same as in Figure 12-3; therefore, we do not show this plot here.

12.4 SIM, BUGS, AND STAN

Why do we need three different programs that all generate random values from the joint posterior distribution of the model parameters? Every software has its pros and cons, which are summarized in the following.

The function `sim` uses a direct sampler. Every generated value is an independent value of the posterior distribution. There is no need to worry about convergence or autocorrelation. It is very fast and it can handle large data sets normally without problems. The high speed and robustness of `sim` is traded off with reduced flexibility, that is, `sim` only works for models that have been fitted by `lm`, `glm`, `lmer`, `glmer`, and a few more. Further, `sim` always uses uniform prior distributions over the range of the possible parameter values when applied to these models. The user can specify the prior distributions only when using the function `bayesglm` from the package arm to fit the model.

WinBUGS, OpenBUGS, and JAGS have the flexibility so that the user can specify a large variety of models using an intuitively understandable language. However, for complex ecological models, for example, the Cormack–Jolly–Seber model presented in Section 14.5, computing time can be very long, such as days or weeks. For large data sets, computing time can make fitting some models infeasible.

Stan can handle large data sets and complex models in less computing time than BUGS. Specifically, the number of effective samples compared to the total number of iterations is substantially higher in Stan compared to BUGS, since Stan uses more efficient MCMC algorithms and is implemented using a language focusing on efficiency and stability. But R users must expend more effort to learn Stan than they need to learn BUGS because the Stan-language differs more from the R language. However, the large collection of model code that is freely available, the detailed documentation, online forum, and active discussion groups make the learning comparatively easy.

Our philosophy is to use the simplest methods and software possible for a given problem. Thus, if we have no prior information and we want to fit one of the classical linear models, we use `sim`. If we have a complex model, we either use BUGS or Stan. If we have a large data set, we use Stan.

FURTHER READING

To learn more about MCMC, we recommend Ntzoufras (2009), who gives a very understandable introduction to the principles of MCMC. If you want to learn how to program MCMC yourself, one of the following books may help. Albert (2007), Rizzo (2008), and Robert and Casella (2009) provide full R code. Stauffer (2008) briefly introduces Gibbs sampling and Metropolis–Hasting algorithms. A comprehensive introduction to MCMC is given by Gilks et al. (1996), Brooks (1998), or Geyer (2011). Geyer (2011) is freely available from www.mcmchandbook.net/HandbookSampleChapters.html. In the same book you will also find a chapter on monitoring convergence (Gelman & Shirley, 2011) and one on Hamiltonian Monte Carlo (Neal, 2011). Hamiltonian Monte Carlo is also introduced in Gelman et al. (2014). Specific MCMC algorithms used for fitting demographic models are given in King et al. (2010).

The developers of WinBUGS, Lunn et al. (2013), have written a comprehensive introduction to WinBUGS. McCarthy (2007) gives an understandable introduction to Bayesian data analysis with example BUGS code for many different classically used methods such as ANOVA. Kéry (2010) and Ntzoufras (2009) give introductions to linear and the latter also nonlinear models using WinBUGS. Kéry and Schaub (2012) introduce Bayesian population models with WinBUGS code. Introduction to Bayesian data analyses with some BUGS examples are provided by Gelman and Hill (2007), Stauffer (2008), Link and Barker (2010), and Kruschke (2011). A large collection of ecological models are presented together with R and BUGS code in Royle and Dorazio (2008).

Stan is pretty new so there is not much literature around about it yet. Gelman et al. (2014) give a short introduction to Stan. There is extensive online material including a comprehensive manual (www.mc-stan.org).

Chapter 13

Modeling Spatial Data Using GLMM

Chapter Outline

13.1 BACKGROUND

The first law of geography says: "Everything is related to everything else, but near things are more related than distant things" (Tobler, 1970). Statisticians call this phenomenon spatial autocorrelation. It can be seen as a simple 2D generalization of temporal autocorrelation, which describes the tendency of two things nearer to each other along the time axis to be more similar than things further apart in time. Legendre (1993) proposed a more formal definition: "The property of random variables taking values, at pairs of locations a certain distance apart, that are more similar (positive autocorrelation) or less similar (negative autocorrelation) than expected for randomly associated pairs of random observations". Some common examples of spatial autocorrelation in ecology are patchiness, gradients, or regular distributions.

Spatial autocorrelation is a general property of most ecological data sets, and can occur at different spatial scales. It has various causes but two of them are particularly common. Habitat heterogeneity is one of them. For example, a species living only in forests will have a patchy distribution because of its environmental requirements. We account for this spatial autocorrelation when we include the relevant habitat variables in the statistical models. Biotic processes such as dispersal, conspecific attraction, competition with another species, or other complex dynamics (e.g., source-sink) can also cause spatial autocorrelation. Such biotic processes are usually much harder to include in our statistical models.

Bayesian Data Analysis in Ecology Using Linear Models with R, BUGS, and Stan
http://dx.doi.org/10.1016/B978-0-12-801370-0.00013-7.

If we fail to include important habitat variables or biotic processes, the spatial autocorrelation will still be present in the residuals of our model and we thus violate one of the key assumptions of most statistical models: independence. This happens often in ecology because it is obviously difficult to include all the important spatially structured covariates and biotic processes in a model.

An observation in close proximity to another does not necessarily increase the information available in the data if it is similar to the one already measured. If this happens, the value of a random variable characterizing a site (e.g., a habitat variable or the local density of an animal species) can thus be partially predicted by the values at neighboring sites. Such measurements only artificially increase sample size without contributing a full unit of independent information. Spatial autocorrelation can thus be described as one of the mechanisms leading to pseudoreplication (Hurlbert, 1984).

Spatial autocorrelation can be seen as a problem as long as we focus only on the statistics, but it can also be very interesting for ecologists. If we detect autocorrelation in the residuals of our model, it tells us that we forgot an important predictor or process or that the relationship is different from the one we modeled (e.g., not linear). We can get some insights about this missing process by looking carefully at how the residuals are structured, as described in Section 6.5. Here, we describe how spatial autocorrelation can be included, and thus accounted for or investigated, in a linear model.

13.2 MODELING ASSUMPTIONS

All the methods we present here assume stationarity and isotropy. Stationarity is a difficult concept to explain but it basically says that the scale and effects of spatial autocorrelation are constant across the whole study area. The stationarity assumption may be unrealistic for large study areas. It is, for example, easily violated when studying dispersal in a region with heterogeneous habitat: the scale of dispersal will typically be different in different habitats (e.g., a floodplain versus a forest). Unfortunately, very few methods are able to deal with nonstationarity. Isotropy means that the process causing spatial autocorrelation is acting similarly in all directions. This assumption will be violated when studying wind dispersal in areas with a prevailing wind direction, or organisms living in rivers with a strong current, for example.

13.3 EXPLICIT MODELING OF SPATIAL AUTOCORRELATION

13.3.1 Starting the Model Fitting

We will use a data set containing counts of common redstart *Phoenicurus phoenicurus* breeding pairs in a part of Switzerland. Each row of the data set contains the coordinates of the sites, the elevation, and the forest cover. Each site has an area of 1 km^2.

We can first try to model the counts using a linear and a quadratic effect of elevation. For simplicity, we log-transform the response variable and use a linear model with a normal error distribution instead of a generalized linear model with a Poisson error distribution (see Chapter 8).

```
data(redstart)
mod <- lm(log(counts+1) ~ elevation + I(elevation^2),
          data = redstart)
```

If we check the standard residual plots (using `plot(mod)`, Chapter 6), we see that the model does not fit the data well. For example, the normality assumption seems to be violated if we look at the QQ plot. And we also detect a spatial pattern in the residuals with the help of a bubble plot: positive and negative residuals seem to be clustered, and large residuals occur more often in the northwestern and southern parts of our study area (Figure 13-1). This probably indicates that we forgot an important covariate in the model.

Furthermore, the semivariance (see Section 6.5) is first increasing with distance and seems to reach some plateau for longer distances (Figure 13-2). Figures 13-1 and 13-2 clearly indicate spatial autocorrelation in the residuals, which is a violation of the independence assumption. Note that the standard variogram assumes isotropy, which means that spatial dependence of the residuals is the same in any direction over the whole study area (i.e., there is no directional dependence of the correlation structure). As we have seen in the introduction, this assumption can be violated. We can easily check that by giving directions to the alpha argument in the `variogram` function.

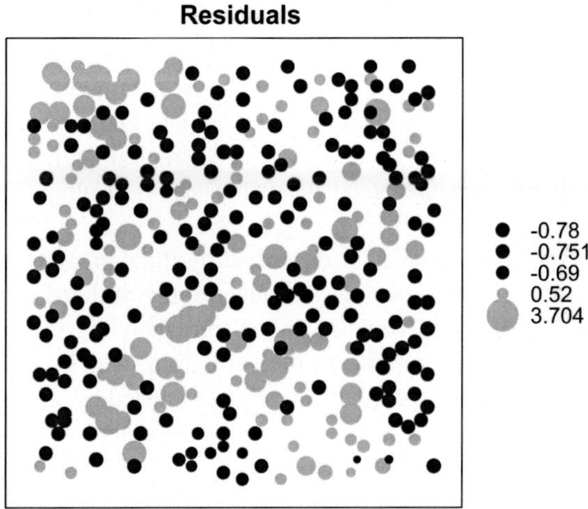

FIGURE 13-1 Bubble plot of the residuals from the redstart initial (nonspatial) model. Orange dots indicate positive, blue negative residuals. The size of the dots corresponds to the absolute value of the residuals. The bubble plot code is described in Chapter 6, Section 6.5.

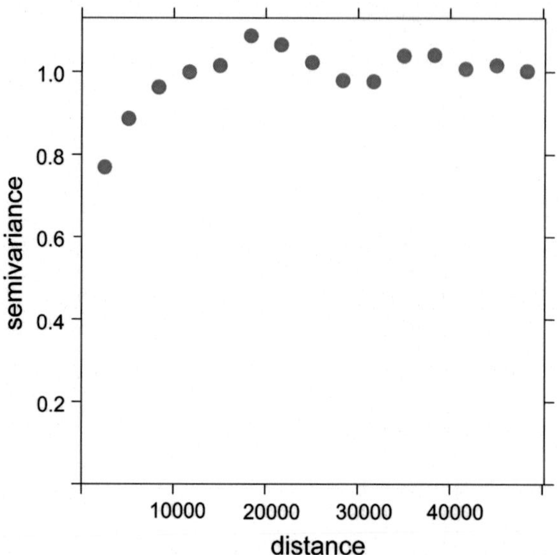

FIGURE 13-2 Semivariogram of the residuals from the redstart initial (nonspatial) model. The variogram code is described in Chapter 6, Section 6.5.

13.3.2 Variogram Modeling

Let's assume that we cannot think of any other predictor that may improve the model fit and that we would like to account for spatial autocorrelation in our model and study it. This may be interesting in relation to biotic processes such as social attraction.

To model spatial autocorrelation, we need to fit a theoretical model to our sample variogram (refer to Figure 13-2). The most common variogram models used by ecologists are the exponential, the Gaussian, and the spherical models (Figure 13-3). They can be fit easily in combination with a normal error distribution using the packages gstat (Pebesma, 2004) or nlme (Pinheiro et al., 2011).

A standard variogram is usually described by three parameters. The range is the distance at which the sites are no longer autocorrelated (the start of the plateau). The sill is the semivariance value corresponding to the range. And lastly, the nugget is a discontinuity of the variogram, which can be present at the origin. It represents the unexplained variation caused by spatial structure at a distance less than the minimum lag or by measurement error.

13.3.3 Bayesian Modeling

In spatial modeling, the Bayesian approach has two main advantages above the frequentist methods. First, it is not based on approximation (like the penalized

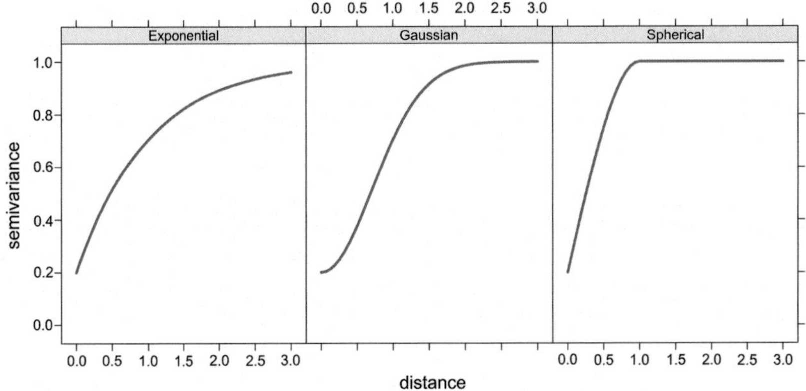

FIGURE 13-3 The exponential, Gaussian, and spherical variogram models (with sill = 1 and nugget = 0.2).

quasilikelihood) and thus provides exact results even for binary responses. Second, it correctly propagates the uncertainties linked to the estimation of the variogram parameters. This is not done by the frequentist approach, which uses a two-step approach: it first estimates the variogram parameters and then assumes them as known and plugs them into the linear model. We thus have an underestimation of uncertainty. A drawback is, however, that the Bayesian method still requires much longer computation time than the frequentist one.

In the previous example we modeled the counts using a transformation of the variable and a simple linear model (with a Gaussian error distribution). This was, of course, not optimal. As we saw in Chapter 8, counts are usually modeled using generalized linear models with a Poisson distribution. We will thus use such models in this section.

If we fit a standard Poisson GLM ignoring spatial autocorrelation, we see that the diagnostics plots are far from optimal. Moreover, the bubble plot and the variogram also show that the residuals are spatially autocorrelated.

We thus need to fit a model that explicitly accounts for spatial autocorrelation. Let's have a closer look at the model we are going to use. We can express our model as a generalized linear mixed model (GLMM) with a Poisson distribution:

$$y_i \sim \mathrm{Pois}(\lambda_i)$$

$$\log(\lambda_i) = \alpha + \beta x_i + w_i$$

This model is actually a GLMM with a site random effect (w_i). This random effect is not a usual random effect but it will vary spatially according to some parametric covariance function. It is drawn from a multivariate normal distribution with a mean of 0 and a variance-covariance matrix which is a function of the between-sites distances. This random effect is also called a

Gaussian process or a Gaussian random field. We use the following general formulation:

$$W \sim \text{MVN}\left(0, \sum\right)$$

$$\sum_{ij} = \sigma^2 \rho\left(\phi, d_{ij}\right)$$

where σ^2 is a variance parameter and ρ is a valid correlation function depending on ϕ, a parameter describing how rapidly the correlation declines with distance ("spatial decay parameter") and on d_{ij}, the distance between sites i and j. In the case of the exponential covariance function, we get:

$$\sum_{ij} - \sigma^2 e^{-\phi d_{ij}}$$

The `spGLM` function in the spBayes package (Finley & Banerjee, 2013) is one possibility to fit such a generalized linear mixed model accounting for spatial autocorrelation in the Bayesian framework. We first standardize the covariates to improve the convergence of the MCMC algorithm. We also rescale the geographic coordinates to avoid problems with the numerical estimation of ϕ.

```
library(spBayes)

# Standardize the covariates
redstart$elevation.z  <- as.numeric(scale(redstart$elevation))
redstart$elevation.z2 <- redstart$elevation.z^2

# Rescale the coordinates
redstart$x <- redstart$x / 1000
redstart$y <- redstart$y / 1000
```

We now fit a simple GLM and use the estimated parameters as starting values for the MCMC algorithm. Finally we generate suitable tuning values (the initial values for the variance of the normal proposal distribution used by the MCMC sampler) by computing the Cholesky factorization of the variance-covariance matrix of the GLM.

```
m0 <- glm(counts ~ elevation.z + elevation.z2, data = redstart,
          family = poisson)

beta.starting <- coefficients(m0)
beta.tuning   <- t(chol(vcov(m0)))
```

We can now fit the spatial GLM. As an example, we use an exponential covariance function but other functions are also available (see `?spGLM`).

```
n.batch <- 1000
batch.length <- 100
n.samples <- n.batch*batch.length

# Fit the spatial model
mod <- spGLM(counts ~ elevation.z + elevation.z2, data = redstart,
    family = "poisson", coords = as.matrix(redstart[,c("x", "y")]),
    starting = list("beta" = beta.starting, "phi" = 0.5, "sigma.sq" = 1,
    "w" = 0), tuning = list("beta" = beta.tuning, "phi" = 0.5, "sigma.sq" = 0.1,
    "w" = 0.01), priors = list("beta.Flat", "phi.Unif" = c(0.01, 50),
    "sigma.sq.IG" = c(2, 1)), amcmc = list("n.batch" = n.batch,
    "batch.length" = batch.length, "accept.rate" = 0.43),
    cov.model = "exponential", verbose = TRUE, n.report = 10)
```

We use an adaptive MCMC algorithm to improve the convergence and the estimation of the parameters. We use 1000 sequential batches of 100 iterations. At the start of each batch, the algorithm will try to improve its tuning and try to reach an acceptance rate of 0.43. A similar algorithm is used by OpenBUGS. We assume flat priors for the effects of the covariates, a uniform prior between 0.01 and 50 for ϕ, and an inverse-gamma prior with shape = 2 and scale = 1 for σ^2. (It is not yet possible to use weakly informative priors as we recommend in Chapter 15.)

There is usually little information in the data about ϕ and σ^2; the priors can therefore have a large effect on the results and we strongly advise to perform a prior sensitivity analysis using different values for the priors (see Chapter 15). One possibility for ϕ is to set the lower bound of the uniform distribution equal to the average nearest neighbor distance between sites and the upper bound equal to half the span of the study area (Swanson et al., 2013).

We visually check the convergence of the MCMC chain and also get parameter estimates based on the posterior distributions. The object returned by the `spGLM` function contains an mcmc object (called p.beta.theta.samples) which can be processed by functions of the coda package (Plummer et al., 2006). For this analysis we use a burn-in equal to half the number of iterations.

```
burn.in <- 0.5*n.samples
plot(window(mod$p.beta.theta.samples, start = burn.in)) # plots not
                                                            shown
summary(window(mod$p.beta.theta.samples, start = burn.in))

Iterations = 50000:1e+05
Thinning interval = 1
Number of chains = 1
Sample size per chain = 50001
```

```
1. Empirical mean and standard deviation for each variable,
   plus standard error of the mean:

                   Mean      SD  Naive SE Time-series SE
(Intercept)    -0.6687 0.25296 0.0011313       0.015941
elevation.z     0.1335 0.16710 0.0007473       0.005149
elevation.z2   -0.2541 0.10897 0.0004873       0.002889
sigma.sq        1.3850 0.35729 0.0015978       0.016525
phi             0.2412 0.07495 0.0003352       0.003043

2. Quantiles for each variable:

                  2.5%      25%     50%     75%    97.5%
(Intercept)    -1.2367 -0.80985 -0.6536 -0.5061 -0.20131
elevation.z    -0.1843  0.01862  0.1337  0.2474  0.46221
elevation.z2   -0.4818 -0.32432 -0.2477 -0.1791 -0.05421
sigma.sq        0.8642  1.14379  1.3259  1.5569  2.23983
phi             0.1078  0.18931  0.2363  0.2867  0.40392
```

The `summary` function gives the posterior means, standard deviations, and quantiles for each model parameter. It also gives the standard errors of the posterior means ignoring the autocorrelation of the chains, called "Naïve SE", and the time-series standard errors accounting for autocorrelation ("Time-series SE"). In this example we see a possible effect of elevation on the counts, but the uncertainty is large.

If the MCMC algorithm does not converge, try to increase the number of iterations. If this doesn't work, then try to change the starting or the tuning values.

This model is a natural way to model spatial autocorrelation for point data. Unfortunately the time needed to fit it can be quite long even for small data sets, and computation time increases as a cubic function of the sample size. This makes this model almost useless for moderately large data sets (more than 1000 data points).

That is why Banerjee et al. (2008) provided an approximation to the spatial part. Instead of modeling the spatial random effect over all the sampled sites, they propose to model it over a reduced set of locations called knots. The values of the random effect for the sites are then computed from the spatial process based on the knots. This technique is called predictive process modeling, and is very efficient, even for large data sets as long as the number of knots is small enough (see Latimer et al., 2009, for a complete but simplified explanation).

To perform this analysis with our data set we first need to define the locations of the knots. It is very important that our knot design adequately covers our study area. The distance between knots also has to be short enough to allow a valid estimation of the spatial decay parameter (ϕ). The design may have a significant effect on the results and we advise performing a small

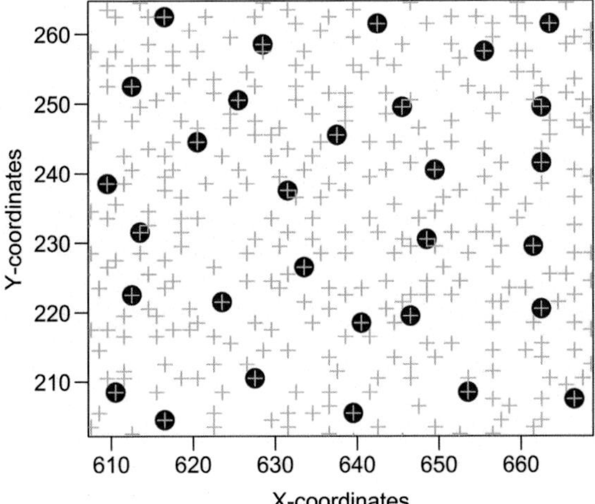

FIGURE 13-4 Spatial distribution of the sampling sites (orange crosses) and the knots (blue dots) in the study area.

sensitivity analysis using different locations and numbers of knots. In this example, we use the `cover.design` function in the fields package (Nychka et al., 2014) to generate a more or less regular grid of 30 knots in our study area (Figure 13-4).

```
library(fields)

# Generate the knots locations to cover the whole landscape
knotlocs <- cover.design(redstart[,c("x", "y")], 30)
knotlocs <- cbind(knotlocs$design[,1], knotlocs$design[,2])

plot(knotlocs, pch = 16, col = "blue", cex = 2)
points(redstart[,c("x", "y")], pch = 3, col = "red")
```

We can now fit the predictive process model using again the `spGLM` function. Note that we now use the knots argument with a matrix containing the coordinates of the knots.

```
mod.pp <- spGLM(counts ~ elevation + elevation2, data = redstart,
  family = "poisson", coords = as.matrix(redstart[,c("x", "y")]),
  knots = as.matrix(knotlocs), starting = list("beta" = beta.starting,
  "phi" = 0.5, "sigma.sq" = 1, "w" = 0), tuning = list("beta" = beta.tuning,
  "phi" = 0.5, "sigma.sq" = 0.1, "w" = 0.01), priors = list("beta.Flat",
  "phi.Unif" = c(0.01, 20), "sigma.sq.IG" = c(1.5, 0.5)),
  amcmc = list("n.batch" = n.batch, "batch.length" = batch.length,
  "accept.rate" = 0.43), cov.model = "exponential", verbose = TRUE,
  n.report = 10)
```

We can then visually check the convergence of the MCMC chain and also get parameter estimates based on the posterior distributions.

```
plot(window(mod.pp$p.beta.theta.samples, start = burn.in))
summary(window(mod.pp$p.beta.theta.samples, start = burn.in))
```

Note that the running time is a lot shorter than for the previous model but the results are (fortunately) similar. Before drawing any conclusions we still need to check if spatial autocorrelation was properly accounted for. We thus compute the residuals of the model and plot the variogram. The spatial structure in the residuals should no longer be recognizable. Here is the code to get the Pearson residuals:

```
# Extract median predictions at sample points (fitted values)
Xb <- cbind(1, redstart$elevation, redstart$elevation2) %*%
     t(mod.pp$p.beta.theta.samples[burn.in:n.samples,1:3])
yhat <- exp(Xb + mod.pp$p.w.samples[,burn.in:n.samples])
yhat.median <- apply(yhat, 1, median)

# Compute the Pearson residuals
resi <- (redstart$counts - yhat.median)/sqrt(yhat.median)
```

13.3.4 OpenBUGS Example

The functions available in the spBayes package are ideal for relatively simple models. But it is, unfortunately, not possible to add more complexity such as a second random effect. In this case we have to implement the desired model in the BUGS language. OpenBUGS provides some functions to fit spatial models. For example, the spatial random effect described earlier can be fit easily using the `spatial.exp` or `spatial.disc` functions, but we have not been able to implement the predictive process model in the BUGS language.

Don't forget that we standardized all the covariates to improve the mixing of the MCMC chains. We also rescaled the coordinates of the sites to avoid numerical problems with the estimation of the spatial parameters. We now use three chains with 10,000 iterations each. This is enough to achieve convergence for this model and this data set.

```
# MCMC settings
ni <- 10000
nt <- 1
nb <- 5000
nc <- 3
```

We define a model that is equivalent to the one we first fitted using the `spGLM` function. The spatial random effect with an exponential covariance function is defined with the spatial.exp function. Its first argument is the mean of the multivariate normal distribution, which equals 0. The second and third arguments are the coordinates of the sites. The fourth one is the spatial precision parameter, which is equal to the inverse of σ^2. The fifth argument is the spatial decay parameter (ϕ). The last argument is the power of the exponentiated term (see the exponential covariance function in Section 13.3.3); it can also be estimated, theoretically, from the data but there is almost never enough information. That is why we advise setting it to 1 *a priori*; this gives the exponential covariance function seen previously. Setting it to 2 gives the Gaussian covariance function, but this can sometimes cause numerical estimation troubles.

Of course, we need to specify prior distributions for all the estimated parameters. Note that we use other priors than the ones that were available with the spGLM function.

```
# Model in BUGS language
model {

    # Likelihood
    for (i in 1:n) {
       y[i] ~ dpois(lambda[i])
       log(lambda[i]) <- alpha + beta1*elev[i] + beta2*elev2[i] +
                        W[i]
       muW[i] <- 0
       }

    # Spatial exponential random effect
    W[1:n] ~ spatial.exp(muW[], xcoord[], ycoord[], tauSp, phi, 1)

    # Priors for the fixed effects
    alpha ~ dnorm(0, 0.01)
    beta1 ~ dnorm(0, 0.04)
    beta2 ~ dnorm(0, 0.04)

    # Priors for the spatial random effect
    tauSp <- pow(sdSp, -2)
    sdSp ~ dunif(0, 5)
    phi ~ dunif(0.01, 5)

}
```

We now create an object to store the needed data and we also specify initial values for some parameters. We strongly advise specifying initial values for the spatial random effect. Experience has shown that a value of zero is often a good solution. Finally, we define the parameters that will be monitored by OpenBUGS.

```
# Bundle data
bugs.data <- list(n = nrow(redstart), y = redstart$counts, xcoord =
  redstart$x, ycoord = redstart$y, elev = redstart$elevation.z,
  elev2 = redstart$elevation.z^2)

# Initial values
inits <- function() {list(alpha = runif(1), beta1 = runif(1),
  beta2 = runif(1), phi = runif(1), W = rep(0, nrow(redstart)))}

# Parameters monitored
params <- c("alpha", "beta1", "beta2", "phi", "tauSp")
```

Because this model takes a long time to fit, we do not use the standard function in the R2OpenBUGS package. We run the model with the bugs.parfit function in the dclone package (Solymos, 2010), which allows us to run the three chains on three different computer cores (parallel processing). The output is an mcmc.list object that can be processed with functions of the coda package.

```
library(dclone)

# Call OpenBUGS from R
cl <- makePSOCKcluster(3)
parallel::clusterExport(cl, varlist = "redstart")
samples <- bugs.parfit(cl, data = bugs.data, params = params,
    model = "SpatialExp.txt", inits = inits, n.chains = nc, n.thin = nt,
    n.iter = ni, n.burnin = nb, DIC = FALSE, program = "openbugs", seed = 1:3,
    save.history = FALSE)
stopCluster(cl)

plot(samples)
summary(samples)
```

The estimated coefficients for the fixed effects are very similar to those produced by the spGLM function. However, there are small differences for the spatial random effect parameters. This is due to the different priors that we used.

FURTHER READING

Dormann et al. (2007) and Beale et al. (2010) are accessible reviews of the models currently available to model residual spatial autocorrelation. However, the predictive process model of Banerjee et al. (2008) is not mentioned. An understandable presentation of this model is given by Latimer et al. (2009); Swanson et al. (2013) provide an applied example.

There are also many textbooks on spatial data analyses. Mathematical details of spatial models are presented in Banerjee et al. (2004) and Diggle and Ribeiro Jr. (2007). Bivand et al. (2008) provides R examples for statisticians already familiar with spatial statistics.

Chapter 14

Advanced Ecological Models

Chapter Outline

14.1 HIERARCHICAL MULTINOMIAL MODEL TO ANALYZE HABITAT SELECTION USING BUGS

Categorical response variables are quite common in ecological studies. For example, when studying habitat selection of animals, the outcome variable often is one of several habitat types in which an animal is observed at different time points. Questions can be about whether the animal uses the different habitat types proportional to their availability in its home range, whether the use of the habitat types differs between the sexes or between young and adult animals, or whether the use of habitat types is affected by any covariate such as weather or age of the animal. The statistical methods used to analyze habitat selection with respect to such questions are manifold. They range from simple preference indices to complicated multivariate methods (see overview in Manly et al., 2002) or compositional analysis (Aebischer & Robertson, 1993).

All these classical methods assume independence among observations. However, particularly in telemetry studies, this is often not the case, because single individuals are followed over time producing repeated measurements per individual. The multinomial model is a linear model with a categorical outcome variable. To account for nonindependent data we use random effects, which we introduced earlier in the framework of generalized linear mixed models.

Bayesian Data Analysis in Ecology Using Linear Models with R, BUGS, and Stan
http://dx.doi.org/10.1016/B978-0-12-801370-0.00014-9.

The multinomial model can be used in any case where the outcome variable is categorical, for example, to study differences in diet composition between the sexes or to analyze choice experiments in behavioral studies where the number of possible choices is larger than two (we could use logistic regression if the number of possible choices is two). One important requirement is that the observations are counts, for example, number of locations in specific habitat types or number of items eaten by an individual. Thus, multinomial models cannot be applied when the outcome variable is a continuous variable such as the proportion of time (measured in seconds) an animal has shown different behaviors or proportion of area (measured in square meters) of different habitat types in a home range. In such cases, multivariate methods or compositional analyses have to be considered, see, for example, Legendre and Legendre (2012).

In the multinomial model, it is assumed that a total of N_i counts of observation i are distributed among K categories with proportions $\mathbf{p}_i$ (the bold letter indicates a vector). N_i is the size parameter of the multinomial distribution and $\mathbf{p}_i$ is a probability vector. The length of $\mathbf{p}_i$ is the number of categories, K. The vector $\mathbf{p}_i$ contains the probabilities of the outcome for each of the categories. The sum of $\mathbf{p}_i$ is 1. The outcome variable $\mathbf{y}_i$ is a vector of length K that contains the number of counts in each category. The sum of $\mathbf{y}_i$ is N_i.

$$\mathbf{y}_i \sim Multinomial(\mathbf{p}_i, N_i)$$

Often, the data contain the outcome as a single observation, the category name, rather than as a vector of numbers per category. Then we can use the categorical distribution.

$$y_i \sim Categorical(\mathbf{p}_i)$$

The categorical distribution corresponds to a multinomial distribution with size parameter $N_i = 1$. In the following sections we assume that the data set contains the outcome as a category $k = 1, \ldots, K$. The probability vector $\mathbf{p}_i$ contains, for observation i, the expected values. These K values are probabilities that the outcome is category k. To model the relationship between the outcome and the predictor variables, for each category k, a separate linear predictor is needed.

$$z_{i,k} = \mathbf{X}_i \boldsymbol{\beta}_k$$

where $\mathbf{X}_i$ is a vector with the values of the predictors for observation i and $\boldsymbol{\beta}_k$ contains the model coefficients for category k. The inverse link function of the value $z_{i,k}$ gives the probability, $p_{i,k}$, that an observation with predictor values as in observation i is category k.

Then, a link function is needed that guarantees that all values of $\mathbf{p}_i$, $p_{i,1}, \ldots, p_{i,K}$, are probabilities (i.e., between 0 and 1) and that the sum of $\mathbf{p}_i$ is 1.

Classically, a multivariate version of the logistic function is used as the link function. Therefore, the multinomial model is sometimes also called the multinomial logistic model, or "mlogit" model (see, e.g., the R package mlogit; Croissant, 2012).

$$p_{i,k} = \frac{e^{z_{i,k}}}{\sum_{k=1}^{K} e^{z_{i,k}}} \text{ which makes that } \sum_{k=1}^{K} p_{i,k} = 1$$

The constraint that the probability vector $\mathbf{p}_i$ sums to 1 has important consequences: if all $p_{i,1}$ through $p_{i,K-1}$ are known, the last element of the probability vector ($p_{i,K}$) equals one minus the sum of the others. If $K - 1$ linear predictors are estimated, the last one is defined by default. Therefore, we fix the $\boldsymbol{\beta}$ vector of one linear predictor (baseline category) to be a vector of zero values.

Multinomial models can be fitted in R using frequentist methods with the function `multinom` from the package nnet or the function `mlogit` from the package mlogit. The first does not allow inclusion of random factors, whereas it is possible with the `mlogit` function. Here, we use OpenBUGS to fit a multinomial model to habitat selection data in a Bayesian framework using two random effects.

Our example data are from Bock et al. (2013). They located little owls *Athene noctua* using radio telemetry during the daytime in winter and noted the type of roosting site: nest box, tree canopy, wood stack, or tree cavity. One important question was whether the choice of the roosting site type depends on ambient temperature. Thus, temperature was the predictor and type of roosting site the outcome variable. We include the individual and family as random factors because individuals were repeatedly located and the individuals were grouped into families. We define the model using BUGS code and save the BUGS code in the file "siteselection.bugs".

BUGS Code for the Habitat Selection Model (file "siteselection.bugs")

```
model {
  ## priors
  for(k in 2:ncat) {
    ## priors for model coefficients (see Chapter 15)
    for(j in 1:2) {
      beta[j,k] ~ dnorm(0, 0.04)
    }
    ## hyperpriors for random effects (see Chapter 15)
    sigmaind[k] ~dt(0,1,2)I(0,)
    sigmafam[k] ~dt(0,1,2)I(0,)
  }
  sigmaind[1] <- 0
  sigmafam[1] <- 0
```

Continued

BUGS Code for the Habitat Selection Model (file "siteselection.bugs")—cont'd

```
## random effects
for(i in 1:nind) {rind[i, 1] <- 0}      # zero for baseline category
for(f in 1:nfam) {rfam[f, 1] <- 0}      # zero for baseline category
for(k in 2:ncat) {
  for(i in 1:nind) {
    rind[i,k]~dnorm(0, 1)               # random individual effect
  }
  for(f in 1:nfam) {
    rfam[f,k]~dnorm(0, 1)               # random family effect
  }
}

## likelihood
for(i in 1:n) {
  roost[i]~dcat(p[i,1:ncat])
  for(k in 1:ncat) {
    ## linear predictors including the random factors
    z[i,k]<-beta[1,k] + beta[2,k]*temp[i] + sigmafam[k]*rfam
          [fam[i],k] + sigmaind[k]*rind[ind[i],k]
    expz[i,k] <- exp(z[i,k])
    p[i,k] <- expz[i,k] / sum(expz[i,1:ncat])  # logit link
  }
}

## constrain coefficients of the baseline category to zero
for(j in 1:2) {
  beta[j,1] <- 0
}

##  predict  site  selection  probabilities  for  different
temperatures
for(k in 1:ncat) {
  for(i in 1:nnew) {
    pnew[i,k] <- expznew[i,k] / sum(expznew[i,1:ncat])
    znew[i,k] <- beta[1,k] + beta[2,k]*newtemp[i]
    expznew[i,k] <- exp(znew[i,k])
    }
  }
}
```

Note that the random effects have to be fixed to zero for the reference category. Further, we specified the random effects by a standard normal distribution (*Norm*(0,1)) and multiply these effects by the between-individual and between-family standard deviation in the linear predictor. This improves the fitting process (Gelman et al., 2014).

In the last part of the model code, we obtain the fitted values, that is, for each roosting site type, we get the probability that the type is used given specific ambient temperature values (provided in the vector "newtemp"). These fitted values are visualized in Figure 14-1. To fit the model, we first bundle the data for OpenBUGS.

```
data(roostingsiteuse)
dat <- roostingsiteuse
dat$roosting.loc <- factor(dat$roosting.loc, levels=c("nest box",
   "tree canopy", "stack of wood", "tree cavity"))
dat$temp.z <- (dat$temp-mean(dat$temp))/sd(dat$temp)
dat$roostingnum <- as.numeric(dat$roosting.loc)

# Prepare data frame for predictions
newdat <- data.frame(temp=seq(min(dat$temp), max(dat$temp),
                              length=100))
newdat$temp.z <- (newdat$temp-mean(dat$temp))/sd(dat$temp)

datax <- list(roost=dat$roostingnum, n=nrow(dat),
              ncat=nlevels(dat$roosting.loc), temp=dat$temp.z,
              fam=dat$familynum, nfam=max(dat$familynum),
              ind=dat$indnum, nind=max(dat$indnum),
              newtemp=newdat$temp.z, nnew=nrow(newdat))
```

Then, we define a function that generates initial values for the betas and the random factors. Parameters that were fixed to zero in the model (baseline category) cannot have any initial value but NAs must be given.

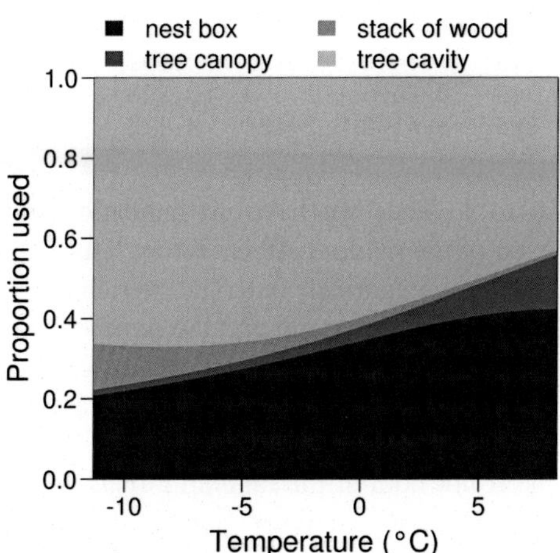

FIGURE 14-1 Proportions of four different roosting site types used by little owls *Athene noctua* during winter in relation to ambient temperature.

```
inits <- function() {
  list(beta=matrix(c(rep(NA,2), runif(2*(datax$ncat-1), -1,1)),
                  ncol=datax$ncat, nrow=2),
        sigmafam=c(NA, runif(datax$ncat-1, 0.1, 2)),
        sigmaind=c(NA, runif(datax$ncat-1, 0.1,2)))
}
```

At last, we have to specify which parameters we want to save.

```
parameters <- c("beta", "sigmaind", "sigmafam","pnew")
```

Then, we can run the MCMC to fit the model:

```
library(R2OpenBUGS)
fit <- bugs(datax, inits, parameters, model.file="siteselection.bugs",
            n.thin=2, n.chains=2, n.burnin=1000, n.iter=10000,
            debug=FALSE)
```

When OpenBUGS has finished, we check the history plots of the Markov chains for indications of nonconvergence and, if all is ok, look at the results:

```
print(head(fit$summary, 12), 3)
```

```
              mean    sd    2.5%    25%    50%     75%  97.5% Rhat n.eff
beta[1,2]   -3.668 2.171 -8.2750 -5.103 -3.516 -2.1107  0.195 1.00  1500
beta[1,3]   -4.380 2.234 -8.8920 -5.934 -4.252 -2.7500 -0.371 1.00  2700
beta[1,4]    0.802 1.806 -2.7741 -0.378  0.795  1.9642  4.449 1.00  3500
beta[2,2]    0.988 1.177 -1.1261  0.176  0.906  1.7100  3.481 1.00  6500
beta[2,3]   -1.325 1.032 -3.6260 -1.936 -1.233 -0.6048  0.431 1.00  9000
beta[2,4]   -0.606 0.957 -2.6081 -1.194 -0.571  0.0318  1.209 1.00  5600
sigmaind[2]  2.717 1.453  0.1535  1.497  2.824  4.0052  4.909 1.00  9000
sigmaind[3]  3.396 1.154  0.7729  2.629  3.592  4.3610  4.943 1.00  6000
sigmaind[4]  2.301 1.453  0.0989  1.006  2.211  3.5595  4.846 1.00  2000
sigmafam[2]  3.288 1.312  0.3452  2.431  3.603  4.3662  4.943 1.00  7500
sigmafam[3]  3.013 1.345  0.2605  2.009  3.228  4.1640  4.912 1.00  9000
sigmafam[4]  3.948 0.896  1.6168  3.485  4.178  4.6390  4.968 1.01  2000
```

All $\widehat{R}$ are close to 1. Thus we have no indication of nonconvergence. However, the precision of the random effects is low. This is typical for random effects in models with categorical outcome variables including logistic regression. Both observations per group and the number of groups need to be high to get precise estimates of between- and within-group variances. In this example, sample size is way too small ($n = 42$) to estimate six additional variance parameters.

Here, at least one if not both of the random effects should be omitted and the model refitted (try this as an exercise). However, even if the uncertainty is high, due to too many random effects, the result is similar to the one obtained by Bock et al. (2013) based on a larger sample size. The second beta values are the estimated temperature effects: the negative values of beta[2,3] and beta[2,4] mean that wood stacks and tree cavities were used less often (compared to nest boxes) with increasing temperature. This is also apparent in Figure 14-1.

The advantage of using simulation techniques for describing the joint posterior distributions of the model parameters is that we can easily obtain a sample of simulated fitted values (of which the means are plotted in Figure 14-1) or other derived parameters. In the little owl example it is helpful to compare the proportion of used roosting sites with the corresponding proportion of available roosting sites, for example, by a preference index such as Jacob's preference index. This is done by simply calculating the Jacob's index for each simulated value of the proportion used; see Bock et al. (2013) for such an example.

14.2 ZERO-INFLATED POISSON MIXED MODEL FOR ANALYZING BREEDING SUCCESS USING STAN

Up until now we have described the outcome variable with a single distribution, such as the normal distribution in the case of linear (mixed) models, and Poisson or binomial distributions in the case of generalized linear (mixed) models. In life sciences, however, quite often the data are actually generated by more than one process.

In such cases the distribution of the data could be the result of two or more different distributions. If we do not account for these different processes our inferences are likely to be biased. In this and the next section, we introduce mixture models that explicitly include two processes that generated the data. The zero-inflated Poisson model is a mixture of a binomial and a Poisson distribution, and the site-occupancy model is a mixture of two binomial distributions. We belief that two (or more)-level models are very useful tools in life sciences because they can help uncover the different processes that generate the data we observe.

Counting animals or plants is a typical example of data that contain a lot of zero counts. For example, the number of nestlings produced by a breeding pair is often zero because the whole nest was depredated or because a catastrophic event occurred such as a flood. However, when the nest succeeds, the number of nestlings varies among the successful nests depending on how many eggs the female has laid, how much food the parents could bring to the nest, or other factors that affect the survival of a nestling in an intact nest. Thus the factors that determine how many zero counts there are in the data differ from the factors that determine how many nestlings there are, if a nest survives. Count data that are produced by two different processes—one produces the zero counts and the other the variance in the count for the ones that were not zero in the first process—are called zero-inflated data. Histograms of zero-inflated data look bimodal, with one peak at zero (Figure 14-2).

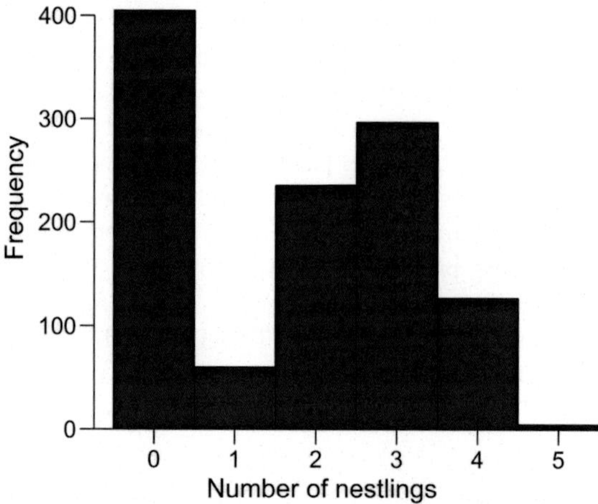

FIGURE 14-2 Histogram of the number of nestlings counted in black stork nests *Ciconia nigra* in Latvia ($n = 1130$ observations of 279 nests). The data were kindly provided by Maris Strazds.

The Poisson distribution does not fit well to such data, because the data contain more zero counts than expected under the Poisson distribution. Mullahy (1986) and Lambert (1992) formulated two different types of models that combine the two processes in one model and therefore account for the zero excess in the data and allow the analysis of the two processes separately.

The hurdle model (Mullahy, 1986) combines a left-truncated count data model (Poisson or negative binomial distribution that only describes the distribution of data larger than zero) with a zero-hurdle model that describes the distribution of the data that are either zero or nonzero. In other words, the hurdle model divides the data into two data subsets, the zero counts and the nonzero counts, and fits two separate models to each subset of the data. To account for this division of the data, the two models assume left truncation (all measurements below 1 are missing in the data) and right censoring (all measurements larger than 1 have the value 1), respectively, in their error distributions. A hurdle model can be fitted in R using the function `hurdle` from the package pscl (Jackman, 2008). See the tutorial by Zeileis et al. (2008) for an introduction.

In contrast to the hurdle model, the zero-inflated models (Mullahy, 1986; Lambert, 1992) combine a Bernoulli model (zero vs. nonzero) with a conditional Poisson model; conditional on the Bernoulli process being nonzero. Thus this model allows for a mixture of zero counts: some zero counts are zero because the outcome of the Bernoulli process was zero (these zero counts are sometimes called structural zero values), and others are zero because their outcome from the Poisson process was zero. The function `zeroinfl` from the package pscl fits zero-inflated models (Zeileis et al., 2008).

The zero-inflated model may seem to reflect the true process that has generated the data closer than the hurdle model. However, sometimes the fit of zero-inflated models is impeded because of high correlation of the model parameters between the zero model and the count model. In such cases, a hurdle model may cause less troubles.

Both functions (`hurdle` and `zeroinfl`) from the package pscl do not allow the inclusion of random factors. The functions `MCMCglmm` from the package MCMCglmm (Hadfield, 2010) and `glmmadmb` from the package glmmADMB (http://glmmadmb.r-forge.r-project.org/) provide the possibility to account for zero-inflation with a GLMM. However, these functions are not very flexible in the types of zero-inflated models they can fit; for example, `glmmadmb` only includes a constant proportion of zero values. A zero-inflation model using BUGS is described in Kéry and Schaub (2012). Here we use Stan to fit a zero-inflated model. Once we understand the basic model code, it is easy to add predictors and/or random effects to both the zero and the count model.

The example data contain numbers of nestlings in black stork *Ciconia nigra* nests in Latvia collected by Maris Stradz and collaborators at 279 nests between 1979 and 2010. Black storks build solid and large aeries on branches of large trees. The same aerie is used for up to 17 years until it collapses. The black stork population in Latvia has drastically declined over the last decades. Here, we use the nestling data as presented in Figure 14-2 to describe whether the number of black stork nestlings produced in Latvia decreased over time. We use a zero-inflated Poisson model to separately estimate temporal trends for nest survival and the number of nestlings in successful nests. Since the same nests have been measured repeatedly over 1 to 17 years, we add nest ID as a random factor to both models, the Bernoulli and the Poisson model. After the first model fit, we saw that the between-nest variance in the number of nestlings for the successful nests was close to zero. Therefore, we decide to delete the random effect from the Poisson model. Here is our final model:

$$z_{it} \sim Bernoulli(\theta_{it})$$

$$\text{logit}(\theta_{it}) = a_1 + a_2\text{year}_t + \varepsilon_{\text{nestID}[i]}$$

$$y_{it} \sim Poisson(\lambda_{it}(1 - z_{it}))$$

$$\log(\lambda_{it}) = b_1 + b_2\text{year}_t$$

$$\varepsilon_{\text{nestID}} \sim Norm(0, \sigma_n)$$

z_{it} is a latent (unobserved) variable that takes the values 0 or 1 for each nest i during year t. It indicates a "structural zero", that is, if $z_{it} = 1$ the number of nestlings y_{it} always is zero, because the expected value in the Poisson model $\lambda_{it}(1 - z_{it})$ becomes zero. If $z_{it} = 0$, the expected value in the Poisson model becomes λ_{it}.

To fit this model in Stan, we first write the Stan model code and save it in a separated text-file with name "zeroinfl.stan".

Stan Code for the Zero-Inflated Poisson Mixed Model

```
data {
    int<lower=0> N;                            // sample size
    int<lower=0> Nnests;                       // number of nests
    int<lower=0> y[N];                      // counts (number of youngs)
    vector[N] year;                            // numeric covariate
    int<lower=0, upper=Nnests> nest[N]; // index of nest
  }

  parameters {
    vector[2] a;                      // coef of linear pred for theta
    vector[2] b;                      // coef of linear pred for lambda
    real<lower=0> sigmanest;          // between nest sd in logit(theta)
    real groupefftheta[Nnests]; // nest effects for theta
  }

  model {
    //transformed parameters (within model to avoid monitoring)
    vector[N] theta;                  // probability of zero youngs
    vector[N] lambda;                 // avg. number of youngs
                                      // in successful nests
          for(i in 1:N){
          // linear predictors with random effect
          theta[i] <- inv_logit(a[1] + a[2] * year[i] +
                    sigmanest * groupefftheta[nest[i]]);
    }
    lambda <- exp(b[1] + b[2] * year);
    // priors
    a[1] ~ normal(0,5);
    a[2] ~ normal(0,5);
    b[1] ~ normal(0,5);
    b[2] ~ normal(0,5);
    sigmanest ~ cauchy(0,5);

    // random effects
    for(g in 1:Nnests) {
      groupefftheta[g] ~ normal(0,1);
    }

    // likelihood
    for (i in 1:N){
      if(y[i] == 0)
```

Stan Code for the Zero-Inflated Poisson Mixed Model—cont'd

```
        increment_log_prob(log_sum_exp(bernoulli_log(1, theta[i]),
                           bernoulli_log(0, theta[i]) +
                           poisson_log(y[i], lambda[i])));
      else
        increment_log_prob(bernoulli_log(0, theta[i]) +
                           poisson_log(y[i], lambda[i]));
    }
  }
```

In the Stan model code we define ε_{nestID} by a standard normal distribution and add the product $\sigma_n \varepsilon_{\ nestID[i]}$ to the linear predictor. This parameterization of a random effect makes the model fit more stable. In the likelihood part, we use the if-else statement and define the likelihood separately for the zero and the nonzero counts. The zero count can be due to the outcome of the Bernoulli model being zero or because the count is zero. Therefore, the density function of the count of zero is the sum of the density function of the Bernoulli distribution for 1 (remember that 1 indicates zero) and the product of the Bernoulli-density for 0 (indicating not a "structural" zero) with the Poisson-density for zero. This is the expression within the `log_sum_exp` function.

The functions `bernoulli_log` and `poisson_log` define the logarithm of the density functions. The `log_sum_exp` function is the logarithm of the sum of the exponents, that is, log(exp(a) + exp(b)). The function `log_sum_exp` does this in a mathematically stable way. Finally, the function `increment_log_prob` is an alternative and flexible way of defining the likelihood in Stan; the code:

```
model{
  y ~ normal(mu, sigma);
}
```

is equivalent to:

```
model{
  increment_log_prob(normal_log(y, mu, sigma));
}
```

There are more efficient ways to code the above model in Stan. For example, the code `y ~ poisson_log(lambda)` includes the logarithm link function so that the exp function for the linear predictor can be skipped. Similarly, the `inv_logit` function can be omitted by using `~ bernoulli_logit` instead of `~ bernoulli`, or `bernoulli_logit_log` instead of `bernoulli_log`. Including the link function in the definition of the model by one of these functions makes the model fitting faster and more stable. However, here, we preferred to code the model so that the structure resembles the notations we have used throughout the book. We recommend the Stan manual to learn programming efficiently.

Then we start R, read and prepare the data, and run Stan to fit the model.

```
data(blackstork)
dat <- blackstork

dat$nest <- factor(dat$nest)                    # define nest as a factor
dat$year.z <- as.numeric(scale(dat$year))  # z-transform year

# Prepare data for Stan
y <- dat$njuvs
N <- nrow(dat)
nest <- as.numeric(dat$nest)
Nnests <- nlevels(dat$nest)
year <- dat$year.z
datax <- c("y", "N", "nest", "Nnests", "year")

# Run Stan
library(rstan)
mod <- stan(file = "STAN/zeroinfl.stan", data=datax, chains=5, iter=1000)
print(mod, c("a", "b", "sigmanest"))

Inference for Stan model: zeroinfl.
5 chains, each with iter=1000; warmup=500; thin=1;
post-warmup draws per chain=500, total post-warmup draws=2500.

           mean se_mean   sd  2.5%   25%   50%   75% 97.5% n_eff Rhat
a[1]      -1.02    0.00 0.12 -1.26 -1.09 -1.01 -0.93 -0.79  2500    1
a[2]       0.56    0.00 0.11  0.35  0.48  0.56  0.63  0.79  2500    1
b[1]       0.89    0.00 0.03  0.84  0.88  0.89  0.91  0.95  2500    1
b[2]      -0.05    0.00 0.02 -0.10 -0.07 -0.05 -0.04 -0.01  2500    1
sigmanest  0.93    0.01 0.17  0.61  0.82  0.93  1.05  1.28   705    1

Samples were drawn using NUTS(diag_e) at Fri Sep 05 11:41:53 2014.
For each parameter, n_eff is a crude measure of effective sample size,
and Rhat is the potential scale reduction factor on split chains (at
convergence, Rhat=1).
```

The “mean” is the average of the simulated values from the marginal posterior distributions for each parameter. The “se_mean” is the standard error of these simulated values, that is, the Monte Carlo uncertainty and *not* the standard error of the parameter estimate! The sd is the sample standard deviation of the simulations. This corresponds to the standard error of the parameter estimates.

After several quantiles of the posterior distribution, the effective sample size n_eff and the potential scale reduction factor $\widehat{R}$ (see Section 12.3) are given. If we decided to draw more simulations, we would set the argument fit=mod to skip the model compilation step which saves time.

Before drawing conclusions from this model, the model fit is assessed using predictive model checking. To do so, we first simulate replicated numbers of nestlings for every year and nest in the data set based on the model fit.

```
nsim <- length(modsims$lp_)  # extract number of simulations
yrep <- matrix(nrow=length(y), ncol=nsim)
Xmat <- model.matrix(~year)
for(i in 1:nsim){
  theta <- plogis(Xmat%*%modsims$a[i,] +
              modsims$groupefftheta[i,nest])
  z <- rbinom(length(y), prob=theta, size=1)
  lambda <- (1-z) *exp(Xmat%*%modsims$b[i,])
  yrep[,i] <- rpois(length(y), lambda=lambda)
}
```

First, we visualize the observed and the first four simulated replicated data sets graphically, by plotting the number of nestlings for each nest and year (Figure 14-3).

The comparison of the simulated data with the observed data shows quite similar patterns, indicating that the model captures the general structure of the data. However, we can recognize that there are fewer zero values observed at the beginning of the study period than assumed by the model (the observed data shows fewer black dots in the lower left part of the panel than the simulated data). We could go back and fit a model allowing for a nonlinear relationship between year and the proportion of successful nests, or we can accept the small lack of fit. The overall proportion of zero values was the same in the replicated data (median 36%) as they were in the observed data (36%).

```
# Proportion of zeros for the observed data
mean(y==0)
[1] 0.3584071

# Proportion of zeros for the nsim simulated data
propzeros <- apply(yrep, 2, function(x) mean(x==0))
quantile(propzeros, prob=c(0.025, 0.5, 0.975))
     2.5%       50%     97.5%
0.3221239 0.3610619 0.4000000
```

The result provides strong evidence that the proportion of failed nests increased over the years, because the estimated effect of year in the zero-model was 0.56 (95% CrI 0.36–0.77). We extract these numbers from the

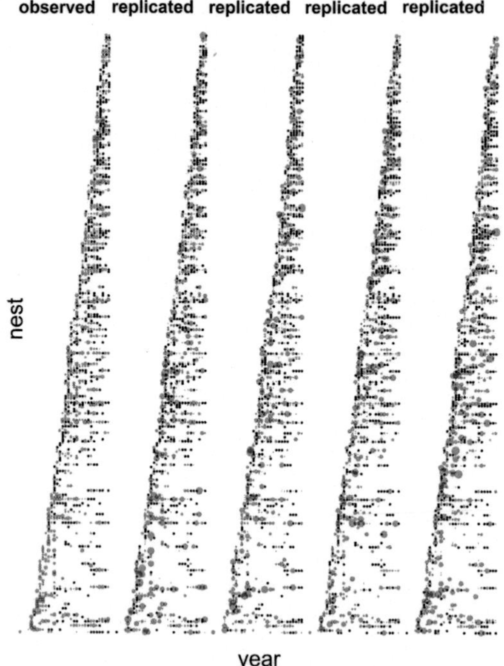

FIGURE 14-3 Numbers of nestlings for every nest in the different years. The smallest red dot corresponds to one, the largest to nine nestlings. The black circles indicate nests with zero nestlings. The nests are sorted according to their first year with a record in the data set. The leftmost panel is the observed data, the other panels are the first four (of 5000) simulated replicated data sets.

object mod1 that contains 5000 simulations from each model parameter using `extract` and `apply`:

```
modsims <- extract(mod1)
apply(modsims$a, 2, quantile, prob=c(0.025, 0.5, 0.957))
              [,1]       [,2]
 2.5%   -1.2670004  0.3556119
 50%    -1.0105234  0.5623700
 95.7%  -0.8169778  0.7650336
```

Further, there is also evidence that the number of nestlings of the successful nests decreased over time because the estimated slope of the regression coefficient in the Poisson model is –0.05 (–0.10 to –0.02).

```
apply(modsims$b, 2, quantile, prob=c(0.025, 0.5, 0.957))
             [,1]         [,2]
 2.5%   0.8430767  -0.09894172
 50%    0.8948438  -0.05489263
 95.7%  0.9402984  -0.01526097
```

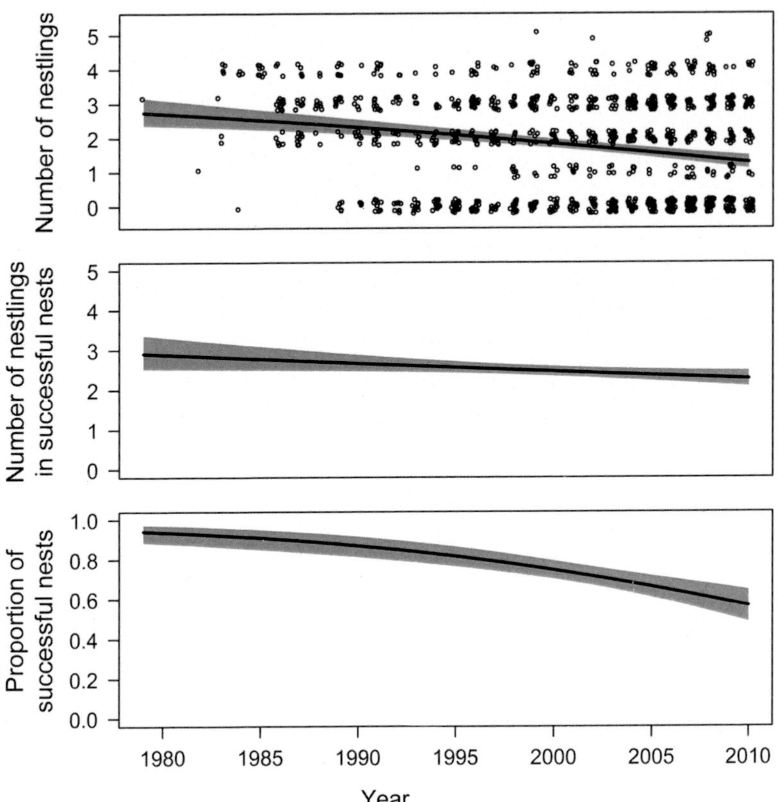

FIGURE 14-4 Fitted regression lines from the zero-inflated model with the number of nestlings as dependent variable: the proportion of successful nests (lowest panel), the number of nestlings given nest success (medium panel), and the average number of nestlings (averaged over all nests; top panel). The shaded area gives the 95% CrI. Sample size was 1130 observations of 279 nests.

To visualize the result more intuitively, we can draw three regression lines: (1) the proportion of successful nests, (2) the number of nestlings from the successful nests, and (3) the average number of nestlings, which is the product of the last two (Figure 14-4). To do so, we proceed as we have learned in the first part of the book: we create a new data frame that contains the predictor variable "year". We transform this variable as we have done with the original variable before fitting the model. We create a new model matrix.

Then, we extract the means of the posterior distributions for the coefficients of the Bernoulli ("ahat") and the Poisson model ("bhat"). We subtract the fitted values of the Bernoulli model from 1 to get the estimated proportions of nests that survived (remember that $z_{it} = 1$ means that the nest has died). We add the fitted value from the Poisson model, which is the average number of nestlings for the successful nests. The average number of nestlings per year (averaged over all nests including the ones that failed) is the product of the proportion of nests that survived and the average number of nestlings for successful nests.

```
newdat <- data.frame(year=1979:2010)
newdat$year.z <- (newdat$year - mean(dat$year))/sd(dat$year)
Xmat <- model.matrix(~year.z, data=newdat)
ahat <- apply(modsims$a, 2, mean)
bhat <- apply(modsims$b, 2, mean)
newdat$propsfit <- 1-plogis(Xmat %*% ahat)
newdat$nnestfit <- exp(Xmat %*% bhat)
newdat$avnnestfit <- newdat$propsfit*newdat$nnestfit
```

Now, repeat the preceding calculations many times (nsim) using each set of model parameters (i.e., a new iteration of the MCMC) once. Finally, we use, for every fitted value, the 2.5% and 97.5% quantiles of the nsim fitted values as lower and upper limits of a 95% CrI.

```
nsim <- length(modsims$lp_)
propsmat <- matrix(ncol=nsim, nrow=nrow(newdat))
nnestmat <- matrix(ncol=nsim, nrow=nrow(newdat))
avnnestmat <- matrix(ncol=nsim, nrow=nrow(newdat))
for(i in 1:nsim){
  propsmat[,i] <- 1-plogis(Xmat %*% modsims$a[i,])
  nnestmat[,i] <- exp(Xmat %*% modsims$b[i,])
  avnnestmat[,i] <- propsmat[,i]*nnestmat[,i]
}
newdat$propslwr <- apply(propsmat, 1, quantile, prob=0.025)
newdat$propsupr <- apply(propsmat, 1, quantile, prob=0.975)
newdat$nnestlwr <- apply(nnestmat, 1, quantile, prob=0.025)
newdat$nnestupr <- apply(nnestmat, 1, quantile, prob=0.975)
newdat$avnnestlwr <- apply(avnnestmat, 1, quantile, prob=0.025)
newdat$avnnestupr <- apply(avnnestmat, 1, quantile, prob=0.975)
```

14.3 OCCUPANCY MODEL TO MEASURE SPECIES DISTRIBUTION USING STAN

When surveying animal or plant populations, it is usually impossible to detect or capture all individuals in the study area. Inevitably, the probability of detecting or capturing an individual will differ between groups of individuals (e.g., between sexes), or may depend on habitat or other factors. If we do not account for these differences in detection or capture probability, the results drawn from such data are likely to be biased. There is a long tradition with a whole range of different models to account for the capture probability in studies that capture and individually mark animals (see the Further Reading section at the end of this chapter).

In contrast, for a long time the issue of imperfect detection (i.e., detection probability < 1) has been neglected in observational studies on animal and plant populations. However, mainly during the last two decades, a lot of effort has been put into the development of models that account for imperfect detection in animal and plant surveys, and nowadays it is becoming the gold standard to account for imperfect detection in most observational surveys.

Occupancy models are an important class of models that account for imperfect detection in animal and plant surveys. Occupancy models deal with the occupancy of a sampling unit by one or more species of interest (MacKenzie et al., 2006). In such studies, a number of sites are usually visited several times by a researcher to infer the presence of a species. Let's say we survey N sites (indexed by i) and visit each site J times (each visit is indexed by j). Our data (let's use the matrix y_{ij} for the data) is then a 1 if a species is observed and a 0 if the species is not observed. We assume that each site is occupied with probability ψ (which we usually call the "probability of occupancy"). Thus, for the true occupancy state x_i, we assume a Bernoulli process:

$$x_i \sim Bernoulli(\psi_i) \tag{14-1}$$

Note, that we can only partly observe whether or not a site is occupied: if we observe the species in a site, we know that this site is occupied, whereas if we do not observe a species in a site the site might not be occupied or we might not have seen the species although the site in fact was occupied. That is why we need to account for the detection process as well. To do so, the main trick is that we visit the site several times within a short period of time. We need to assume that the true occupancy state x_i does not change between the visits. Such an assumption is very common in ecological models that account for the observation process and is usually termed the "closed population assumption". To account for the observation process we assume a second Bernoulli process conditional on the true occupancy state x_i:

$$y_{ij} \sim Bernoulli(x_i p_{ij}) \tag{14-2}$$

Thus, during a visit we observe the species in site i with probability p only if the site is actually occupied by the species ($x_i = 1$). If the site is not occupied ($x_i = 0$), independent of the value of p_{ij}, we cannot observe the species. Therefore, the detection probability p_{ij} is defined as the probability of observing a species during a visit given the site was actually occupied. The Equations 14-1 and 14-2 form the basic structure of an occupancy model.

We now aim to analyze data from the amphibian monitoring in the canton of Aargau in Switzerland. For this monitoring, each of 572 water bodies was visited two times to infer the presence of yellow-bellied toad *Bombina variegata*. The presence of toads is mostly indicated by acoustic cues. Since the calling activity is likely to change over the breeding season, we expect detection probability of yellow-bellied toads to depend on the day when the survey was conducted. Thus, we add the day of the year as a linear and quadratic covariate of the detection probability p_{ij}:

$$\text{logit}(p_{ij}) = \beta_0 + \beta_1 * \text{DAY}_{ij} + \beta_2 * \text{DAY}_{ij} * \text{DAY}_{ij}$$

And we assume that occupancy probability is equal among all sites.

$$\text{logit}(\psi_i) = \alpha_0$$

We can now use Stan to obtain estimates of the parameters. We write the Stan model code and save it in a text file with the name "occupancy.stan".

Stan Code for the Site-Occupancy Model (file "occupancy.stan")

```
data {
  int<lower=0> N;
  int<lower=2> J;
  int<lower=0, upper=1> y[N, J];
  int<lower=0, upper=1> x[N];
  real DAY[N, J];
}

parameters {
  real a0;
  real b0;
  real b1;
  real b2;
}

transformed parameters {
  real<lower=0,upper=1> psi[N];
  real<lower=0,upper=1> p[N, J];
  for(i in 1:N) {
    psi[i] <- inv_logit(a0);  //You may add covariates here
    for(j in 1:J) {
      p[i, j] <- inv_logit(b0 + b1*DAY[i, j] +
                             b2*DAY[i, j]*DAY[i, j]);
    }
  }
}
```

Stan Code for the Site-Occupancy Model (file "occupancy.stan")—cont'd

```
model {
  // Priors (see Chapter 15)
  a0 ~ normal(0, 5);
  b0 ~ normal(0, 5);
  b1 ~ normal(0, 5);
  b2 ~ normal(0, 5);

// likelihood
for(i in 1:N) {
  if(x[i]==1) {
    1 ~ bernoulli(psi[i]);
    y[i] ~ bernoulli(p[i]);
  }
  if(x[i]==0) {
    increment_log_prob(log_sum_exp(log(psi[i]) + log1m(p[i,1]) +
                          log1m(p[i,2]), log1m(psi[i])));
    }
  }
}
```

In Stan, the code usually looks a bit more complex than when using BUGS. This is mainly because we have to strictly define parameters and data structures. Furthermore, discrete latent variables such as x_i are allowed in BUGS but not (yet) in Stan. In Stan we need to find a way around it. We do so by using the if-statement and separately formulating the likelihood for sites where the species has been observed at least once, and sites where the species has not been observed.

Now, we are ready to estimate the parameters using the yellow-bellied toad data. We load the data, assign shorter names, and calculate some additional data that are needed in the Stan model.

```
data(yellow_bellied_toad)
y <- yellow_bellied_toad$y              # Observational data
x <- as.integer(apply(y, 1, sum)>0) # Observed (naive) occupancies
N <- dim(y)[1]                          # Number of sites
J <- dim(y)[2]                          # Number of visits
DAY <- yellow_bellied_toad$DAY          # Julian days of observations
DAY <- (DAY-150)/10                 # scale DAY for easier convergence
```

We need to bundle the data and run the model using Stan.

```
datax <- c("y", "N", "J", "x", "DAY")
fit <- stan(file = "occupancy.stan", data = datax, iter = 2000,
          chains = 2)
```

A summary of the results is printed to the R console with the function `print`.

```
print(fit, c("a0", "b0", "b1", "b2"))
Inference for Stan model: occupancy.
2 chains, each with iter=2000; warmup=1000; thin=1;
post-warmup draws per chain=1000, total post-warmup draws=2000.

    mean se_mean   sd  2.5%   25%   50%   75% 97.5% n_eff Rhat
a0 -1.16    0.00 0.11 -1.38 -1.23 -1.16 -1.09 -0.96   863    1
b0  1.53    0.01 0.29  0.97  1.33  1.52  1.73  2.10   772    1
b1 -0.77    0.01 0.20 -1.17 -0.90 -0.77 -0.63 -0.40   704    1
b2 -0.50    0.00 0.09 -0.68 -0.57 -0.50 -0.44 -0.34   638    1

Samples were drawn using NUTS(diag_e) at Fri Sep 05 12:07:34 2014.
For each parameter, n_eff is a crude measure of effective sample size,
and Rhat is the potential scale reduction factor on split chains (at
convergence, Rhat=1).
```

All $\widehat{R}$ are close to 1. Thus, convergence is likely to be fine. To graphically assess convergence, the Markov chains can be plotted using the function `historyplot` (package blmeco) or `traceplot` (rstan).

```
historyplot(fit, "a0")     # figure not shown
```

Now, we are ready to draw inferences. First, we can make a figure of the change in detection probability over the season (Figure 14-5). We calculate fitted values for detection probability from 1 May to 1 July, which corresponds to day 120 to day 181 of the year. We extract the simulated values from the posterior distribution of β_0, β_1, and β_2, and for each of these simulated values

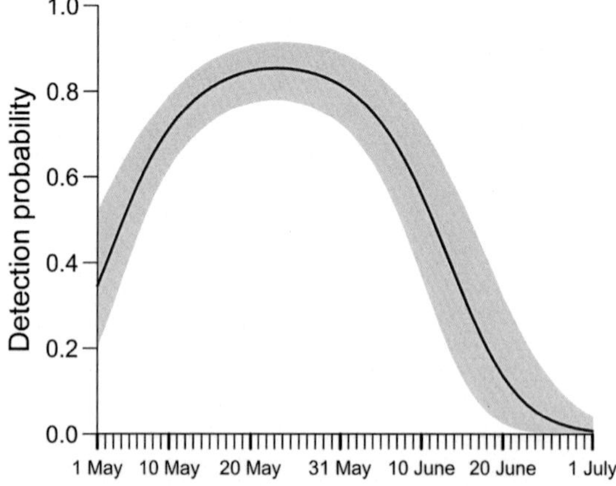

FIGURE 14-5 Fitted detection probability of yellow-bellied toads *Bombina variegata* over the breeding season. The shaded area gives the 95% CrI. Sample size was 572 sites.

and each day between 1 May and 1 July we calculate the fitted value of the detection probability and store them in the matrix fitp.

```
new.DAY <- 120:181
new.trDAY <- (new.DAY-150)/10
modsims <- extract(fit)
nsim <- dim(modsims$b0)[1]
fitp <- array(dim=c(nsim, length(new.DAY)))
for(i in 1:nsim)
  fitp[i,] <- plogis(modsims$b0[i] + modsims$b1[i]*new.trDAY +
                     modsims$b2[i]*new.trDAY^2)
```

Now, we can extract the mean and the lower and upper limits of the 95% CrI and make the figure.

Apparently, detection probability of yellow-bellied toads peaks in the last week of May: a researcher visiting an occupied site during the end of May will detect the species with a probability > 0.8. Since the presence of the species is usually inferred from calling males, we believe that the pattern of detection probability over the season strongly mirrors the calling activity of yellow-bellied toads. Acoustic signaling of many species can strongly change over the course of the day or within a short time period during the breeding season, and detection probability will change accordingly. The yellow-bellied toad data set is thus a good example to show the importance of accounting for detection probability in observational studies: if one does not account for detection probability, visiting the different sites at different dates would bias the results.

Now, we aim to draw inferences on the proportion of sites occupied by yellow-bellied toads, which is the main parameter of interest. In our occupancy model we implemented the true occupancy on the logit scale. Therefore, to draw inferences about occupancy probability we need to back-transform the estimates of α_0:

```
modsims <- extract(fit)
quantile(plogis(modsims$a0), prob=c(0.025, 0.5, 0.957))
     2.5%       50%     95.7%
0.2046530 0.2384540 0.2737034

# observed (naive) occupancy
mean(x)
[1] 0.220279
```

Thus, when accounting for detection probability in our toad example, we estimate that 23.8% (95% CrI: 20.5–27.4%) of the sites were truly occupied. This is not much different from the 22% of sites where we actually observed toads. Such a high detection rate could only be achieved because most of the visits were conducted during the period of highest detection probability and repeated visits were done, which increases the per-year detection probability compared to the per-visit detection probability shown in Figure 14-5.

14.4 TERRITORY OCCUPANCY MODEL TO ESTIMATE SURVIVAL USING BUGS

Estimates of survival and other demographic parameters are usually obtained from data on individually marked animals (Section 14.5). However, a major drawback of these methods is that they are usually invasive because individuals need to be captured and marked. This can be stressful for the animals and time- and labor-intensive for researchers. In contrast, noninvasive observational data from yearly population counts are rarely used to estimate survival because the individuals are not individually marked. Obtaining survival estimates from data of unmarked individuals is currently a very active field of research (e.g., Roth & Amrhein, 2010; Zipkin et al., 2014). The R package unmark is entirely devoted to the statistical analysis of data from surveys of unmarked animals (Fiske & Chandler, 2011).

Our example is from a 10-year population study on nightingales *Luscinia megarhynchos*. Each year during the breeding season, 55 nightingale territories are surveyed several times to check whether they are occupied by male nightingales (Amrhein et al., 2002). Territories were relatively stable across years irrespective of the identity of the territory holder, because nightingales frequently use the edges of bushes, paths, or rivers as territory borders. From these territory occupancy data we aimed to extract information about the survival of the individual male nightingales that occupied the territories.

The estimation of individual survival is complicated because of two main problems: first, territories that are occupied during two consecutive years might not be occupied by the same individual both years. Second, it might be possible that the territory holder was not detected during the surveys and, thus, some occupied territories might be assumed to be unoccupied.

We suggest that the territories that were surveyed for the presence of nightingales might be regarded as the single sites of site occupancy models that are surveyed for the presence of a species (Section 14.3). We thus decided to adapt a multiseason site occupancy model (MacKenzie et al., 2006) to account for the fact that some territory holders might switch territories from year to year, while some other territory holders may not survive to the next year and the territory might be occupied by a new territory holder. We denote the observed nightingale territory occupancy data with y_{itj}: it contains a 1 if a singing nightingale was heard in territory i during year t and visit j and if no nightingale was observed it contains a 0. We assume that three main processes can describe these data.

The first is the survival of individual nightingales from year $t-1$ to year t and returning to the same territory (denoted as Φ). Further, a territory can only be occupied by the same individual as in the previous years if the individual has been present in the previous year, that is, $x_{it\text{-}1} = 1$.

$$z_{it} \sim Bernoulli(x_{it-1}\Phi)$$

The variable z_{it} is a state variable that indicates whether a territory is occupied by the same nightingale as in the previous year. The second process is the colonization of an unoccupied territory by a new individual. The colonization probability is *r*.

$$x_{it} \sim Bernoulli(z_{it} + r[1 - z_{it}])$$

The state variable x_{it} indicates whether a territory i is occupied in year t either by a new individual or by the same one from the previous year. We believe that a territory is occupied by a nightingale if the territory owner from the previous year occupies the territory again (i.e., $z_{it} = 1$) or if the territory is occupied by a new territory owner with probability *r*.

Finally, since a nightingale is not always singing, we might not detect the nightingale in an occupied territory, which is the third important process influencing our data. Consequently, we need to account for this observation process as well. We assume that we observe the territory owner of an occupied territory with probability *p*.

$$y_{itj} \sim Bernoulli(x_{it}p)$$

We now have described how we think our data (y_{itj}) have been generated. The entire model thus includes three Bernoulli distributions with three different parameters (Φ, r, p). Note, however, that z_{it} depends on the territory occupancy of the previous year ($x_{it\text{-}1}$), for which we have no data during the first year. Thus we need yet another model to describe the true occupancy of nightingales during the first year (x_{i1}) and we then start to estimate survival and colonization probability from the second year. We included the additional parameter Ω which is the probability that a territory is occupied during the first study year.

$$x_{i1} \sim Bernoulli(\Omega)$$

Now, everything is defined and we can use OpenBUGS to obtain the posterior distributions of the model parameters. We write the BUGS model code and save it in a text file with name "territoryoccupancy.txt".

BUGS Code for the Territory-Occupancy Model (file: "territoryoccupancy.txt")

```
model{
# priors
omega ~ dunif(0,1)
phi ~ dunif(0,1)
r ~ dunif(0,1)
p ~ dunif(0,1)
```

Continued

BUGS Code for the Territory-Occupancy Model (file: "territoryoccupancy.txt")—cont'd

```
# likelihood
for(i in 1:N){
  x[i,1] ~ dbern(omega)
  for(t in 2:T) {
    mu.phi[i,t] <- x[i,t-1]*phi
    z[i,t] ~ dbern(mu.phi[i,t])
    mu.r[i,t] <- z[i,t] + (1-z[i,t])*r
    x[i,t] ~ dbern(mu.r[i,t])
    for(j in 1:J) {
      mu.p[i,t,j] <- x[i,t]*p
      y[i,t,j] ~ dbern(mu.p[i,t,j])
    } # close j
  } # close t
} # close i
}
```

Note that during the first study year we only estimate which of the territories are occupied. Only from the second year (the *t*-loop starts with "2") can we estimate Φ and *r*. Now, let's see whether we can estimate the parameters using the nightingale data. We load the data using

```
data(nightingales)
```

As usual, we assign a shorter name and look at the data.

```
y <- nightingales
class(y)
```

```
[1] "array"
```

```
dim(y)
```

```
[1] 55 10 8
```

The data comes in the form of a three-dimensional array: the first dimension contains the 55 territories, the second dimension contains the 10 years of the study (2000–2009), and the third dimension contains the eight visits.

To fit the model, we need to calculate the number of territories, the number of study years, and the number of yearly visits, and then bundle the data for OpenBUGS.

```
N <- dim(y)[1]          # Number of territories
T <- dim(y)[2]          # Number of study years
J <- dim(y)[3]          # Number of yearly visits
datax <- list(y=y, N=N, T=T, J=J)
```

Then, we define a function that generates initial values for the four parameters (Ω, Φ, r, p). We should also give some sensible starting values for x, which helps to avoid OpenBUGS from getting stuck from time to time.

```
inits <- function() {
  x <- array(NA, dim = c(N,T))
  for(t in 1:T) x[,t] <- as.integer(apply(y[,t,], 1, sum)>0)
  list(omega = runif(1,0,1),
       phi = runif(1,0,1),
       r = runif(1,0,1),
       p = runif(1,0,1),
       x=x)
}
```

As always, we have to specify which parameters we want to save.

```
parameters <- c("phi", "r", "p", "omega")
```

Then, we can run the MCMC to fit the model:

```
bugsmod <- bugs(datax, inits, parameters,
  model.file="territoryoccupancy.txt",
  n.thin=2, n.chains=2, n.burnin=1000, n.iter=5000,
  working.directory=bugsworkingdir)
```

When OpenBUGS has finished, we check the history plots of the Markov chains for indications of nonconvergence.

```
print(bugsmod$summary[,c("mean", "2.5%", "97.5%", "Rhat")], 3)
              mean      2.5%     97.5% Rhat
  phi        0.591     0.507     0.670    1
  r          0.256     0.202     0.315    1
  p          0.487     0.466     0.508    1
  omega      0.797     0.684     0.891    1
  deviance 3189.221  3171.000  3224.000    1
```

All $\widehat{R}$ are close to 1. Thus, there is no indication of nonconvergence. However, before drawing conclusions from this model, we do some predictive model checking to see how well our model fits the data. To do so, we simulate replicated territory occupancy data for each of the simulated parameter values in the model fit. We simulate nsim replicated sets of territory occupancy data and summarize each set of territory occupancy data using four different test statistics.

The first (nobstot) is the total number of times a territory was observed to be occupied, the second (meanobster) is the average number of years the territory was occupied, the third (obsy1) is the number of observations during the first visit, and the fourth (obsy5) is the number of observations during the fifth visit. First, we need to prepare the vectors where we can store the results of the summary statistics:

```
nsim <- bugsmod$n.sims
nobstot <- integer(nsim)
meanobster <- integer(nsim)
obsy1 <- integer(nsim)
obsy5 <- integer(nsim)
```

Then we start to simulate replicated territory occupancy data and calculate the summary statistics using the following code:

```
for(i in 1:nsim){
  x <- array(dim = c(N, T))
  z <- array(dim = c(N, T))
  yrep <- array(dim = c(N, T, J))
  x[,1] <- rbinom(N, 1, bugsmod$sims.matrix[, "omega"][i])
  for(j in 1:J) {
    yrep[,1,j] <- rbinom(N, 1, x[,1]*bugsmod$sims.matrix[, "p"][i])
  }
  for(t in 2:T) {
    z[,t] <- rbinom(N, 1, x[,t-1]*bugsmod$sims.matrix[, "phi"][i])
    x[,t] <- rbinom(N, 1, z[,t] + (1-z[,t])*
                       bugsmod$sims.matrix[, "r"][i])
    for(j in 1:J) {
       yrep[,t,j] <- rbinom(N, 1, x[,t]*bugsmod$sims.matrix[, "p"][i])
    }
  }
  nobstot[i] <- sum(yrep)
  nobster <- rep(0,N)
  for(t in 1:T) {
     nobster <- nobster + as.integer(apply(yrep[,t,], 1, sum)>0)
  }
  meanobster[i] <- mean(nobster)
  obsy1[i] <- sum(yrep[,,1])
  obsy5[i] <- sum(yrep[,,5])
}
```

Now, we are ready to compare the summary statistics from the simulated values with the summary statistics calculated from the real data. For the summary statistics of the simulated values we calculate 1% and 99% quantiles.

```
quantile(nobstot, probs = c(0.01, 0.99))
      1%      99%
  875.98  1345.01
sum(y)
[1] 1120
```

During the surveys we observed a territory-holding nightingale 1120 times. This is well within the 1% and 99% quantiles of the replicated data (876 and 1345). In this respect, our model is doing a good job. So let's have a look at the average number of years the territories were occupied.

```
quantile(meanobster, probs = c(0.01, 0.99))
      1%        99%
4.127273  6.200000
nobster <- rep(0,N)
for(t in 1:T) {
  nobster <- nobster + as.integer(apply(y[,t,], 1, sum)>0)
}
mean(nobster)
[1] 5.2
```

Again the 5.2 years the territories were, on average, occupied during the study period is well within the 1% and 99% quantiles of the replicated data. Again, we can be satisfied with our model. So we can do the last check. What if we compare only the number of observations during the first visit or only the observations during the fifth visit to each of the territories?

```
quantile(obsy1, probs = c(0.01, 0.99))
     1%     99%
    105     174
sum(y[,,1])
  [1] 60
quantile(obsy5, probs = c(0.01, 0.99))
   1%    99%
  105    175
sum(y[,,5])
[1] 145
```

Now, we have detected an aspect that might be of concern: we only made 60 observations of territory-holding nightingales during the first visit, which is well below the number of observations in the replicated data. In contrast, during the fifth visit the replicated data again fit well with the observations. So what went wrong during the first visit? We assumed constant detection probability during each visit; however, this might be unrealistic as singing activity of nightingales is lower during the beginning of the breeding season than later on in the breeding season (Roth et al., 2012). Thus, we might considerably improve our model by fitting survey-specific detection probabilities. We encourage the reader to do this as an exercise.

Nevertheless, we draw some inferences using the model with constant detection probability. Our estimate for the probability that an individual survives from one year to the next and settles in the same territory again is 0.59 (95% CrI: 0.50–0.67). This estimate is very similar to the survival we estimated from a subsample of individually marked nightingales of the same study population using mark-recapture analyses (Roth & Amrhein, 2010). However, the territory occupancy model allowed including a larger number of individuals in the study because unmarked individuals could also be used.

Therefore, conclusions can be made for a broader population. Of course, the individual-level information about survival is weaker for a sample of unmarked individuals than for the sample of marked individuals. Thus, a higher sample size is needed to obtain the same precision of the survival estimate. If both marked and unmarked individuals exist in a population, an elegant solution would be to combine marked and unmarked individuals in the same model. Using BUGS, it is straightforward to expand the previous model to such a combined model. To learn more about how to combine data from different data sources, we suggest scanning the recent literature about "integrated population models" (e.g., Schaub et al., 2007).

14.5 ANALYZING SURVIVAL BASED ON MARK-RECAPTURE DATA USING STAN

To study the underlying factors that cause animal populations to increase or decrease, individual animals are usually marked and subsequently recaptured to infer information about the demographic parameters such as individual survival or population size. The difficulty with mark-recapture data is that not all marked individuals are recaptured during each capture occasion. If an individual has not been captured, it may have died, emigrated from the study area, or it may have escaped being captured. For the study of survival it is important to disentangle mortality and emigration from capture probability. To do so, Cormack (1964), Jolly (1965), and Seber (1965) developed a statistical model, the Cormack–Jolly–Seber model (CJS), which became the method used by myriad population biologists.

The CJS model explicitly models the capture process (the observation process) conditional on a survival model (the biological process). By including an observation process model, the recapture probability and thus the proportion of individuals still alive and present but not recaptured is estimated and taken into account while estimating the probability that an individual survives and stays in the study area, the so-called apparent survival. The individuals do not need to be recaptured physically, an observation of an individual is sufficient to know that the individual is still alive. In such cases, resighting probabilities instead of recapture probabilities are estimated. To formulate the CJS model, we use a partially observed variable z_{it} that indicates whether individual i is alive and present at time t. The model assumes a Bernoulli process for z_{it}:

$$z_{it} \sim Bernoulli(z_{it-1}\Phi_{it})$$

In other words, an individual i is alive and in the study area at time t ($z_{it} = 1$) with probability Φ_{it}, if it was alive and in the study area at time $t - 1$ ($z_{it\text{-}1} = 1$), or 0, if it was dead (or had permanently emigrated from the study area) at time $t - 1$ ($z_{it\text{-}1} = 0$). The parameter Φ_{it} is the apparent survival probability. The name "apparent" indicates that mortality and emigration from the study area cannot be separated from capture-recapture data (from one study area). A linear predictor for the logit (or another link function) of Φ_{it} can be added to study factors affecting apparent survival.

$$\text{logit}(\Phi_{it}) = \mathbf{X}_{it}\boldsymbol{\beta}$$

This part of the model looks like an autoregressive logistic regression. However, the variable z is only partly observed. It is known that an individual is alive ($z_{it} = 1$) when it has been recaptured or resighted at time t or at any time point later ($y_{it} = 1$). When an individual is not recaptured ($y_{it} = 0$) during any subsequent capture occasion, we do not know whether it has died or whether it is still alive but has not been captured or resighted. The capture process is added to the model as a second Bernoulli process.

$$y_{it} \sim Bernoulli(z_{it}p_{it})$$

where p_{it} is the probability that individual i that is alive and within the study area is captured during capture occasion t. Also, a linear predictor can be added for the logit of p_{it} to study factors affecting the capture probability.

In the original CJS both Φ and p were fully time dependent, that is, for each capture occasion t, independent Φ and p values were estimated. In this fully time dependent model, the Φ and p are not estimable for the last capture occasion, only their product Φp is estimable. Including linear predictors for apparent survival probability or recapture probability makes the two parameters estimable also for the last capture occasion. But they add the constraint that apparent survival probability or recapture probability is constant when all the covariates are constant. This resembles the complete pooling case (Section 7.1.2): the parameter is estimated as an overall mean assuming no differences between the capture occasions. On the other hand, the fully time dependent variant corresponds to the no pooling case, because for each capture occasion an independent parameter is estimated.

Often the reality is something between the two cases, and given we remember what we learned in Section 7.1.2, we use random effects. The advantage of random effects in mark-recapture models was recognized several years ago (Burnham & White, 2002; Royle & Link, 2002). The possibility to use capture occasion as a random effect has been implemented in standard mark-recapture software such as the program MARK (Burnham & White, 2002). The inclusion of individual-level random effects has become feasible only recently with improved computer capacities and MCMC algorithms (Pledger et al., 2003; Royle, 2008, Gimenez & Choquet, 2010; Ford et al., 2012).

However, individual-level random effects are still not widely used, mainly because the fit of such models in user-friendly software such as MARK or BUGS is extremely time consuming. For example, to fit the model we present in the following in BUGS took several days. With Stan we obtained a similar number of effective samples from the posterior distributions in only about three hours. Though this comparison does not meet scientific standards, it is an encouraging observation and motivation for describing here how to fit a CJS model with random effects in Stan.

The example data are from a four-year radio-telemetry study on the post-fledging survival of barn swallows *Hirundo rustica* (Grüebler & Naef-Daenzer, 2008; Grüebler & Naef-Daenzer, 2010). We selected data for the first brood only. The data can be loaded using

```
data(survival_swallows)
```

For convenience, we give a shorter name and look at the data.

```
datax <- survival_swallows
str(datax)
List of 8
$ CH      : int [1:322, 1:18] 1 1 1 1 1 1 1 1 1 1 ...
  ..- attr(*, "dimnames")=List of 2
  .. ..$ : NULL
  .. ..$ : chr [1:18] "day0" "day1" "day2" "day3" ...
$ I       : int 322
$ K       : int 18
$ carez   : num [1:322] -0.584 -0.584 -0.584 -0.584 -0.584 ...
$ year    : num [1:322] 1 1 1 1 1 1 1 1 1 1 ...
$ agec    : num [1:18] -8.5 -7.5 -6.5 -5.5 -4.5 -3.5 ...
$ family  : num [1:322] 5 5 5 5 5 1 1 1 4 4 ...
$ nfam    : num 72
```

The data comes in the format of a list as it is used for BUGS. The element "CH" is a matrix. Every row of the matrix is a capture history of one individual. Every column corresponds to one day after fledging. The first column is the day of fledging. Only individuals that have fledged have been studied. Therefore, the first column contains a 1 for every individual. The entries in columns 2 to 18 indicate whether the individuals have been observed alive at day $t = 1, \ldots, 17$. "I" is the total number of individuals, that is, the number of rows of "CH", and "K" is the number of occasions (here including the day of fledging). "carez" is a family-level covariate that indicates how long the parents have cared for the fledglings after fledging. This variable was z-transformed. "year" contains the numbers 1 to 4, indicating the year of the study. "agec" is the day after fledging (correlating with the age of the individuals) centered around 0. This variable is used as a covariate for the detection probability. The centering eliminates the correlation between the estimate for the intercept and the slope and therefore improves the model-fitting process (better mixing of the MCMC, higher number of effective samples in less time). "family" indicates the family the individual belongs to and "nfam" is the total number of families.

In this exercise we would like to estimate, for every day after fledging, an independent daily survival probability averaged over the four years and corrected for the effort the parents put into the care of their fledglings. We further include family as a random effect because individuals from the same family are not independent, for example, they received the same parental care.

The linear predictor in the survival model contains one intercept per day t, a linear effect of carez, and a year- and family-level modulation of the intercept:

$$\text{logit}(\Phi_{it}) = \alpha_{0t} + \alpha_1 \text{carez}_i + \varepsilon_{\text{year}[i]} + \gamma_{\text{family}[i]}$$

We assume a normal distribution for the year and family random effects.

$$\varepsilon_{\text{year}} \sim Norm(0, \sigma_{\Phi\text{year}})$$

$$\gamma_{\text{family}} \sim Norm(0, \sigma_{\Phi\text{family}})$$

In the linear predictor of the capture probability (in this case we should call this an encounter probability), we include time since fledging (agec) as a covariate because the behavior of the birds is gradually changing as they get older. Because different types of radio transmitters, which differed in coverage, were used in the four different years, the effect of age on observation probability may be different in the four years. We, therefore, included year-specific intercepts and slopes without partial pooling (i.e., agec, year, and their interaction are fixed effects). As a random effect, we add family, because the individuals are moving in family flocks. Thus, if one individual is detected, the probability of also detecting the other members of this family is very high.

$$\text{logit}(p_{it}) = \beta_{0\text{year}[i]} + \beta_{1\text{year}[i]} \text{agec}_i + \delta_{\text{family}[i]}$$

$$\delta_{\text{family}} \sim Norm(0, \sigma_{p\text{family}})$$

In the Stan code, we specify the random effects as standard normal distributions (*Norm*(0,1)) and multiply the random effects with the sigma-parameter within the linear predictor. This parameterization results in a faster and more reliable fit when using MCMC. An important difference between Stan and BUGS is that Stan cannot (yet) handle discrete latent (= unobserved) variables. In the CJS model, after the last recapture of individual i, z_{it} (for $t > last_i$) is an unobserved discrete variable. The problem is circumvented by discarding all observations after the last recapture of each individual by running the likelihood in the model code only until the last observation (`k in 1:last[i]` instead of `k in 1:K`). Then, we add the logarithm of the probability that the individual is never recaptured between its last recapture and the end of the study to the log-likelihood in a separate step. The variable $last_i$ is created in the "transformed data" block. The probability that individual i is never recaptured between its last recapture and the end of the study is defined in the "transformed parameters" block, and named "chi".

Stan Code for the CJS Model with Random Effects (file "CJS_swallows-stan")

```
data {
  int<lower=2> K;                     // capture events
  int<lower=0> I;                     // number of individuals
  int<lower=0,upper=1> CH[I,K];       // CH[i,k]: individual i
                                      // captured at k
  int<lower=0> nfam;                  // number of families
  int<lower=0, upper=nfam> family[I];// index of group variable
  vector[I] carez;         // duration of parental care, z-trans.
  int<lower=1,upper=4> year[I];       // index of year
  vector[K] agec;                    // age of fledling, centered
}
transformed data {
  int<lower=0,upper=K+1> last[I];   // last[i]: ind i last
                                    //capture
  last <- rep_array(0,I);
  for (i in 1:I) {
    for (k in 1:K) {
      if (CH[i,k] == 1) {
        if (k > last[i]) last[i] <- k;
      }
    }
  }
}
parameters {
  real b0[4];                         // intercepts per year for p
  real b1[4];                     // slope for age per year for p
  real a[K-1];                        // intercept of phi
  real a1;                            // coef of phi
  real<lower=0> sigmaphi;  // between family sd in logit(phi)
  real<lower=0> sigmayearphi;  // between-year sd in logit(phi)
  real<lower=0> sigmap;        // between family sd in logit(p)
  real fameffphi[nfam];               // family effects for phi
  real fameffp[nfam];                 // family effects for p
  real yeareffphi[4];                 // year effect on phi
}
transformed parameters {
  real<lower=0,upper=1>p[I,K];        // capture probability
  real<lower=0,upper=1>phi[I,K-1];  // survival probability
```

Stan Code for the CJS Model with Random Effects (file "CJS_swallows-stan")—cont'd

```
  real<lower=0,upper=1>chi[I,K+1]; // probability that an
        // individual is never recaptured after its last capture
  {
    int k;
    for(ii in 1:I){
      for(tt in 1:(K-1)) {
        // linear predictor with random effect for phi:
        // add fixed and random effects here
        phi[ii,tt] <- inv_logit(a[tt] +a1*carez[ii] +
          sigmayearphi * yeareffphi[year[ii]] +
          sigmaphi*fameffphi[family[ii]]);
      }
    }
  for(i in 1:I) {
    // linear predictor with random effect
    // for p: add fixed and random effects here
    p[i,1] <- 1; // first occasion is marking occasion
    for(kk in 2:K)
      p[i,kk] <- inv_logit(b0[year[i]] + b1[year[i]]*agec[kk]+
        sigmap*fameffp[family[i]]);

    // probability that an individual is never recaptured after its
    // last capture
    chi[i,K+1] <- 1.0;
    k <- K;
    while (k > 1) {
      chi[i,k] <- (1 - phi[i,k-1]) + phi[i,k-1] * (1 - p[i,k]) *
        chi[i,k+1];
      k <- k - 1;
    }
    chi[i,1] <- (1 - p[i,1]) * chi[i,2];
    }
  }
}
model {
  // priors
  for(j in 1:4){
    b0[j] ~ normal(0, 5);
    b1[j] ~ normal(0, 5);
  }
  for(v in 1:(K-1)){
    a[v]~normal(0,5);
```

Continued

Stan Code for the CJS Model with Random Effects (file "CJS_swallows-stan")—cont'd

```
    }
    a1~normal(0,5);
    sigmaphi ~ student_t(2,0,1);
    sigmayearphi ~ student_t(2,0,1);
    sigmap ~ student_t(2,0,1);

    // random effects
    for(g in 1:nfam) {
      fameffphi[g] ~ normal(0, 1);
      fameffp[g] ~ normal(0,1);
    }
    for(ye in 1:4){
      yeareffphi[ye] ~ normal(0, 1);
    }
    // likelihood
    for (i in 1:I) {
      if (last[i]>0) {
        for (k in 1:last[i]) {
          if(k>1) 1 ~ bernoulli(phi[i, k-1]);
          CH[i,k] ~ bernoulli(p[i,k]);
        }
      }
      increment_log_prob(log(chi[i,last[i]+1]));
    }
  }
```

As for the zero-inflated model in Section 14.2, we could code the model more efficiently, for example, by omitting the `inv_logit` function by using `bernoulli_logit`. But here we use a less efficient code that allows for recognition of the model structure more easily. The model code is saved in the text file "CJS_swallows.stan" and the model is fitted to the data using Stan:

```
library(rstan)
mod <- stan(file = "CJS_swallows.stan", data=datax, chains=4,
            iter=5000)
```

A summary of the results is printed to the R console with the function `print`.

```
print(mod, c("a", "a1", "b0", "b1","sigmayearphi", "sigmaphi",
      "sigmap"))
Inference for Stan model: CJS_swallows_neu.
4 chains, each with iter=5000; warmup=2500; thin=10;
post-warmup draws per chain=250, total post-warmup draws=1000.
```

```
             mean se_mean   sd  2.5%   25%   50%   75% 97.5% n_eff Rhat
a[1]         3.94    0.02 0.55  2.81  3.61  3.94  4.29  5.00   633 1.00
a[2]         3.86    0.02 0.57  2.67  3.55  3.87  4.20  4.95   512 1.01
a[3]         3.24    0.02 0.49  2.12  2.97  3.27  3.56  4.09   588 1.01
a[4]         3.27    0.02 0.52  2.17  2.97  3.29  3.58  4.25   634 1.01
a[5]         3.43    0.02 0.54  2.27  3.12  3.44  3.76  4.44   679 1.00
a[6]         3.95    0.02 0.65  2.73  3.56  3.96  4.31  5.31   806 1.00
a[7]         3.47    0.02 0.58  2.13  3.13  3.48  3.85  4.53   554 1.00
a[8]         3.86    0.03 0.70  2.52  3.42  3.82  4.28  5.25   774 1.00
a[9]         3.23    0.02 0.58  1.95  2.88  3.21  3.61  4.36   652 1.00
a[10]        2.26    0.02 0.48  1.14  2.01  2.30  2.55  3.11   577 1.00
a[11]       10.21    0.16 5.10  3.78  6.19  9.27 13.08 22.89   966 1.01
a[12]        2.11    0.02 0.50  0.94  1.83  2.14  2.42  2.95   512 1.01
a[13]        2.24    0.02 0.54  1.01  1.94  2.26  2.56  3.27   687 1.01
a[14]        4.69    0.13 3.73  1.84  2.80  3.49  4.66 17.05   851 1.00
a[15]        4.27    0.13 4.07  1.16  2.25  2.86  4.14 18.28   915 1.00
a[16]        1.29    0.02 0.54  0.14  0.97  1.30  1.65  2.27   604 1.00
a[17]        2.24    0.14 2.96  0.26  1.12  1.61  2.21 10.35   420 1.01
a1           0.72    0.00 0.13  0.46  0.63  0.71  0.81  0.98  1000 1.01
b0[1]        2.18    0.01 0.36  1.47  1.93  2.18  2.42  2.92   890 1.00
b0[2]        1.97    0.01 0.22  1.53  1.82  1.95  2.10  2.40  1000 1.00
b0[3]        3.43    0.01 0.27  2.93  3.24  3.41  3.61  3.97  1000 1.00
b0[4]        3.38    0.01 0.39  2.66  3.13  3.37  3.64  4.15  1000 1.00
b1[1]        0.04    0.00 0.04 -0.03  0.02  0.04  0.07  0.13  1000 1.00
b1[2]       -0.31    0.00 0.03 -0.37 -0.33 -0.31 -0.29 -0.26  1000 1.00
b1[3]       -0.41    0.00 0.04 -0.49 -0.44 -0.41 -0.38 -0.34  1000 1.00
b1[4]       -0.07    0.00 0.05 -0.18 -0.11 -0.07 -0.03  0.03   899 1.01
sigmayearphi 0.70    0.02 0.50  0.13  0.35  0.55  0.89  2.09   767 1.00
sigmaphi     0.48    0.01 0.16  0.11  0.37  0.48  0.59  0.76   878 1.00
sigmap       0.89    0.00 0.14  0.66  0.78  0.87  0.97  1.20   981 1.00

Samples were drawn using NUTS(diag_e) at Sun Sep 14 13:46:47 2014.
For each parameter, n_eff is a crude measure of effective sample size,
and Rhat is the potential scale reduction factor on split chains (at
convergence, Rhat=1).
```

We first check whether the MCMCs have converged. Then, we assess model fit and, lastly, we draw conclusions. The $\widehat{R}$ seem to be close to 1. Thus there is no indication of nonconvergence. Also, the Monte Carlo error (se_mean) is close to 0 at least for the parameters belonging to the first 10 days, where sample size is high. To graphically assess convergence, the Markov chains can be plotted within the same plot. To plot the whole chains (inclusive of the values discarded due to thinning, alternatively with or without the warm-up phase), use

```
traceplot(mod, "sigmap")     # figure not shown
```

To look at only the saved values to base inferences on, use, for example, the function `historyplot`.

```
historyplot(mod, "sigmap")     # figure not shown
historyplot(mod, "sigmaphi")   # figure not shown
historyplot(mod, "a")          # figure not shown
```

To assess whether the model fits the data, we simulate replicated data from the model (posterior predictive model checking). To do so, we use the 1000 values from the posterior distributions of Φ and p, here called Φ^*_{rit} and p^*_{rit}, where r is the index for the simulation and the star is added to distinguish these values from the theoretical (unknown) true Φ_{it} values and from the estimated parameter values $\widehat{\Phi}_{it}$. The Φ^*_{rit} and p^*_{rit} values can be accessed by the function extract.

```
modsims <- extract(mod)
```

First, we define arrays for the replicated observations (y^{rep}_{rit}) and the replicated state variable (z^{rep}_{rit}).

```
nsim <- dim(modsims$phi)[1]  # extract the number of simulations
yrep <- array(dim=c(nsim, datax$I, datax$K))
zrep <- array(dim=c(nsim, datax$I, datax$K))
```

Second, the state and observation at the first capture occasion (day of fledging) is filled with 1s for every individual i and simulation r.

```
yrep[,,1] <- 1     # first occasion is 1 for all individuals
zrep[,,1] <- 1
```

Third, we go through all subsequent capture occasions and simulate states z^{rep}_{rit} based on $z^{rep}_{rit\text{-}1}$ and Φ^*_{rit}, and new observations y^{rep}_{rit} based on z^{rep}_{rit} and p^*_{rit}.

```
for(j in 1:(datax$K-1)){
  zrep[,,j+1] <- rbinom(nsim*datax$I, size=zrep[,,j],
                        prob=modsims$phi[,,j])
  yrep[,,j+1] <- rbinom(nsim*datax$I, size=zrep[,,j+1],
                        prob=modsims$p[,,j])
}
```

The yrep object contains 1000 new capture histories for the 322 individuals in the data set. All of these data sets reflect possible observations assuming that the structure of the model is true but taking into account the uncertainty in the estimated model parameters.

In the last step, we compare the replicated data sets to the observed data. There are many ways to do this comparison. When the interest is in an overall mean survival probability, we may compare the average time until the last capture between the observed and simulated data; but when the interest is in temporal patterns of survival we may do better by visualizing each individual capture history. As an example here, we assess whether the between-family variance assumed in the model fits the data. To do so, we count the number of observations per individual and calculate the mean, minimum, and maximum of this number per family. We do this first for the observed data.

```
nobspind <- apply(datax$CH, 1, sum)
     # number of observations per individual
mpfam <- tapply(nobspind, datax$family, mean)
minpfam <- tapply(nobspind, datax$family, min)
maxpfam <- tapply(nobspind, datax$family, max)
```

Then, we do the same for the 1000 replicated data sets.

```
repnpind <- apply(yrep, c(1,2), sum)
repmpfam <- matrix(nrow=nsim, ncol=datax$nfam)
repminpfam <- matrix(nrow=nsim, ncol=datax$nfam)
repmaxpfam <- matrix(nrow=nsim, ncol=datax$nfam)

for(f in 1:nsim) {
  repmpfam[f,] <- tapply(repnpind[f,], datax$family, mean)
  repminpfam[f,] <- tapply(repnpind[f,], datax$family, min)
  repmaxpfam[f,] <- tapply(repnpind[f,], datax$family, max)
}
```

The observed data are then compared with the simulated data graphically. In Figure 14.6 the average number of observations is given per family and ordered increasingly (bold black line). In addition the familywise minimum and maximum number of observations per individual are given (thin black lines). The blue lines give the same statistics as for the observed data, but averaged over the 1000 simulated data sets. We see that the replicated data from the model resemble the observed data quite well, particularly for families with an average number of observations. Maybe, for families with low numbers of observations, the model predicts slightly higher numbers of observations whereas for families with many observations, the model predicts slightly lower numbers. This may indicate that the model slightly underestimates the between-family variance either in recapture probability or in survival probability. This predictive model checking is not exhaustive. See Chapter 10 for more on predictive model checking.

After we have decided that our model may adequately describe the process that generated the data, the results can be extracted from the modsims object (obtained by the function `extract`). The estimates for daily survival probabilities are directly calculated from the parameter α_0 (Figure 14-7).

```
ests <- plogis(apply(modsims$a, 2, mean))
ests.lwr <- plogis(apply(modsims$a, 2, quantile, prob=0.025))
ests.upr <- plogis(apply(modsims$a, 2, quantile, prob=0.975))
```

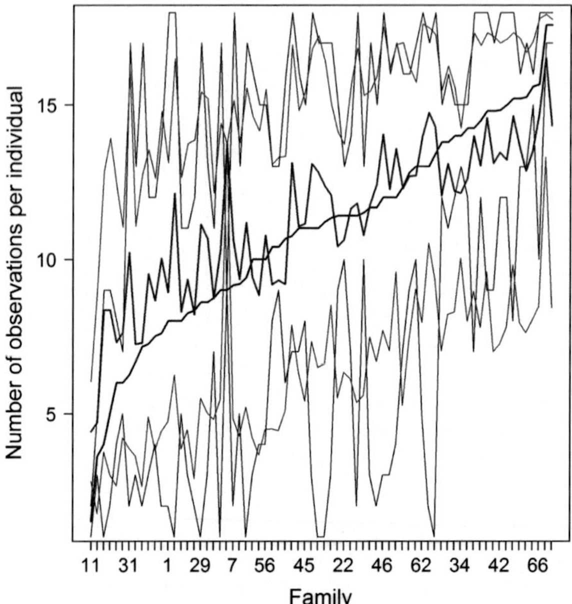

FIGURE 14-6 Number of times an individual has been observed during the study. Shown is the average per family (bold black line) in ascending order. For every family, the observed minimum and the maximum is given in thin black lines. The blue lines give the same numbers averaged over 1000 simulated data sets (i.e., mean of the posterior predictive distribution for the specific statistic).

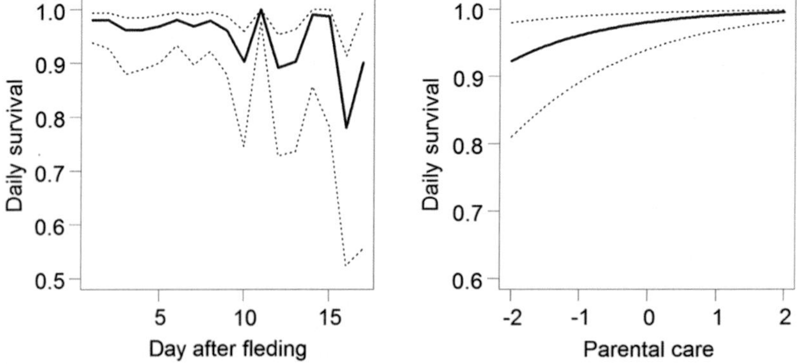

FIGURE 14-7 Estimated daily survival probability for every day after fledging (left) and effect of parental care on daily survival of the fledglings (right). Bold lines = mean of the posterior distribution; dotted lines = 2.5% and 97.5% quantiles of the posterior distribution.

To visualize the effect of parental care on survival, we average the intercept over the first 12 days and extract the slope for carez from the modsim object. Then we obtain the fitted values in the same way as we calculate the regression line in a logistic regression.

```
ma <- apply(modsims$a[,1:12], 1, mean) # mean over the first 12 days
newdat <- data.frame(carez=seq(-2, 2, length=100))
b <- c(mean(ma), mean(modsims$a1))      # model coefficients
Xmat <- model.matrix(~carez, data=newdat)
newdat$fit <- plogis(Xmat %*% b)
nsim <- nrow(modsims$a)
fitmat <- matrix(ncol=nsim, nrow=nrow(newdat))
for(i in 1:nsim) fitmat[,i] <- plogis(Xmat %*%
                                    c(ma[i], modsims$a1[1]))
newdat$lwr <- apply(fitmat, 1, quantile, prob=0.025)
newdat$upr <- apply(fitmat, 1, quantile, prob=0.975)
```

We see that duration of parental care has a positive effect on daily survival of the fledglings (Figure 14-7). The posterior probability that parental care has a positive effect on fledgling survival is the proportion of positive values among the simulated values from the posterior distribution of the parameter α_1, which is larger than 0.999.

```
mean(modsims$a1>0)
[1] 1
```

All 1000 simulated values from the posterior distribution of α_1 are positive. Because we do not know whether the 1001st simulation would have produced a negative number (which would have led to a posterior probability of 0.999001), we can report our results as $\Pr(\alpha_1>0|y) > 0.999$ and conclude that we found a well supported positive effect of parental care on fledgling survival.

FURTHER READING

An informative and well-written chapter about multinomial models in WinBUGS is given by Ntzoufras (2009).

In the book about analysis of count data, Cameron and Trivedi (2013) give an overview of models for data with excess zeros. Zuur et al. (2012) is an applied textbook on zero-inflated models for ecologists. Poulsen et al. (2011) apply a zero-inflated Poisson model to disentangle different effects on animal communities. Lecomte et al. (2013) use zero-inflated models in spatial analyses of biomass data.

Royle and Dorazio (2008) introduce a variety of different ecological models that explicitly model the observation process conditional on the biological process. In their recent book, they expand these models to allow for studying the spatial movements of animals (Royle et al., 2014). Link and Barker (2010) give an introduction to the Bayesian theory with applications in ecology.

The territory occupancy model is discussed in Roth and Amrhein (2010).

Besides the original literature on CSJ models (Cormack, 1964; Jolly, 1965; Seber, 1965), a multitude of authors have discussed variations of these models. A seminal review is Lebreton et al. (1992). Practical introductions for biologists are given by White and Burnham (1999) (software Program MARK), Royle and Dorazio (2008) (R and WinBUGS), and Kéry and Schaub (2012) (WinBUGS). Theoretical background is found in King et al. (2010) and King (2014).

Chapter 15

Prior Influence and Parameter Estimability

Chapter Outline

15.1 HOW TO SPECIFY PRIOR DISTRIBUTIONS

In Bayesian data analysis, the prior distribution is an inherent part of the model, as are the error distribution and the link function. It is, therefore, important to choose meaningful prior distributions given the study at hand.

In the first part of this book, we did not deal much with prior distributions ("priors") at all, which is not good practice. In the introductory chapters, we exclusively used `sim` to obtain samples from the posterior distributions based on `lm`, `glm`, `lmer`, or `glmer` objects. In these cases, the function `sim` uses improper priors, $p(\beta) \propto 1$ (a horizontal line at 1) for the model coefficients and $p(\sigma^2) \propto 1/\sigma^2$ for the variance parameters. These priors are called improper because they are not proper probability distributions: their density functions do not integrate to 1.

For the type of models used, these priors produce results that are mostly equal to results one would obtain using frequentist methods. Thus, all information in the results stems from the data. There are, however, special situations where prior distributions will have a marked influence on the results. But in such situations, classical linear models cannot be applied. For example, when the outcomes of a binary variable are completely separated with respect to a covariate (no overlap in the covariate between the two outcome groups) the function `glm` (and any other frequentist method used to fit the model) fails to fit a logistic regression. In such cases, the model becomes estimable when proper prior distributions are used. Proper prior distributions are priors that are not based on the data and that integrate to 1. We use proper priors in the second part of the book (from Chapter 12 onwards). When using proper prior distributions, we need to know how to choose such a prior, and we need to assess whether and how this choice influences our conclusions.

Bayesian Data Analysis in Ecology Using Linear Models with R, BUGS, and Stan
http://dx.doi.org/10.1016/B978-0-12-801370-0.00015-0.

Prior distributions provide the opportunity to formally include previous knowledge in a statistical analysis. If available, it is usually a good idea to use such previous knowledge. In almost every study, there exists some information, at least about the range of reasonable parameter values. If this information is used to construct the prior, the parameter estimates are kept within the range of reasonable values. In addition, such informative priors help making MCMC algorithms stable.

Weakly informative prior distributions are priors that are constructed based on the range of reasonable parameter values, but without compiling information from existing studies. These priors contain some information, but obviously much less than what would be possible if the relevant information from the literature would be gathered (Gelman et al., 2014). Weakly informative priors can be defined, for example, from the model structure. In a logistic regression with a z-transformed numeric predictor, a slope of 10 would mean that the probability changes from `plogis(−5)` $= 0.007$ to `plogis(5)` $=$ 0.993 when the predictor variable is increased by 1 standard deviation. This is a huge effect and unrealistic in most situations. In this case, we might use a normal distribution with mean 0 and standard deviation 5 ($Norm$(0,5)) as the prior; such a prior assigns most of its mass to slope values from -10 to $+10$. This means that, before looking at the data we know that the effect can be something between a huge negative and a huge positive effect; we do not know much, but we give implausible parameter values (less than -10 or larger $+10$) very low probabilities. In contrast, a flat prior, such as $Norm$(0,100), would give similar probabilities to both implausible and plausible parameter values.

We try to avoid the term *noninformative* prior (or, sometimes, *uninformative*) because every prior distribution contains information. For example, an improper prior $\propto 1$ says that values close to 0 are equally likely as very large values. Often, a seemingly noninformative prior becomes highly informative when the parameter is transformed, for example, when a link function is used. The Jeffreys prior is such a clever prior that is invariant to transformations. These priors have some nice mathematical properties. Gelman et al. (2014) call such priors "reference" priors. Such reference priors, or flat priors are handy for a quick analysis. But because they can have a large influence on the results, the results should be checked carefully and the prior influence should be assessed.

In general, informative priors are chosen to keep the posterior distribution within a range of reasonable values and to stabilize MCMC algorithms, and, in case of strong priors, to base the conclusions not only on the data at hand but also on previous knowledge. Using informative priors is a way to integrate different data sets into the same analysis.

What prior distributions should be chosen for which parameters? So-called conjugate priors have the mathematical property that the posterior distribution has the same parameterization as the prior distribution. If you would like to derive the posterior distribution analytically (instead of by using `sim` or an

MCMC sampler), the calculations are much easier when conjugate priors are chosen. Neither `sim`, BUGS, nor Stan require that conjugate priors are specified. But, sometimes, it is helpful to be able to quickly assess by hand how a posterior distribution looks like. Of course, this is only possible for simple problems.

For example, an ornithologist has installed a data logger at the entrance of a woodpecker's breeding hole to record the visits of the bird at the nest. After installation he or she tests whether every visit is correctly recorded by visual observations done in tandem. We assume that the error rate is constant over time and that only false negatives are a problem, that is, the device sometimes misses visits but never records a visit when there was in fact none. The ornithologist observed 10 visits and the comparison with the data logger shows that it has recorded 10. Thus there was no error in 10 visits. How sure can we be that the logger will work correctly for the following 200 visits? The number of errors can be modeled as a binomial variable $y \sim Binom(p, 10)$ and the error probability can be estimated. The function `glm` cannot fit this model, because the outcome variable is 0 for all observations. However, after specifying a prior distribution, it is possible to fit this model using BUGS or Stan. But, because it is such a simple case, we can also obtain the posterior by hand using a conjugate prior distribution.

From Table 15-1 we see that the conjugate prior distribution for the binomial data model is the beta distribution, and we see what the posterior looks like, given the data and the prior. Thus, we only have to choose parameters a and b for the beta prior distribution. Let's assume that the ornithologist heard from colleagues that some of these loggers work very well (error rate close to 0) whereas others do not work at all (error rate close to 1). Therefore, we use $Beta(1,1)$, which is a uniform distribution between 0 and 1,

TABLE 15-1 Conjugate Prior Distributions for Four Types of Models with Corresponding Posterior Distribution

Data model	Conjugate prior	Posterior
$y \sim Binom(p,N)$	$p(p) \sim Beta(a,b)$	$p(p \mid y) \sim Beta(\alpha + y, b + N - y)$
$y \sim Pois(\lambda)$	$p(\lambda) \sim Gamma(a,b)$	$p(\lambda \mid y) \sim Gamma(a + n\overline{y}, b + n)$ n: sample size, $\overline{y}$: sample mean
$y \sim Norm(\mu,\sigma^2)$ with variance known	$p(\mu) \sim Norm(\mu_0,\tau_0^2)$	$p(\mu \mid y) \sim Norm(\mu_n,\tau_n^2)$ with $\mu_n = \frac{\frac{1}{\tau_0^2}\mu_0 + \frac{n}{\sigma^2}\overline{y}}{\frac{1}{\tau_0^2} + \frac{n}{\sigma^2}}$ and $\frac{1}{\tau_n^2} = \frac{1}{\tau_0^2} + \frac{n}{\sigma^2}$
$y \sim Norm(\mu,\sigma^2)$ with variance unknown	$p(\mu,\sigma^2) \sim N\text{-}Inv\text{-}\chi^2$	$p(\mu,\sigma^2 \mid y) \sim N\text{-}Inv\text{-}\chi^2$

Note: The parameterization of the N-inverse-χ^2 distribution is rather complex and is not easily done by hand. We only added this model for completeness (see Gelman et al., 2014 for more detailed information).

as prior. The posterior, then, is *Beta*(1 + 0, 1 + 10). From this distribution we can extract the mean, which is 1/11= 0.09, and the 95% quantile, which is `qbeta(0.95, 1, 11)` = 0.24. Thus, the ornithologist can be 95% sure that the number of visits that are missed is lower than 24%.

But, if it is known that such loggers have error rates between 0 and 10%, we can use a *Beta*(2,40) as prior distribution. This prior distribution has a mean of 0.05 and a 95% quantile of 0.11. And, the posterior is *Beta*(2 + 0, 40 + 10), which has a mean of 2/50 = 0.04 and a 95% quantile of `qbeta(0.95, 2, 50)` = 0.09.

The Beta Distribution

The beta distribution is a continuous probability distribution for values between 0 and 1. It is, therefore, useful for describing the distribution of probabilities *p*. Its density function is

$$p(p) = \frac{\Gamma(a+b)}{\Gamma(a)\Gamma(b)} p^{a-1}(1-p)^{b-1},$$

where $\Gamma(x) = (x - 1)!$ is the gamma function. The gamma function has some exciting mathematical properties that are not important for us practitioners, except that it is used to define the density functions of some useful statistical distributions such as the beta distribution. The parameters *a* and *b* can have any positive value (>0). The mean of a beta distribution is $a/(a + b)$ and the variance is $\frac{ab}{(a+b)^2(a+b+1)}$. The higher the sum $a + b$, the smaller the variance (Figure 15-1).

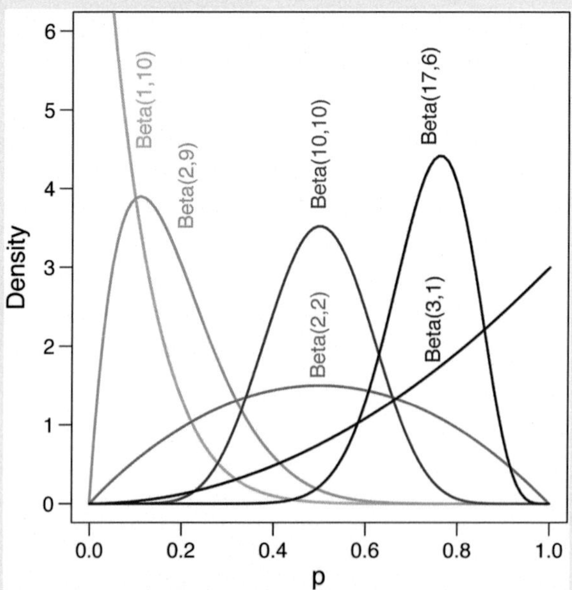

FIGURE 15-1 Density functions of different beta distributions.

How do we find the correct parameter values for a prior distribution? For example, if we need a prior for a probability as in the preceeding example, and we choose a beta distribution, what values should we use for the parameters a and b? If the prior knowledge can be expressed as an estimate and an uncertainty measure, say 0.05 ± 0.02 (estimate $\pm$ standard error), then the estimate and the square of the standard error (= variance of the parameter) can be inserted in the formulas for the mean and variance of the beta distribution (see box), and the equations can be solved for a and b. This is done by the function `shapeparameter` from the package carcass. In other cases, the "trial and error" approach is also valuable. The functions `qgamma`, `qbinom`, and `qnorm` are helpful for this purpose.

Prior distributions for variance parameters are discussed in the next section.

15.2 PRIOR SENSITIVITY ANALYSIS

In cases of flat or weakly informative priors we should make sure that the choice of the prior does not markedly influence the conclusions. And, when we have strong priors, we may want to report the relative contribution of the prior and the data to the results, respectively. In this section, we present three examples to show how the influence of priors can be assessed.

We start with the normal linear regression that we first introduced in Chapter 4 and then used to introduce BUGS and Stan in Chapter 12. The data consists of 50 pairs of x and y measurements, and we are interested in the slope of the regression line of y on x. In Chapter 12, we fit the following model with weakly informative priors:

$$
\begin{aligned}
y_i &\sim Norm(\mu, \sigma) \\
\mu &= \beta_0 + \beta_1 X_i \\
\beta_k &\sim Norm(0, 5) \\
\sigma &\sim Cauchy(0, 5)[0,]
\end{aligned}
$$

To measure the influence of the choice of *Norm*(0,5) for the prior of the slope parameter, we fit 20 different models using different priors. The used priors vary from very strongly informative, *Norm*(0,0.01), to flat, *Norm*(0,150). The 20 different models can be fitted in one go, for example, using Stan. To do so, we include in the data a vector that we call "sdprior", in addition to y, x, and n. This vector contains the 20 standard deviations of the 20 different prior normal distributions that we would like to use for the slope parameter "beta1". Then, we run through the 20 different regressions and estimate 20 different intercepts, slopes, and sigma parameters. We use a truncated Cauchy distribution as prior for sigma (see below).

Stan Code for 20 Different Linear Regressions, Each with a Different Prior Distribution for the Slope Parameter

```
data {
  int<lower=0> n;
  vector[n] y;
  vector[n] x;
  vector[20] sdprior;
}
parameters {
  vector[20] beta0;
  vector[20] beta1;
  real<lower=0> sigma[20];
}
model {
  //priors
  beta0 ~ normal(0,5);
  sigma ~ cauchy(0,5);
  for(k in 1:20){
    beta1[k] ~ normal(0,sdprior[k]);
    // likelihood
    y ~ normal(beta0[k] + beta1[k] * x, sigma[k]);
  }
}
```

From the results, we see that priors with standard deviations smaller than about 0.8 have an influence on the result, whereas priors with larger variances do not influence the result markedly (Figure 15-2). More informative priors speed up MCMC and Hamiltonian Monte Carlo algorithms. Thus, it is recommended to use priors that are as informative as possible, as long as they do not noticeably affect the posterior distribution. Sensitivity analyses, as the one we did here, help in finding such prior distributions.

For a variance parameter, such as the sigma in the linear regression, or a between-group variance in a mixed model, it is often difficult to find an appropriate prior. A prior for a variance must be nonnegative and continuous. Commonly used priors are long-tailed inverse-gamma distributions or uniform distributions spanning a range of positive values. However, inverse-gamma distributions can result in improper posterior distributions (Gelman, 2006) and could, therefore, cause troubles in the model fitting process. For this reason, in many practical Bayesian data analyses, uniform priors have been preferred. However, uniform priors tend to overestimate the

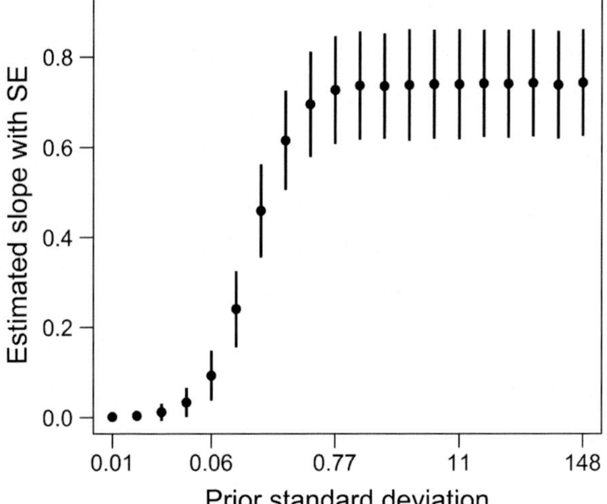

FIGURE 15-2 Estimated slope parameter of a linear regression fitted to the same data ($n = 50$) but using different normal priors for the slope parameter. Values on the x-axis are the standard deviations of the normal prior distributions used for the slope parameter. The vertical lines are standard errors.

variance parameter, particularly when sample size (or the number of groups) is small.

Further, uniform priors are very unnatural in that they assume that variances up to a threshold value are, *a priori*, all equally likely, and values above the threshold are impossible. A more natural prior for a variance parameter would be one with a large mass in a range of likely values with an upper tail that gradually becomes smaller and approaches zero for unrealistically large values. Folded t-, half-Cauchy, and inverse-gamma distributions have such shapes (Gelman, 2006).

To assess prior sensitivity we use 20 different prior distributions for the variance parameter σ of the linear regression from Chapter 4, and for the between-nest variance "sigmatheta" in the zero-inflated Poisson mixed model from Chapter 14.

The 20 priors used are shown in Figure 15-3. The folded t-distribution is the distribution of the absolute value of a t-distributed variable with a mean of zero. This distribution has been recommended as prior distribution for variance parameters in hierarchical models (Gelman, 2006). We construct the folded t-distribution by using a zero-truncated t-distribution (retain only nonnegative values). The folded t-distribution has three parameters: the degrees of freedom (ν), the mean, and the standard deviation. The larger the ν the closer the t-distribution looks to a normal distribution. A t-distribution with $\nu = 1$ and $\sigma = 1$ is a standard Cauchy distribution.

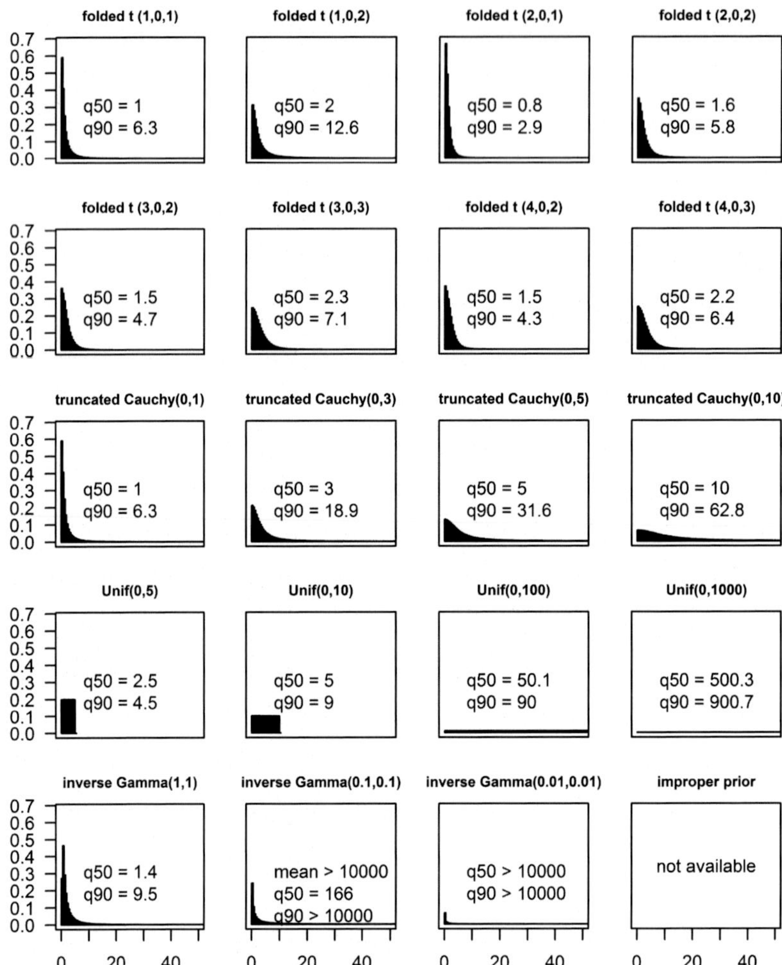

FIGURE 15-3 The different prior distributions used to assess the sensitivity of the variance parameter estimate to the prior choice. The parameters of the folded *t*-distribution are the degrees of freedom (ν), the mean, and the standard deviation. The parameters of the Cauchy distributions are the location and the scale. For both, the folded *t*- and the truncated Cauchy distributions, we used absolute values of the *t*- and Cauchy distribution respectively, so that distributions are restricted to nonnegative values. The parameters of the uniform distributions are the lower and the upper limit. The parameters of the inverse-gamma distributions are the shape and the inverse scale. For each distribution, the median (q50) and the 90% quantile (q90) are given. The scale on the *x*- and *y*-axes are equal in all plots to ease comparison between the distributions. Long tails are not shown.

The Cauchy distribution describes the distribution of the ratio between two independent normal variables. It is a symmetric distribution with long tails. A half-Cauchy distribution is the upper half (the positive values if the location is zero) of the Cauchy distribution (shown in Figure 15-3). It has a large mass at low positive values, and then the density thins out with larger values. Therefore, the Cauchy distribution is another candidate for priors of variance parameters.

We also look at the commonly used (but not recommended) uniform and inverse-gamma distributions, and at the improper prior used by `sim`. The BUGS and Stan codes we used to specify these prior distributions are given in Table 15-2.

In the two examples we used here, the prior does not seem to have a marked influence on the estimate of the variance (Figure 15-4). Very strong priors such as *Unif*(0,5) obviously affect the posterior when they differ from the prior, as was the case in the linear regression, where the mean of the posterior of σ actually should have been around 5.

We expected that, compared to the linear regression, the prior would have a stronger influence on the between-nest variance parameter of the zero-inflated model because the between-nest variance is estimated based on an unobserved (latent) and binary variable (the indicator variable of "structural" zero values). However, the prior influence was unexpectedly weak. One important reason may be that the estimate coincided very well with most of the priors, and, furthermore, because sample size (measured as the number of nests and the observations per nest) is very large in this data set. In other data sets the variance parameters may be more strongly influenced by the choice of the prior, and a sensitivity analysis as suggested here may be more critical.

15.3 PARAMETER ESTIMABILITY

The more complex a model becomes, particularly when latent variables are included, the more difficult it becomes to assess whether parameters are estimable at all. Parameter estimability is, particularly in multilevel ecological models, an important issue. Some parameters in a model may not be estimable for two different reasons. First, the parameter is mathematically not identifiable because of the model structure. This is called "intrinsic" nonestimability. For example, the recapture probability and the survival probability for the last time period of a mark-recapture study are only estimable as a product. Thus, the survival and recapture parameters for the last time period are intrinsically not estimable. Second, the parameter may theoretically be identifiable, but there is too little information in the data to obtain a reliable estimate. This is called "extrinsic" nonestimability.

Whether and which parameters are intrinsically estimable in a model can be unambiguously assessed using numeric algebra. We do not deal with these

TABLE 15-2 Candidate Prior Distributions for Variance Parameters with Corresponding BUGS and Stan Code

Prior	BUGS	Stan

Folded t for σ
For $\nu = 1$, 90% of mass is below 6.3
For $\nu = 2$, 90% of mass is below 2.9
	sigma ~ dt(0,1,nu)I(0,)
	The value ν has to be equal or larger than 1 when using OpenBUGS or Jags, and equal or larger than 2 when using WinBUGS.
		parameters {
		  ...
		  real<lower=0> sigma;
		}
		model {
		  sigma ~ student_t(nu,0,1);
		  ...
		}

Truncated Cauchy for σ
$\tau = 2.5$, 90% of mass is below 7.7
$\tau = 5$, 90% of mass is below 15.4
	No preprogrammed Cauchy distribution available	parameters {
		  ...
		  real<lower=0> sigma;
		}
		model {
		  sigma ~ cauchy(0, tau);
		  ...
		}

Uniform for σ
υ is the upper limit of the uniform distribution
	`sigma~dunif(0,nu)`	`parameters {`
		`  ...`
		`  real<lower=0, upper=nu> sigma;`
		`}`
		`model {`
		`  sigma ~ uniform(0, nu);`
		`  ...`
		`}`

Inverse Gamma for σ^2
$(\alpha, \beta)=(1, 1)$: 90% of mass is below 9.2
$(\alpha, \beta) = (0.1, 0.1)$, 90% of mass is below ~1'600'000'000
	`tau <- dgamma(alpha, beta)`
	`sigma <- pow(tau, -0.5)`
		`parameters {`
		`  ...`
		`  real<lower=0> invsigma2;`
		`}`
		`transformed parameters{`
		`  real<lower=0> sigma;`
		`  sigma <- 1/sqrt(invsigma2);`
		`}`
		`model {`
		`  invsigma2 ~ gamma(alpha, beta);`
		`  ...`
		`}`

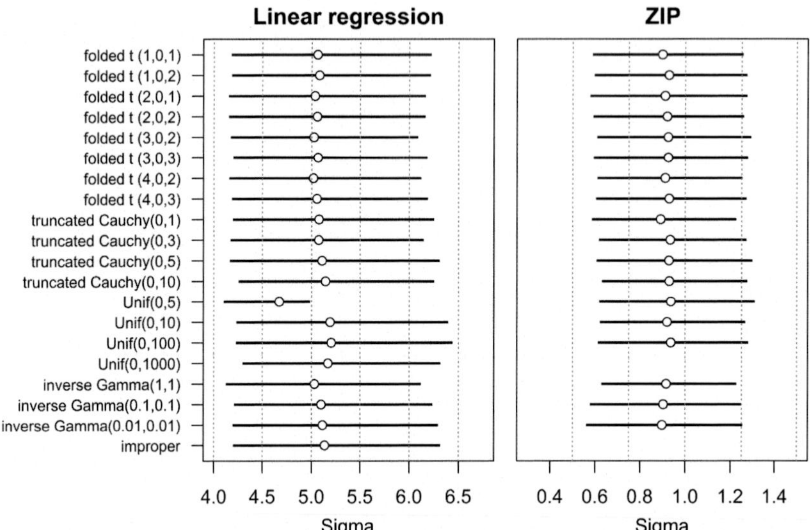

FIGURE 15-4 Estimated variance parameters based on different prior distributions. The "sigma" in the linear regression from Chapter 4 is the standard deviation of the residuals, and the "sigma" in the zero-inflated Poisson model (ZIP) is the between-nest variance of the logit of nest survival (black stork example of Chapter 14). The MCMC algorithm failed to sample for the zero-inflated Poisson (ZIP) model with a Unif(0,1000) prior. All models were fit in Stan except the ones with improper priors, for which we used the R functions `lm` and `sim`. These functions cannot fit ZIP mixed models.

methods here because in this book all parameters are intrinsically identifiable in most cases. See Further Reading for more information about the algebraic methods to assess parameter identifiability.

Extrinsic nonestimability is less straightforward to define than intrinsic nonestimability. As the information in the data lessens, it becomes more difficult to estimate a parameter. In Bayesian analyses, the posterior distribution for a weakly identifiable parameter will resemble the prior distribution. The closer the posterior distribution is to the prior distribution, the weaker is the information in the data, and the less estimable is a parameter. Therefore, one possibility to assess how well a parameter is informed by the data is to measure the overlap between the prior and the posterior distributions.

To illustrate how this is done in practice, we use the territory occupancy model from Chapter 14. There, we estimate survival probability ϕ, territory colonization rate r, detection probability p, and the probability that a territory is occupied during the first year Ω. All these parameters are probabilities. The true state of a territory is only partially observable, therefore, parameter estimability may be an issue. From the BUGS output, we see that the Markov chains seem to have converged and we have enough samples to draw inference.

```
round(bugsmod$summary, 2)
       mean    sd  2.5%   25%   50%   75% 97.5% Rhat n.eff
phi    0.59  0.04  0.50  0.56  0.59  0.62  0.67 1.00  1000
r      0.26  0.03  0.20  0.24  0.26  0.28  0.31 1.00  1600
p      0.49  0.01  0.47  0.48  0.49  0.49  0.51 1.00  6300
omega  0.79  0.05  0.68  0.76  0.80  0.83  0.89 1.00  8000
dev    3188    13  3171  3182  3184  3194  3218 1.01   220
```

The posterior distributions of the model parameters are visualized in Figure 15-5 together with the flat prior (*Unif*(0,1)). The gray area is the overlap between the prior and the posterior.

The function `overlap` from the package birdring calculates the overlap between two distributions (e.g., the prior and the posterior) using approximate density functions and Monte Carlo simulation. The posterior distribution is given to the function `overlap` as a sample of simulated values. In case of a *Unif*(0,1) prior, we can use the argument prior = "unif01". Alternatively, the prior distribution can also be given as a sample of random values.

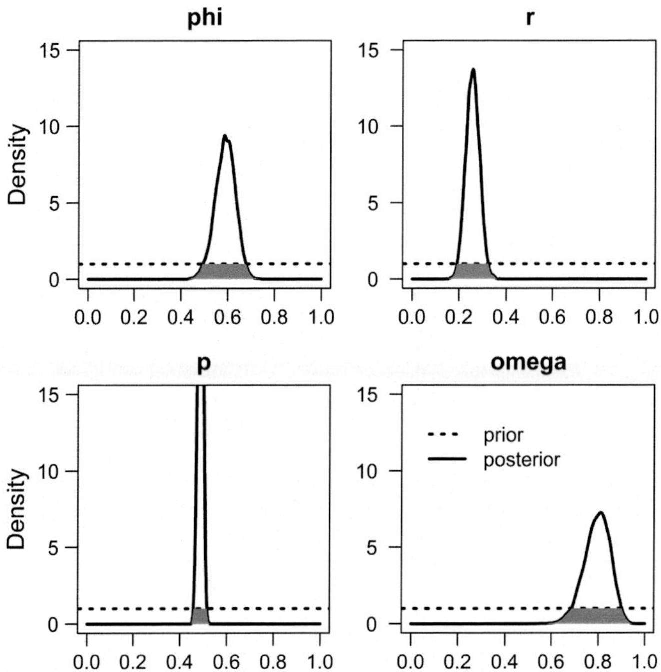

FIGURE 15-5 Posterior distributions (solid lines) and prior distributions (broken lines) of the four model parameters in the territory occupancy model. The overlap between the prior and the posterior is shaded.

```
library(birdring)
overlap(bugsmod$sims.list$phi, prior="unif01")
[1] 0.2143415
overlap(bugsmod$sims.list$r, prior="unif01")
[1] 0.1524065
overlap(bugsmod$sims.list$p, prior="unif01")
[1] 0.06212017
overlap(bugsmod$sims.list$omega, prior="unif01")
[1] 0.261083
```

The overlap between the prior and the posterior is smallest for the estimate of the detection probability p, and larger for survival probability ϕ and the occupancy during the first study year Ω. Note that these values are not exact because the function overlap uses approximations and Monte Carlo simulations. Nevertheless, they give the degree of the overlap, which helps us to see which parameter estimates are well informed by the data. Values close to 1 would mean that there is no or only very weak information about the parameter in the data. This is also true for informative prior distributions. The results shown indicate that there is substantial information in the data regarding all four parameters.

FURTHER READING

Gelman (2006) shows that the inverse-gamma distribution for a variance parameter can cause problems and proposes the folded t-distribution as prior for sigma-values. Chapter 5, Section 5.7 in Gelman et al. (2014) explains the mathematical properties of the uniform, inverse-Gamma, half-Cauchy, and folded t-distributions as prior distributions for variance parameters in hierarchical models. Appendix A of Gelman et al. (2014) gives the density function, the expected value, the variance, the mode, and a short description on the interpretation and application for 26 different probability distributions.

Stauffer (2008) contains a short and concise chapter on prior distributions.
Catchpole and Morgan (1997) present an algebraic method that allows identifying model parameters that are not estimable due to the model structure. Applications and further discussion of parameter identifiability can be found in Catchpole et al. (2001), Schaub and Lebreton (2004), and Gimenez et al. (2009).

The quantification of the estimability of parameters by measuring the overlap between the prior and the posterior distribution has been discussed, for example, by Garrett and Zeger (2000) and Gimenez et al. (2009).

Chapter 16

Checklist

Chapter Outline

16.1 DATA ANALYSIS STEP BY STEP

This checklist can give some guidance for data analysis. However, the list is not complete. For specific studies, a different order of the steps may make more sense or further data structures not considered here may need to be checked. We usually repeat steps 2 to 7 until we find one or a set of models that fit the data well and that are realistic enough to be useful for the intended purpose. Data analysis is always a lot of work and, often, the following steps have to be repeated many times until we find a useful and robust model. There is a danger with this: we may find interesting results that answer different questions than we asked originally.

We can report such findings, but we should state that they appeared (more or less by chance) during the data exploration and model fitting phase, and we have to be aware that the estimates may be biased because the study was not optimally designed with respect to these findings. It is important to always keep the original aim of the study in mind. Do not adjust the study question according to the data. We also recommend reporting what the model started with at the first iteration and describing the strategy and reasoning behind the model development process.

Bayesian Data Analysis in Ecology Using Linear Models with R, BUGS, and Stan
http://dx.doi.org/10.1016/B978-0-12-801370-0.00016-2.

Step 1: Plausibility of Data

Check graphically, or via summary statistics, whether all the data are plausible. Prepare the data so that errors (typos, etc.) are minimal, for example, by double-checking the entries.

Step 2: Relationships

Think about the direct and indirect relationships among the variables of the study.

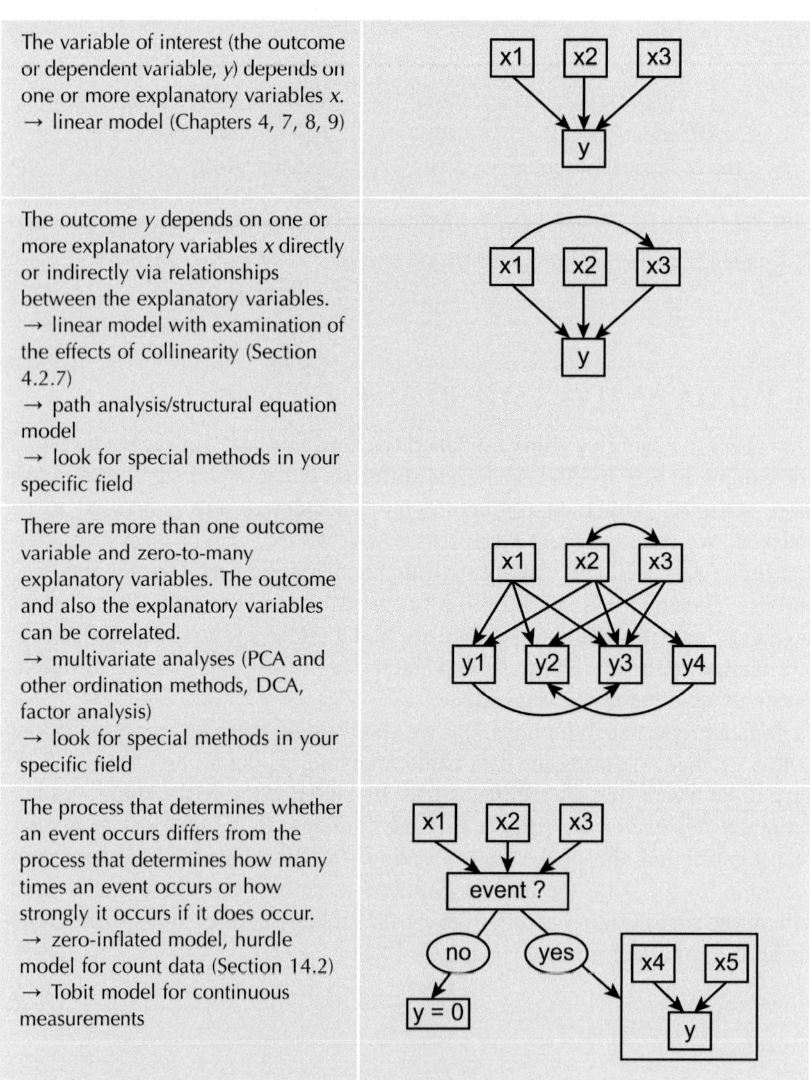

Relationship
The variable of interest (the outcome or dependent variable, *y*) depends on one or more explanatory variables *x*. → linear model (Chapters 4, 7, 8, 9)
The outcome *y* depends on one or more explanatory variables *x* directly or indirectly via relationships between the explanatory variables. → linear model with examination of the effects of collinearity (Section 4.2.7) → path analysis/structural equation model → look for special methods in your specific field
There are more than one outcome variable and zero-to-many explanatory variables. The outcome and also the explanatory variables can be correlated. → multivariate analyses (PCA and other ordination methods, DCA, factor analysis) → look for special methods in your specific field
The process that determines whether an event occurs differs from the process that determines how many times an event occurs or how strongly it occurs if it does occur. → zero-inflated model, hurdle model for count data (Section 14.2) → Tobit model for continuous measurements

—cont'd	
The variable of interest z is not fully observed, for example, because not all individuals are detected during a count. The observation y is biased because of the observation process. → two-levels ecological model (the biological and the observation process is modeled separately, Sections 14.3, 14.4, 14.5)	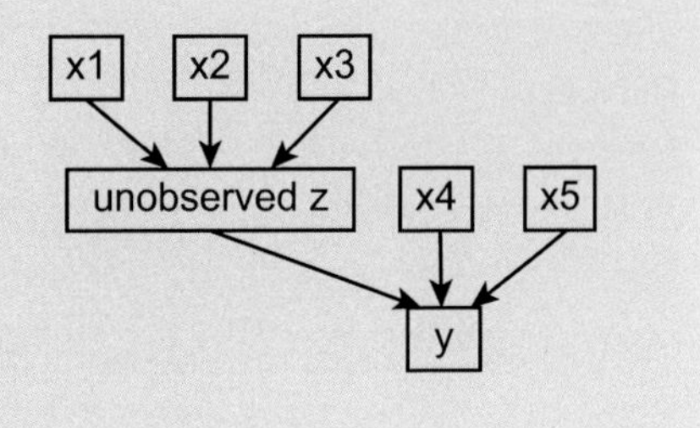

...and there are many more statistical models.

We normally start a data analysis by drawing a sketch of the model including all explanatory variables and interactions that may be biologically meaningful.

We will most likely repeat this step after having looked at the model fit. To make the data analysis transparent we should report every model that was considered. A short note about why a specific model was considered and why it was discarded helps make the modeling process reproducible.

Step 3: Error Distribution

What is the nature of the variable of interest (outcome, dependent variable)?

Number of outcome variables	Type of outcome variable	Characteristics of outcome variable	Error distribution/ model type that (probably) is appropriate
One	Numeric	Range: −inf to +inf or positive values not too close to zero	Gaussian (= normal) (Chapters 4 and 7)
		Number of successful trials among a defined number of trials	Binomial (Sections 8.2 and 9.1)
		Binary (0/1, presence/ absence, yes/no)	Binomial (Bernoulli) (Section 8.3)
		Count	Poisson (Sections 8.4 and 9.2)

Continued

—cont'd

Number of outcome variables	Type of outcome variable	Characteristics of outcome variable	Error distribution/ model type that (probably) is appropriate
One	Numeric	Rate or density (e.g., number of events within a time span, number of observations in a given area)	Poisson with an offset (Sections 8.4.5 and 9.2.5)
		Time to event (e.g., survival times, durations, time until an event happens or if it does not happen until the end of the study)	Survival analysis; for example, Klein and Moeschberger (2003)
		Composition based on (independent) counts (e.g., how many times an animal was observed in different habitats)	Multinomial (Section 14.1)
	Categorical	Ordered	Cumulative logit or proportional odds model (see Agresti 2007 for theory, and Lunn et al. 2013 for BUGS code)
		Nominal	Multinomial (Section 14.1)
>1	Numeric	Normal distribution	Multivariate linear models (Manova) PCA, DCA, factor analysis, cluster analysis More specialized methods (e.g. Borcard et al. 2011)
	Nonnumeric		Ordination methods, such as correspondence analysis Cluster analysis, decision trees, classification methods, etc.; for example, Legendre and Legendre (2012)

Step 4: Preparation of Explanatory Variables

1. *Look at the distribution (histogram) of every explanatory variable:* Linear models do not assume that the explanatory variables have any specific distribution. Thus there is no need to check for a normal distribution! However, very skewed distributions result in unequal weighting of the observations in the model. In extreme cases, the slope of a regression line is defined by one or a few observations only. We also need to check whether the variance is large enough, and to think about the shape of the expected effect. The following four questions may help with this step:
   - Is the variance (of the explanatory variable) big enough so that an effect of the variable can be measured?
   - Is the distribution skewed? If an explanatory variable is highly skewed, it may make sense to transform the variable.
   - Does it show a bimodal distribution? Consider making the variable binary.
   - Is it expected that a change of 1 at lower values for x has the same biological effect as a change of 1 at higher values of x? If not, a transformation (e.g., log) could linearize the relationship between x and y.
2. *Centering:* Centering (x_centered $= x -$ mean(x)) is a transformation that produces a variable with a mean of 0. With centered predictors, the intercept and main effects in the linear model are better interpretable (they are measured at the center of the data instead of at the covariate value = 0 which may be far off; Section 4.2.6); the model fitting algorithm converges faster and better.
3. *Scaling:* To make the estimate of the effect sizes comparable between variables, the variables can be scaled, that is, x_scaled = x/sd(x). The unit of the scaled variable is then 1 standard deviation. Gelman and Hill (2007, p. 55 f) propose to scale the variables by two times the standard deviation (x_scaled = x/(2*sd(x))) to make effect sizes comparable between numeric and binary variables. Scaling can be important for model convergence, especially when polynomials are included. Consider the use of orthogonal polynomials (Section 4.2.9).
4. *Collinearity:*
   - Look at the correlation among the explanatory variables (pairs plot or correlation matrix).
   - If the explanatory variables are correlated, go back to step 2 and see Section 4.2.7 for more details about collinearity.
5. *Are interactions and polynomial terms needed in the model?* If not already done in step 2, think about the relationship between each explanatory variable and the dependent variable.
   - Is it linear or do polynomial terms have to be included in the model? If the relationship cannot be described appropriately by polynomial terms, think of a nonlinear model or a generalized additive model (GAM).
   - May the effect of one explanatory variable depend on the value of another explanatory variable (interaction)?

Step 5: Data Structure

After having taken into account all of the (fixed effect) terms from step 4: are the observations independent or grouped/structured?

Structure in residuals	**Type of model**
All observations are independent	`lm` or `glm` (Chapters 4 and 8)
The observations are grouped (e.g., repeated measurements on the same subject, block-design, some individuals are from the same nest)	mixed models: `lmer` or `glmer` (Chapters 7 and 9)
Temporal correlation (has to be considered when measurements are taken over time)	Include the specific correlation structure in the model, see Chapter 6
Spatial correlation	Include the specific correlation structure in the model, see Chapters 6 and 13

Step 6: Fit the Model

Fit the model as described in Chapters 4, 7, 8, 9, or 14.

Step 7: Check Model Assumptions, Fit, and Sensitivity

We assess model fit by graphical analyses of the residuals (Chapter 6), by predictive model checking (Section 10.1), or by sensitivity analysis (Chapter 15). The following is a nonexhaustive list of aspects that can be looked at in a residual analysis and posterior predictive model checking.

Model type	**Possible residual plots and residual checks**	**Possible plots/test statistics in posterior predictive checks**
Normal linear model (LM, LMM)	Residuals vs. fitted: is the mean zero across the range of fitted values? (in mixed models the mean of the residuals can have a trend due to the shrinkage effect) QQ plot of residuals: are they normally distributed? sqrt(abs(residuals)) vs. fitted values: homogeneity of variance? Residuals vs. leverage: influential observations? Residuals vs. every explanatory variable: linear trend, homogeneity of variance? For LMM: QQ plot of the random effects: are they normally distributed?	Plots: spatial distribution, histogram Test statistics: quantiles, quantiles of residuals, range, skewness parameter

—cont'd

Model type	**Possible residual plots and residual checks**	**Possible plots/test statistics in posterior predictive checks**
Binomial model (GLM, GLMM)	Residuals vs. fitted values: is the mean zero across the range of fitted values? (in mixed models the mean of the residuals can have a trend due to the shrinkage effect) QQ plot of deviance-residuals: are they normally distributed? Residuals vs. every explanatory variable: linear trend, homogeneity of variance? Observations (or classwise means) vs. fitted values. If number of trials > 1: compare residual deviance with residual df (overdispersion) For glmer objects: QQ plot of the random effects; Is the mean of the random effects zero? If not, the model fitting algorithm failed! Check the results using, e.g., the function `glmmPQL` from the package MASS	Plots: spatial distribution, histograms Test statistics: when number of trials > 1: proportion of zeros, quantiles, range Bernoulli model (number of trials = 1): mean length of zero-rows, number of switches between 0 and 1
Poisson model (GLM, GLMM)	Residuals vs. fitted: is the mean zero across the range of fitted values? (in mixed models the mean of the residuals can have a trend due to the shrinkage effect) QQ plot of deviance-residuals: Are they normally distributed? Residuals vs. every explanatory variable: linear trend, homogeneity of variance? Compare residual deviance with residual df (overdispersion) For glmer - objects: QQ plot of the random effects: normal distribution of the random effects? Is the mean of the random effects zero? (nonzero values mean that the model fitting algorithm failed!) check results using the function `glmmPQL` from the package MASS	Plots: histogram, spatial distribution Test statistics: range, quantiles, proportion of zero values

For non-Gaussian models it is often easier to assess model fit using posterior predictive checks (Chapter 10) rather than residual analyses. Posterior predictive checks usually show clearly in which aspect the model failed so we can go back to step 2 of the analysis. Recognizing in what aspect a model does not fit the data based on residual plots improves with experience. Therefore, the following table lists some patterns that can appear in residual plots together with what these patterns possibly indicate. We also indicate what could be done in the specific cases.

Pattern	Possible reasons and what can be done
The mean of the residuals along the fitted values or along any explanatory variable is not constantly zero	The relationship is not linear: include polynomial terms, use generalized additive models (GAM; Wood, 2006), transform the variables.
Residuals are not normally distributed	In the case of an LM or LME, this is of concern: transform the outcome variable (use e.g., Tukey's first aid transformations or Box–Cox transformations). In the case of a GLM or GLMM, this is probably not a problem. Compare with residuals from fitting the model to simulated data. If the residuals from the observed data have a different distribution than the ones from simulated data, do additional posterior predictive model checking to find out whether and why the model does not fit the data.
For LM or LME: variance is not homogeneous	Sometimes variance increases with the mean. Then a log-transformation might help or the use of a Poisson model (if the outcome is a count). Sometimes the variance differs between different groups in the data or the variance depends on other variables: account for this heteroscedasticity in the model (Chapter 6).
There are influential observations (a.k.a. outliers)	There could be a mistake in the data file: check it! The model does not capture the real variance in the data, i.e., the data distribution has heavier tails compared to what the model assumes: Include an additional variance parameter in the model, or use robust methods (e.g., assume *t*-distribution instead of normal error distribution), or show the results with and without the influential observations.
For Poisson models and binomial models with the number of trials > 1: there is overdispersion in the data	An important predictor variable might be missing. An important structure of the data is not accounted for in the model. If overdispersion remains: Include an overdispersion term (an observation-level random factor) in the model, or use quasi-Poisson, quasibinomial, or negative-binomial models.

Step 8: Model Uncertainty

If, while working through steps 1 to 7, possibly repeatedly, we came up with one or more models that fit the data reasonably well, we then turn to the methods presented in Chapter 11 to draw inference from more than one model. If we have only one model, we proceed to step 9.

Step 9: Draw Conclusions

Simulate values from the joint posterior distribution of the model parameters (`sim`, BUGS, Stan). Use these samples to present parameter uncertainty, to obtain posterior distributions for predictions, probabilities of specific hypotheses, and derived quantities.

FURTHER READING

The paper by Zuur et al. (2010) presents a similar checklist in a frequentist framework.

Chapter 17

What Should I Report in a Paper

Chapter Outline

17.1 HOW TO PRESENT THE RESULTS

Sometimes, we find it helpful to write the results before the methods, especially when data analyses were extensive. Once the results are written it is often easier to distinguish between the important modeling steps that go into the main text from the ones that are supplied in an (electronic) appendix only.

The results are the information extracted from the data that can be used to (partially) answer the research question. All results that are necessary to understand the conclusions should be presented. In the case of linear models, a results section contains at least two elements: (1) the estimate of the effect size with a standard error or a credible interval, and (2) information about the location (e.g., the mean) and variance in the data. The first shows how large and how certain an effect is, whereas the comparison of the effect size with the variance in the data informs the reader about the relevance of the measured effect.

The order of these two points does not matter. Sometimes it is preferable to describe the data (point 2) before presenting the effect size (point 1). For example, a results section in a paper on the effect of a treatment of captured prairie dogs with an insecticide that should free the animals from fleas could look like this:

"The mean animal weight was 9.84 kg (SD 0.26 kg) in the control group and 10.22 kg (SD 0.23 kg) in the treatment group. Animals in the treatment group were, on average, 0.38 kg (95% CrI 0.14–0.55 kg) heavier than animals in the control group." Sample size also belongs to the description of the data.

From this description of the results, we read that based on the data and given the model assumptions, we can be 95% sure that the insecticide treatment is associated with an increase in the average weight of a prairie dog of between 0.14 and 0.55 kg. Further, we see that this increase is likely to be

Bayesian Data Analysis in Ecology Using Linear Models with R, BUGS, and Stan
http://dx.doi.org/10.1016/B978-0-12-801370-0.00017-4.

relevant: the estimated effect (0.38 kg) is even larger than the standard deviations for each group (the variance in the data).

In most studies, there is more than one parameter of interest. In these cases, the results may be displayed in a table. Such a table usually contains the estimate with a 95% CrI or a standard error and, optionally, a posterior probability of the estimate being larger than zero or larger than a relevant threshold. Such a table may look like Table 17-1. Because the variance parameters are not given in the table, we describe them in the main text, for example, "The between-brood standard deviation was 0.25 (95% CrI: 0.22–0.33) and the between-individual (within brood) standard deviation was 0.19 (0.16–0.21). The residual standard deviation was 0.61 (0.57–0.67)." The unit should be given somewhere. Here, we give indicate it in the header of Table 17-1.

We did not include the variance parameters in the table because the last column of the table, the posterior probability of the hypothesis that the parameter is larger than zero, is not meaningful for variance parameters. There were only three variance parameters, thus it was feasible to state them in the main text. If the legend does not become too long, it would also be possible to give them in the table legend. However, in models with a larger number of variance parameters it may be preferable to present them in their own table or in a subtable. Sometimes, people highlight statistically

TABLE 17.1 Estimates of Effects of an Implant (corticosterone or placebo), Days Since Implant ("before", "2", and "20"), and Interaction Implant x Day on Corticosterone Concentration in Blood of Barn Owl Nestlings

Parameter	Mean [log(ng/ml)]	2.5% [log(ng/ml)]	97.5% [log(ng/ml)]	$P(\beta > 0)$
Intercept	1.95	1.78	2.13	1
Implant (placebo)	−0.1	−0.35	0.15	0.2
Days (2)	1.64	1.39	1.90	1
Days (20)	0.26	0.01	0.51	0.98
Implant (placebo) x days (2)	−1.69	−2.04	−1.33	0
Implant (placebo) x days (20)	−0.08	−0.43	0.28	0.33

Note: This is an example from Section 7.8; sample size is 151 individuals. Shown are estimated model coefficients (fixed effects) of the linear mixed model for the logarithm of the corticosterone concentration. See text for the estimated variance parameters. Given are the mean, the 2.5%, the 97.5% quantiles of the posterior distribution, and the posterior probability of the hypothesis that the parameter is larger than zero (last column).

significant results in bold or with an asterisk. This may direct the reader to important results.

Table 17-1 should be seen as a suggestion. Depending on the focus of the manuscript, such tables can (and should) look quite different. Instead of the actual model parameters, biologically more meaningful derived parameters may be presented. In the example here (Table 17-1), the main interest is in the effect of a corticosterone implant before, after 2, and after 20 days since implantation in comparison with a placebo implant. Thus, we could give the difference between the corticosterone-treated and placebo-treated individuals for the days "before", 2, and 20 directly (Table 17-2). In the table, the result is seen more quickly and it is more understandable than in Table 17-1. We give the residual standard deviation in the table footnote so that the effect of corticosterone (1.59) can be compared to the natural between-individual variance of corticosterone levels (0.61) showing that the corticosterone implant indeed seems to have a biologically relevant effect.

The convenience of Table 17-2 is traded for a loss of information; for example, there is no information about the absolute corticosterone levels in the two groups since the intercept is missing, and we also do not see how the corticosterone levels change over time because the main effect of days is not given. Therefore, we suggest that a table like Table 17-1 could be given in an appendix or electronic supplement of a paper.

An informative and concise presentation of the results from the model presented in Table 17-1 could be Table 17-2 in combination with Figure 7-4. In Figure 7-4, the information that is missing in Table 17-2 is given; that is, the fitted values per treatment-day combination, as well as the raw data.

TABLE 17.2 Mean Differences, Estimated by a Linear Mixed Model, between Corticosterone- and Placebo-Treated Individuals (effect of corticosterone) in Blood Corticosterone Concentration (log-transformed) of Barn Owls. Means, 2.5% and 97.5% Quantiles of the Posterior Distributions are given

Day	Effect of corticosterone [log(ng/ml)]	Baseline-corrected effect of corticosterone [log(ng/ml)]
Before	0.1 (−0.15;0.35)	0
2	1.69 (1.33;2.04)	1.59 (1.05;2.13)
20	0.08 (−0.28;0.43)	−0.03 (−0.59;0.53)

Note: The complete list of estimates for the model parameters is given in Table 17-1, where "before" is the measurement immediately before the application of the treatment. Therefore, this difference is the natural difference between the two groups (baseline). Because the corticosterone-treated group may have had slightly higher baseline corticosterone levels, we also give baseline-corrected corticosterone effects (i.e., the corticosterone effect after having subtracted the baseline difference).

A figure is often a more appealing way to present statistical results than a table. Using a table, the reader has to construct the corresponding picture in her or his head. To make all readers see the picture, or result, in a way we think is most appropriate, we usually prefer to provide the results graphically. To help with interpretation, we depict the result on the original, natural scale (using back-transformations if needed), and we add, if appropriate, the raw data points to the plot.

It is our responsibility to draw the figures in such a way that the results are represented correctly without giving a misleading impression. For example, an increase in a population by 10 individuals from 1000 to 1010 looks huge when the *y*-axis spans 1000 to 1010, but when the *y*-axis spans 0 to 1500, the difference between the two measurements may hardly be detectable. Thus, the scaling of the *y*-axis needs to be chosen so that the represented effect sizes are in line with the biological relevance. If effects are compared relative to the absolute measurement to assess biological relevance, for example, when studying population sizes, then the *y*-axis should usually start at zero.

When effect sizes are compared to the variance in the data to assess biological relevance, the *y*-axis should at least span the range of the data. How to reproduce statistical results graphically is a science of its own. In the box, we summarize the points that we think are important. For more advice on how to draw figures in statistics see Further Reading at the end of this chapter.

Short Checklist for Presenting Statistical Results in Figures

(1) The figure should represent the answer to the study question. Often, the classical types of plots such as a scatterplot, a bar plot, or an effects plot are sufficient. However, in many cases, adding a little bit of creativity can greatly improve readability or the message of the figure.

(2) Label the *x*- and *y*-axes. Make sure that the units are indicated either within the title or in the figure legend.

(3) Start *y*-axis at zero if the reference to zero is important for the interpretation of effects. The argument `ylim=c(0, max(dat$y))` in R is used for this purpose.

(4) Scale the axes so that all data are shown. Make sure that sample size is indicated either in the figure or in the legend (or, at least, easy to find in the text).

(5) Use interpretable units. That means, if the variable on the *x*-axis has been *z*-transformed to fit the model, back-transform the effects to the original scale.

(6) Give the raw data whenever possible. Sometimes, a significant effect cannot be seen in the raw data because so many other variables have an influence on the outcome. Then, you may prefer showing the effect only (e.g., a regression line with a credible interval) and give the residual standard deviation in the figure legend. Even then, we think it is important to show the raw data graphically somewhere else in the paper or in the supplementary material. A scatterplot of the data can contain structures that are lost in summary statistics.

Short Checklist for Presenting Statistical Results in Figures—cont'd

(7) Draw the figures as simply as possible. Avoid 3D graphics. Delete all unnecessary elements.

(8) Reduce the number of different colors to a minimum necessary. A color scale from orange to blue gives a gray scale in a black-and-white print. `colorRampPalette(c("orange", "blue"))(5)` produces five colors on a scale from orange to blue. Remember that around 8% of the northern European male population have difficulties distinguishing red from green but it is easier for them to distinguish orange from blue.

In addition to the standard results, often more information is needed to make your research transparent and reproducible. Such information may be comprehensive tables with parameter estimates of complex models, residual plots, results from posterior predictive model checks, results from sensitivity analyses, or results from preliminary or alternative analyses. Most readers and editors are happy if you provide all this information in the supplementary material.

As mentioned, effect sizes are often only meaningful in relation to the variance in the data. Therefore, it is important to report the variance in the data. Alternatively, it has been suggested that standardized effect sizes should be reported rather than raw effect sizes to ease future meta-analyses and comparison between different studies. However, such comparisons assume that the overall variance in the data is comparable between the studies, which we think is an unrealistic assumption given the large variety of spatial and temporal scales of studies. Therefore, we prefer to give the estimates that most directly answer the questions of the study at hand. These can be standardized or raw estimates. However, in both cases, it is important to give all information necessary to transform the given estimate into the other type of estimate—that is, to standardize a raw estimate, or to transform the standardized estimate into a raw estimate.

The information needed to do these transformations is the standard deviations of all variables in the data and/or residual variances of all models presented. Therefore, it is valuable to add a table with summary statistics of all variables in the results section or in a supplement. When standardized effects are presented, a detailed description of how these effects were standardized is essential. This will make your study comparable with other studies, for example in meta-analyses.

17.2 HOW TO WRITE UP THE STATISTICAL METHODS

The methods section of a paper contains all of the information needed to reproduce the study. It contains a detailed description of the study design, how and what data you collected, and how you analyzed them. Here, we only discuss what should be reported in the statistical analysis part of the methods section.

The analysis part usually contains the following seven points if we used one of the classical linear models (LM, GLM, LMM, GLMM) in a Bayesian framework:

(1) The error distribution and a declaration of the outcome variable
(2) The link function
(3) The linear predictor and the description of the explanatory variables
(4) Description of the random structure of the model
(5) The prior distributions
(6) The technique used to obtain the posterior distribution
(7) How we draw inference

Avoid R, BUGS, or Stan code within the main text. Such code is only meaningful for people familiar with these languages. In the main text, we use English or algebra to describe a model. However, it is highly recommended to provide R, BUGS, or Stan code together with the data in an electronic supplement to make the analyses reproducible.

For some classical types of linear models it is not necessary to describe the error distribution and link function because these are defined by the name of the model (Table 17-3). For all other types of linear models it is necessary to define all points just listed. For example, it is insufficient to write "We used GLMMs to analyze the data." Let's describe the GLMM we used in Section 9.2.5 to analyze the number of breeding territories of the whitethroat in wildflower fields in relation to the age of the field.

First, we declare the outcome and the error distribution, for example "We modeled the number of breeding territories per field using a Poisson linear mixed model."

Second, we define the link function. "We used the natural logarithm link function."

TABLE 17.3 Classical Linear Models That Are Defined by Only Their Names

Name	Error distribution	Link function	Linear predictor
Ordinary linear regression	Gaussian	Identity	One numeric variable
Analysis of variance (ANOVA)	Gaussian	Identity	One or more categorical variables
Analysis of covariance (ANCOVA)	Gaussian	Identity	Numeric and categorical variables
Multiple regression	Gaussian	Identity	Numeric variables
Logistic regression	Bernoulli	logit	One numeric variable

Third, to describe the linear predictor, we name all explanatory variables and explain how these were transformed. For example, "The size and the age of the field were used as numeric explanatory variables. Size was z-transformed (so that its mean was 0 and its standard deviation 1); for age, we used orthogonal polynomials up to the third degree. As an additional explanatory variable, we added the z-transformed year to account for a potential population trend over the years of the study." In this model, we had an offset because the fields were of different size. Thus, we add: "The size (in ha, log-transformed) of the field was used as an offset to account for the different field sizes."

Fourth, we describe the random structure of the model. "Field id and year were included as random factors to account for repeated measures of the same fields, and to account for between-year variance in population density not captured by the fixed effects, respectively."

Fifth, describe the prior distributions. In cases where we used `lm`, `glm`, `lmer`, or `glmer` together with `sim`, this may look like: "We used improper prior distributions, namely $p(\beta) \propto 1$ for the coefficients, and $p(\sigma) \propto \frac{1}{\sigma}$ for the variance parameters." If we fitted our model using BUGS or Stan, of course, we describe the respective prior distributions.

Sixth, the technique used to obtain the posterior distribution could be analytical or by simulation. The latter could be a direct simulation, that is, a combination of analytics and simulation, or an MCMC technique. The function `sim` uses direct simulation, thus we could write: "To obtain the posterior distribution we directly simulated 5000 values from the joint posterior distribution of the model parameters using the function `sim` of the package arm (Gelman and Hill, 2007)." If we used MCMC, it is common to give more information. Depending on the journal, this information could be moved to the supplementary material, because it inflates the methods section without adding information relevant for understanding the study. The information includes the software and type of algorithm, the number of iterations, the length of the burn-in (or warm-up), the thinning, and how convergence was assessed. The number of effective samples, $\widehat{R}$, and Monte Carlo errors can be given in the results section with their respective parameter estimates, or it can be stated in the methods section that they all were below or above a specific threshold value.

Finally, we describe how we used the simulated values from the joint posterior distribution of the model parameters to draw inference. For example: "The means of the simulated values from the joint posterior distributions of the model parameters were used as estimates, and the 2.5% and 97.5% quantiles as lower and upper limits of the 95% credible intervals." If we used derived parameters such as fitted values or predictions, we may write: "Posterior distributions of fitted values were obtained by calculating 2000 fitted values each with a different set of model parameters from the posterior distribution. Again, the mean and the 2.5% and 97.5% quantiles of these 2000 values were used as the estimate with a 95% credible interval."

If we use posterior probabilities of specific hypotheses, we explain: "Posterior probabilities of the hypothesis A (e.g, that the effect is larger than 0.3) were calculated by the proportion of simulated values from the posterior distribution being A (e.g., larger than 0.3)." If we use the word "significant" in the results section, we define here what we mean. For example, "We declare an effect to be significant if its posterior probability of being positive (or larger than a biologically relevant threshold) is larger than 0.99, or if its posterior probability of being negative (or smaller than a relevant threshold) is larger than 0.99."

To describe more complex ecological models, for example, those in Chapter 14, we follow the same principles, but, of course, a description of the model structure is needed prior to the details of the error distributions, link functions, linear predictors, and prior distributions. To do so, it is sometimes easier to use algebraic notation instead of prose. Further, such models often make crucial model assumptions, such as that the occupancy state of a site does not change between the repeated observations during one season (see occupancy model in Section 14.3). Such assumptions need to be declared in the methods section.

FURTHER READING

Kruschke (2011) summarizes arguments for Bayesian data analyses and also contains a chapter on how to write up methods and results from a Bayesian data analysis.

Murrell (2006) and Mittal (2011) are practical guides on graphics in R.

Schielzeth (2013) discusses different ways of standardizing effect sizes in linear models.

There are plenty of papers and books explaining how to write a paper; Magnusson (1996) is one of the most concise papers about writing a paper.

References

Aebischer, N.J., Robertson, P.A., 1993. Compositional analysis of habitat use from animal radio-tracking data. Ecology 74, 1313–1325.

Agresti, A., 2007. An introduction to categorical data analysis. Wiley, Hoboken, NJ.

Aitkin, M., Francis, B., Hinde, J., Darnell, R., 2009. Statistical modelling in R. Oxford University Press, Oxford.

Akaike, H., 1974. A new look at the statistical model identification. IEEE Transactions on Automatic Control 19, 716–723.

Albert, J., 2007. Bayesian computation with R. Springer, Berlin.

Almasi, B., Roulin, A., Jenni-Eiermann, S., Breuner, C.W., Jenni, L., 2009. Regulation of free corticosterone and CBG capacity under different environmental conditions in altricial nestlings. General and Comparative Endocrinology 164, 117–124.

Almasi, B., Roulin, A., Korner-Nievergelt, F., Jenni-Eiermann, S., Jenni, L., 2012. Coloration signals the ability to cope with elevated stress hormones: effects of corticosterone on growth of barn owls are associated with melanism. Journal of Evolutionary Biology 25, 1189–1199.

Amrhein, V., Korner, P., Naguib, M., 2002. Nocturnal and diurnal singing activity in the nightingale: correlations with mating status and breeding cycle. Animal Behaviour 64, 939–944.

Anderson, J.A., 1974. Diagnosis by logistic discriminant function: Further practical problems and results. Journal of Applied Statistics 23, 397–404.

Anderson, D.R., 2008. Model based inference in the life sciences: A primer on evidence. Springer, New York.

Anderson, D.R., Burnham, K.P., Thompson, W.L., 2000. Null hypothesis testing: Problems, prevalence, and an alternative. Journal of Wildlife Management 64, 912–923.

Anthes, N., Werminghausen, J., Lange, R., 2014. Large donors transfer more sperm, but depletion is faster in a promiscuous hermaphrodite. Behavioural Ecology and Sociobiology 68, 477–483.

Armagan, A., Zaretzki, R.L., 2010. Model selection via adaptive shrinkage with *t* priors. Computational Statistics 25, 441–461.

Arnold, T.W., 2009. Uninformative parameters and model selection using Akaike's information criterion. Journal of Wildlife Management 74, 1175–1178.

Banerjee, S., Carlin, B.P., Gelfand, A.E., 2004. Hierarchical modeling and analysis for spatial data. Chapman & Hall/CRC, Boca Raton.

Banerjee, S., Gelfand, A.E., Finley, A.O., Sang, H., 2008. Gaussian predictive process models for large spatial data sets. Journal of the Royal Statistical Society: Series B (Statistical Methodology) 70, 825–848.

Bartoń, K., 2011. Package 'MuMIn' for multimodel inference. http://stat.ethz.ch/CRAN/.

Bayes, R.T., 1763. An essay towards solving a problem in the doctrine of chances. Philosophical Transactions of the Royal Society of London 53, 370–418.

Beale, C.M., Lennon, J.J., Yearsley, J.M., Brewer, M.J., Elston, D.A., 2010. Regression analysis of spatial data. Ecology Letters 13, 246–264.

Berkson, J., 1938. Some difficulties of interpretation encountered in the application of the chi-square test. Journal of the American Statistical Association 33, 526–536.

Betancourt, M., 2013. Generalizing the no-u-turn sampler to Riemannian manifolds. arXiv 1304.1920, pp. 1–8.

Betancourt, M., Girolami, M., 2013. Hamiltonian Monte Carlo for hierarchical models. arXiv 1312.0906v1.

Bezanson, J., Karpinski, S., Shah, V.B., Edelman, A., 2012. Julia: A fast dynamic language for technical computing. arXiv 1209.5145, pp. 1–27.

Bivand, R.S., Pebesma, E.J., Gómez-Rubio, V., 2008. Applied spatial data analysis with R. Springer, New York.

Bock, A., Naef-Daenzer, B., Keil, H., Korner-Nievergelt, F., Perrig, M., Grüebler, M.U., 2013. Roost site selection by little owls *Athene noctua* in relation to environmental conditions and life-history stages. Ibis 155, 847–856.

Bolker, B.M., Brooks, M.E., Clark, C.J., Geange, S.W., Poulsen, J.R., Stevens, H.H., et al., 2008. Generalized linear mixed models: a practical guide for ecology and evolution. TREE 24, 127–135.

Borcard, D., Gillet, F., Legendre, P., 2011. Numerical Ecology with R. Springer, New York.

Box, G.E.P., 1979. Robustness in scientific model building. In: Launer, R.L., Wilkinson, G.N. (Eds.), Robustness in statistics. Academic Press, New York, pp. 201–236.

Brémaud, P., 1999. Markov chains, Gibbs fields, Monte Carlo simulations, and queues. Springer, New York.

Brooks, S.P., 1998. Markov chain Monte Carlo method and its application. Journal of the Royal Statistical Society, Series D (The Statistician) 47, 69–100.

Brooks, S., Gelman, A., 1998. General methods for monitoring convergence of iterative simulations. Journal of Computational Graphical Statistics 7, 434–455.

Brown, P.J., Vannucci, M., Fearn, T., 2002. Bayes model averaging with selection of regressors. Journal of the Royal Statistical Society, Series B 64, 519–536.

Buckland, S.T., Burnham, K.P., Augustin, N.H., 1997. Model selection: An integral part of inference. Biometrics 53, 603–618.

Burnham, K.P., Anderson, D.R., 2002. Model selection and multimodel inference, a practical information-theoretic approach, 2nd ed. Springer, New York.

Burnham, K.P., Anderson, D.R., 2014. *P* values are only an index to evidence: 20th- vs. 21st-century statistical science. Ecology 95, 627–630.

Burnham, K.P., White, G.C., 2002. Evaluation of some random effects methodology applicable to bird ringing data. Journal of Applied Statistics 29, 245–264.

Cameron, A.C., Trivedi, P.K., 2013. Regression analysis of count data, 2nd ed. Cambridge University Press, Cambridge.

Cameron, A.C., Windmeijer, F.A.G., 1997. An R-squared measure of goodness of fit for some common nonlinear regression models. Journal of Econometrics 77, 329–342.

Carlin, B.P., Chib, S., 1995. Bayesian model choice via Markov chain Monte Carlo methods. Journal of the Royal Statistical Society B 57, 473–484.

Carlin, B., Louis, T., 2000. Bayes and empirical Bayes methods for data analysis. Chapman & Hall/CRC, New York.

Catchpole, E.A., Kgosi, P.M., Morgan, B.J.T., 2001. On the near-singularity of models for animal recovery data. Biometrics 57, 720–726.

Catchpole, E.A., Morgan, B.J.T., 1997. Detecting parameter redundancy. Biometrika 84, 187–196.

Chambers, J.M., 1998. Programming with data. A guide to the S language. Springer-Verlag, New York.

Chambers, J.M., 2008. Software for data analysis, programming with R. Springer, New York.
Chambert, T., Rotella, J.J., Higgs, M.D., 2014. Use of posterior predictive checks as an inferential tool for investigating individual heterogeneity in animal population vital rates. Ecology and Evolution 4, 1389–1397.
Chang, W., 2012. R graphics cookbook. O'Reilly Media, Inc., Sebastopol, CA.
Chevan, A., Sutherland, M., 1991. Hierarchical partitioning. The American Statistician 45, 90–96.
Christensen, R., Johnsen, W., Branscum, A., Hanson, T.E., 2011. Bayesian ideas and data analysis, an introduction for scientists and statisticians. Chapman & Hall, Boca Raton.
Claeskens, G., Hjort, N.L., 2003. The focused information criterion. Journal of the American Statistical Association 98, 900–945.
Claeskens, G., Hjort, N.L., 2008. Model selection and model averaging. Cambridge University Press, Cambridge.
Cohen, J., 1994. The earth is round ($p < 0.05$). American Psychologist 49, 997–1003.
Cormack, R.M., 1964. Estimates of survival from the sighting of marked animals. Biometrika 51, 429–438.
Crawley, M.J., 2005. Statistics: An introduction using R. John Wiley & Sons, Chichester.
Crawley, M.J., 2007. The R book. John Wiley & Sons, Chichester.
Croissant, Y., 2012. Estimation of multinomial logit models in R: The mlogit package. http://cran.r-project.org/web/packages/mlogit/vignettes/mlogit.pdf.
Dalgaard, P., 2008. Introductory statistics with R. Springer, New York.
Davidson, R.R., Solomon, D.L., 1974. Moment-type estimation in the exponential family. Communications in Statistics 3, 1101–1108.
Davis, M.J., 2010. Contrast coding in multiple regression analysis: strengths, weaknesses, and utility of popular coding structures. Journal of Data Science 8, 61–73.
Davison, A.C., Snell, E.J., 1991. Residuals and diagnostics. In: Hinkley, D.V., Reid, N., Snell, E.J. (Eds.), Statistical theory and modelling. In Honour of Sir David Cox. Chapman & Hall, FRS. London.
Dekking, F.M., Kraaikamp, C., Lopuhaä, H.P., Meester, L.E., 2005. A modern introduction to probability and statistics; understanding why and how. Springer, London.
Diggle, P.J., Ribeiro Jr, P.J., 2007. Model based geostatistics. Springer, New York.
Dormann, C.F., Elith, J., Bacher, S., Buchmann, C., Carl, G., Carré, G., et al., 2013. Collinearity: A review of methods to deal with it and a simulation study evaluating their performance. Ecography 36, 27–46.
Dormann, C.F., McPherson, J.M., Araújo, M.B., Bivand, R., Bolliger, J., Carl, G., et al., 2007. Methods to account for spatial autocorrelation in the analysis of species distributional data: a review. Ecography 30, 609–628.
Draper, N.R., Smith, H., 1998. Applied regression analysis, 3rd ed. Wiley, New York.
Duane, S., Kennedy, A.D., Pendleton, B.J., Roweth, D., 1987. Hybrid Monte Carlo. Physics Letters B 195, 216–222.
Ellenberg, H., 1953. Physiologisches und oekologisches Verhalten derselben Pflanzenarten. Berichte der Deutschen Botanischen Gesellschalft 65, 350–361.
Ellenberg, H., 1954. Über einige Fortschritte der kausalen Vegetationskunde. Plant Ecology 5/6, 199–211.
Engqvist, L., 2005. The mistreatment of covariate interaction terms in linear analysis of behavioural and evolutionary ecology studies. Animal Behaviour 70, 967–971.
Faraway, J.J., 2005. Linear models with R. Chapman & Hall/CRC, Boca Raton.
Faraway, J.J., 2006. Extending the linear model with R, generalized linear, mixed effecs and nonarametric regression models. Chapman & Hall/CRC, Boca Raton.

Fielding, A.H., Bell, J.F., 1997. A review of methods for the assessment of prediction errors in conservation presence/absence models. Environmental Conservation 24, 38–49.

Finley, A.O., Banerjee, S., 2013. spBayes: Univariate and multivariate spatial-temporal modeling. R package version 0. 3–8. http://CRAN.R-project.org/package=spBayes.

Fisher, R.A., 1925. Statistical methods for research workers. Oliver and Boyd, Edinburgh.

Fiske, I.J., Chandler, R.B., 2011. Unmarked: An R package for fitting hierarchical models of wildlife occurrence and abundance. Journal of Statistical Software 43, 1–23.

Ford, J.H., Bravington, M.V., Robbins, J., 2012. Incorporating individual variability into mark-recapture models. Methods in Ecology and Evolution 3, 1047–1054.

Forstmeier, W., Schielzeth, H., 2011. Cryptic multiple hypotheses testing in linear models: overestimated effect sizes and the winner's curse. Behavioural Ecology and Sociobiology 65, 47–55.

Fox, J., Weisberg, S., 2011. An R companion to applied regression. Sage Publications, London.

Garrett, E.S., Zeger, S.L., 2000. Latent class model diagnosis. Biometrics 56, 1055–1067.

Gelfand, A., Hills, S., Racine-Poon, A., Smith, A., 1990. Illustration of Bayesian inference in normal data models using Gibbs sampling. Journal of the American Statistical Association 85, 972–985.

Gelfand, A.E., Smith, A.F.M., 1990. Sampling-based approach to calculating marginal densities. Journal of American Statistical Association 85, 398–409.

Gelman, A., Hill, J., Yajima, M., 2012. Why we (usually) don't have to worry about multiple comparisons. Journal of Research on Educational Effectiveness 5, 189–211.

Gelman, A., 2006. Prior distributions for variance parameters in hierarchical models. Bayesian analysis 1, 515–533.

Gelman, A., Carlin, J.B., Stern, H.S., Dunson, D.B., Vehtari, A., Rubin, D.B., 2014. Bayesian data analysis, 3rd ed. CRC Press, Boca Raton.

Gelman, A., Hill, J., 2007. Data analysis using regression and multilevel/hierarchical models. Cambridge University Press, Cambridge.

Gelman, A., Hill, J., Yajima, M., 2012. Why we (usually) don't have to worry about multiple comparisons. Journal of Research on Educational Effectiveness 5, 189–211.

Gelman, A., Pardoe, I., 2006. Bayesian measures of explained variance and pooling in multilevel (hierarchical) models. Technometrics 48, 241–251.

Gelman, A., Rubin, D.B., 1995. Avoiding model selection in Bayesian social research. Sociological Methodology 25, 165–173.

Gelman, A., Rubin, D.B., 1999. Evaluating and using statistical methods in the social sciences: A discussion of "A critique of the Bayesian information criterion for model selection". Sociological Methods & Research 27, 403–410.

Gelman, A., Shirley, K., 2011. Inference from simulations and monitoring convergence. In: Brooks, S., Gelman, A., Jones, G.L., Meng, X.-L. (Eds.), Handbook of Markov chain Monte Carlo. Chapman & Hall/CRC, Boca Raton.

Geman, S., Geman, D., 1984. Stochastic relaxation, Gibbs distributions and the Bayesian restoration of images. IEEE Transactions on Pattern Analysis and Machine Intelligence 6 (72), 721–741.

Geyer, C.J., 2011. Introduction to Markov chain Monte Carlo. In: Brooks, S., Gelman, A., Jones, G.L., Meng, X.-L. (Eds.), Handbook of Markov chain Monte Carlo. Chapman & Hall/CRC, Boca Raton.

Gilks, W.R., Richardson, S., Spiegelhalter, D.J., 1996. Markov chain Monte Carlo in practice. Chapman & Hall/CRC, Boca Raton.

Gimenez, O., Bonner, S.J., King, R., Parker, R.A., Brooks, S.P., Jamieson, L.E., et al., 2009. WinBUGS for population ecologists: Bayesian modelling using Markov chain Monte Carlo methods. In: Thomson, D.L., Cooch, E.G., Conroy, M.J. (Eds.), Modeling demographic processes in marked populations. Springer, New York, pp. 883–915.

Gimenez, O., Brooks, S.P., Morgan, B.J.T., 2009. Weak identifiability in models for mark-recapture-recovery data. In: Thomson, D.L., Cooch, E.G., Conroy, M.J. (Eds.), Modelling demographic processes in marked populations. Springer, New York, pp. 1055–1067.

Gimenez, O., Choquet, R., 2010. Individual heterogeneity in studies on marked animals using numerical integration: capture-recapture mixed models. Ecology 91, 951–957.

Gimenez, O., Grégoire, A., Lenormand, T., 2009. Estimating and visualizing fitness surfaces using mark-recapture data. Evolution 63, 3097–3105.

Glantz, S.A., Slinker, B.K., 2001. Primer of applied regression and analysis of variance, 2nd ed. McGraw-Hill, New York.

Gonick, L., Smith, W., 1993. The cartoon guide to statistics. HarperPerennial, New York.

Gonick, L., Smith, W., 2005. The cartoon guide to statistics, 240 pages, HarperCollins, ISBN: 9780062731029.

Gottschalk, T., Ekschmitt, K., Wolters, V., 2011. Efficient placement of nest boxes for the little owl (*Athene noctua*). The Journal of Raptor Research 45, 1–14.

Graham, M.H., 2003. Confronting multicollinearity in ecological multiple regression. Ecology 84, 2089–2815.

Grueber, C.E., Nakagawa, S., Laws, R.J., Jamieson, I.G., 2011a. Multimodel inference in ecology and evolution: challenges and solutions. Journal of Evolutionary Biology 24, 699–711.

Grueber, C.E., Nakagawa, S., Laws, R.J., Jamieson, I.G., 2011b. Corrigendum. Journal of Evolutionary Biology 24, 1627.

Grüebler, M.U., Korner-Nievergelt, F., von Hirschheydt, J., 2010. The reproductive benefits of livestock farming in barn swallows *Hirundo rustica*: quality of nest site or foraging habitat? Journal of Applied Ecology 47, 1340–1347.

Grüebler, M.U., Naef-Daenzer, B., 2008. Fitness consequences of pre-and post-fledging timing decisions in a double-brooded passerine. Ecology 89, 2736–2745.

Grüebler, M.U., Naef-Daenzer, B., 2010. Survival benefits of post-fledging care: experimental approach to a critical part of avian reproductive strategies. Journal of Animal Ecology 79, 334–341.

Guttman, I., 1967. The use of the concept of a future observation in goodness-of-fit problems. Journal of the Royal Statistical Society B 29, 83–100.

Hadfield, J.D., 2010. MCMC methods for multi-response generalized linear mixed models: The MCMCglmm R package. Journal of Statistical Software 33, 1–22.

Hastings, W.K., 1970. Monte Carlo sampling methods using Markov chains and their applications. Biometrika 57, 97–109.

Hector, A., von Felten, S., Hautier, Y., Weilenmann, M., Bruelheide, H., 2012. Effects of dominance and diversity on productivity along Ellenberg's experimental water table gradients. PlosOne, 7, e43358.

Hector, A., von Felten, S., Schmid, B., 2010. Analysis of variance with unbalanced data: an update for ecology & evolution. Journal of Animal Ecology 79, 308–316.

Henningsen, A., 2013. censReg: Censored regression (Tobit). models. R package version 0.5-20. http://CRAN.R-project.org/package=censReg.

Hoeting, J.A., 2009. The importance of accounting for spatial and temporal correlation in analyses of ecological data. Ecological Applications 19, 574–577.

Hoeting, J.A., Madigan, D., Raftery, A.E., Volinsky, C.T., 1999. Bayesian model averaging: a tutorial (with discussion). Statistical Science 14, 382–417.

Hoffman, M.D., Gelman, A., 2012. The no-u-turn sampler: Adaptively setting path lengths in Hamiltonian Monte Carlo. arXiv:1111.4246v1, pp. 1–30.

Hooten, M.B., Hobbs, N.T., 2015. A guide to Bayesian model selection for ecologists. Ecological Monographs 85, 3–28.

Hurlbert, S.H., 1984. Pseudoreplication and the design of ecological field experiments. Ecological Monographs 54, 187–211.

Jackman, S., 2008. pscl: Classes and methods for R developed in the Political Science Computational Laboratory, Stanford University. Department of Political Science, Stanford University, Stanford, California.

Jenni, L., Winkler, R., 1989. The feather-length of small passerines: a measurement for wing-length in live birds and museum skins. Bird Study 36, 1–15.

Johnson, J.B., Omland, K.S., 2004. Model selection in ecology and evolution. Trends in Ecology and Evolution 19, 101–108.

Jolly, G., 1965. Explicit estimates from capture-recapture data with both death and immigration-stochastic model. Biometrika 52, 225–247.

Kadane, J.B., Lazar, N.A., 2004. Methods and criteria for model selection. Journal of the American Statistical Association 99, 279–290.

Kass, R.E., Carlin, B.P., Gelman, A., Neal, R.M., 1998. Markov chain Monte Carlo in practice: A roundtable discussion. The American Statistician 52, 93–100.

Kéry, M., 2010. Introduction to WinBUGS for ecologists. Academic Press, London.

Kéry, M., Schaub, M., 2012. Bayesian population analysis using WinBUGS. Elsevier, Amsterdam.

King, R., 2014. Annual review of statistics and its application. Statistical Ecology 1, 401-U983. http://dx.doi.org/10.1146/annurev-statistics-022513-115633.

King, R., Brooks, S.P., 2002. Bayesian model discrimination for multiple strata capture-recapture data. Biometrika 89, 785–806.

King, R., Morgan, B.J.T., Gimenez, O., Brooks, S.P., 2010. Bayesian analysis for population ecology. CRC Press, London.

Kleiber, C., Zeileis, A., 2008. Applied econometrics with R. Springer-Verlag, New York.

Klein, J.P., Moeschberger, M.L., 2003. Survival analysis, techiques for censored and truncated data. Springer, New York.

Korner-Nievergelt, F., Korner-Nievergelt, P., Baader, E., Fischer, L., Schaffner, W., Kestenholz, M., 2007. Jahres- und tageszeitliches Auftreten von Singvögeln auf dem Herbstzug im Jura (Ulmethöchi, Kanton Basel-Landschaft). [Seasonal and daily occurrence of passerines on autumn migration in the Jura mountains (Ulmethöchi, northern Switzerland).] Der Ornithologische Beobachter 104, 101–130.

Korner-Nievergelt, F., Liechti, F., Thorup, K., 2014. A bird distribution model for ring recovery data: where do the European robins go? Ecology and Evolution 4, 720–731.

Kruschke, J.K., 2011. Doing Bayesian data analysis, a tutorial with R and BUGS. Academic Press, Boston.

Lambert, D., 1992. Zero-inflated Poisson regression, with an application to defects in manufacturing. Technometrics 34, 1–14.

Latimer, A.M., Banerjee, S., Sang, H., Mosher, E.S., Silander, J.A., 2009. Hierarchical models facilitate spatial analysis of large data sets: A case study on invasive plant species in the northeastern United States. Ecology Letters 12, 144–154.

Lebreton, J.-D., Burnham, K.P., Clobert, J., Anderson, D.R., 1992. Modelling survival and testing biological hypotheses using marked animals: a unified approach with case studies. Ecological Monographs 62, 67–118.

Lecomte, J.B., Benoît, H.P., Etienne, M.P., Bela, L., Parenta, E., 2013. Modeling the habitat associations and spatial distribution of benthic macroinvertebrates: A hierarchical Bayesian model for zero-inflated biomass data. Ecological Modelling 265, 74–84.

Legendre, P., 1993. Spatial autocorrelation: trouble or new paradigm? Ecology 74, 1659–1673.

Legendre, P., Legendre, L., 2012. Numerical ecology. Elsevier, Amsterdam.

Link, W.A., Barker, R.J., 2010. Bayesian inference with ecologial applications. Academic Press, London.

Link, W.A., Eaton, M.J., 2012. On thinning of chains in MCMC. Methods in Ecology and Evolution 3, 112–115.

Link, W.A., Sauer, J.R., 2002. A hierarchical analysis of population change with application to Cerulean Warblers. Ecology 83, 2832–2840.

Little, R.J.A., Rubin, D.B., 2002. Statistical analysis with missing data. John Wiley & Sons, New York.

Logan, M., 2010. Biostatistical design and analysis using R–a practical guide. Wiley-Blackwell, Chichester.

Lunn, D., Jackson, C., Best, N., Thomas, A., Spiegelhalter, D., 2013. The BUGS book–a practical introduction to Bayesian analysis. Taylor & Francis Group, Boca Raton.

Lunn, D., Spiegelhalter, D., Thomas, A., Best, N., 2009. The BUGS project: Evolution, critique and future directions. Statistics in Medicine, Published online in Wiley InterScience, pp. 1–19. www.interscience.wiley.com. http://dx.doi.org/10.1002/sim.3680.

MacKenzie, D.I., Nichols, J.D., Lachman, G.B., Droege, S., Royle, J.A., Langtimm, C.A., 2002. Estimating site occupancy rates when detection probabilities are less than one. Ecology 83, 2248–2255.

MacKenzie, D.I., Nichols, J.D., Royle, J.A., Pollock, K.H., Bailey, L.L., Hines, J.E., 2006. Occupancy estimation and modelling. Inferring patterns and dynamics of species occurrence. Elsevier, Boston.

MacLehose, R.F., Dunson, D.B., Herring, A.H., Hoppin, J.A., 2007. Bayesian methods for highly correlated exposure data. Epidemiology 18, 199–207.

Magnusson, W.E., 1996. How to write backwards. Bulletin of the Ecological Society of America 77, 88.

Manly, B.F., 1994. Multivariate statistical methods, a primer, 2nd ed. Chapman & Hall, London.

Manly, B.F.J., McDonald, L.L., Thomas, D.L., McDonald, T.L., Erickson, W.P., 2002. Resource selection by animals, statistical design and analysis for field studies. Springer Netherlands, Dordrecht.

Mansfield, E.R., Helms, B.P., 1982. Detecting multicollinearity. The American Statistician 36, 158–160.

Marescot, L., Chapron, G., Chadès, I., Fackler, P.L., Duchamp, C., Marboutin, E., Gimenez, O., 2013. Complex decisions made simple: a primer on stochastic dynamic programming. Methods in Ecology and Evolution 4, 872–884.

Martin, A.D., Quinn, K.M., Park, J.H., 2011. MCMCpack: Markov chain Monte Carlo in R. Journal of Statistical Software 42, 1–21.

Matthews, R., 2000. Storks deliver babies ($p = 0.008$). Teaching Statistics 22, 36–38.

Mazerolle, M.J., 2014. AICcmodavg: Model selection and multimodel inference based on (Q) AIC(c). R package version 2.00. http://CRAN.R-project.org/package=AICcmodavg.

McCarthy, M.A., 2007. Bayesian methods for ecology. Cambridge University Press, Cambridge.

McCullagh, P., Nelder, J.A., 1989. Generalized linear models, 2nd ed. Chapman & Hall, London.

McCulloch, C.E., Neuhaus, J.M., 2011. Prediction of random effects in linear and generalized linear models under model misspecification. Biometrics 67, 270–279.

Menard, S., 2000. Coefficients of determination for multiple logistic regression analysis. American Statistician 54, 17–24.

Metropolis, N., Rosenbluth, A.W., Rosenbluth, M.N., Teller, A.H., Teller, E., 1953. Equations of state calculations by fast computing machine. J. Chem. Phys. 21, 1087–1091.

Millar, R.B., 2009. Comparison of hierarchical Bayesian models for overdispersed count data using DIC and Bayes' factors. Biometrics 65, 962–969.

Miller, R.G., 1980. Simultaneous statistical inference, 2nd ed. Springer-Verlag, New York.

Mittal, H.V., 2011. R Graphs Cookbook. Packt Publishing, Birmingham.

Mullahy, J., 1986. Specification and testing of some modified count data models. Journal of Econometrics 33, 341–365.

Murrell, P., 2006. R Graphics. Chapman & Hall/CRC, Boca Raton.

Nakagawa, S., Schielzeth, H., 2013. A general and simple method for obtaining R^2 from generalized linear mixed-effects models. Methods in Ecology and Evolution 4, 133–142.

Neal, R.M., 1994. An improved acceptance procedure for the hybrid Monte Carlo algorithm. Journal of Computational Physics 111, 194–203.

Neal, R.M., 2011. MCMC using Hamiltonian Dynamics. In: Brooks, S., Gelman, A., Jones, G.L., Meng, X.-L. (Eds.), Handbook of Markov chain Monte Carlo. Chapman & Hall/CRC, Boca Raton.

Ntzoufras, I., 2009. Bayesian modelling using WinBUGS. Wiley, Hoboken.

Nychka, D., Furrer, R., Sain, S., 2014. Fields: Tools for spatial data. R package version 7.1. http://CRAN.R-project.org/package=fields.

O'Hara, R.B., Sillanpää, M.J., 2009. A review of Bayesian variable selection methods: What, how and which. Bayesian Analysis 4, 85–118.

Pearce, J., Ferrier, S., 2000. Evaluating the predictive performance of habitat models developed using logistic regression. Ecological Modelling 133, 225–245.

Pearl, J., 2000. Causality: Models, reasoning and inference. Cambridge University Press, Cambridge.

Pearson, K., 1901. On lines and planes of closest fit to systems of points in space. Philosophical Magazine 2, 559–572.

Pebesma, E.J., 2004. Multivariable geostatistics in S: The gstat package. Computers & Geosciences 30, 683–691.

Pebesma, E.J., Bivand, R.S., 2005. Classes and methods for spatial data in R. R News 5, 9–13.

Pinheiro, J.C., Bates, D.M., 2000. Mixed-effects models in S and S-Plus. Springer, New York.

Pinheiro, J., Bates, D., DebRoy, S., Sarkar, D., 2011. NLME: Linear and nonlinear mixed effects models. R package, version 3. 1–102. http://cran.r-project.org.

Pledger, S., Pollock, K.H., Norris, J.L., 2003. Open capture-recapture models with heterogeneity: I. Cormack–Jolly–Seber Model. Biometrics 59, 786–794.

Plummer, M., Best, N., Cowles, K., Vines, K., 2006. CODA: Convergence, diagnosis and output analysis for MCMC. R News 6, 7–11.

Poulsen, J.R., Clark, C.J., Bolker, B.M., 2011. Decoupling the effects of logging and hunting on an Afrotropical animal community. Ecological Applications 21, 1819–1836.

Quinn, G.P., Keough, M.J., 2009. Experimental design and data analysis for biologists. Cambridge University Press, Cambridge.

R Core Team, 2015. R: A language and environment for statistical computing. R Foundation for Statistical Computing. Austria, Vienna.

Rice, W.R., 1989. Analyzing tables of statistical tests. Evolution 43, 223–225.

Ridley, J., Kolm, N., Freckelton, R.P., Gage, M.J.G., 2007. An unexpected influence of widely used significance thresholds on the distribution of reported P-values. Journal of Evolutionary Biology 20, 1082–1089.

Ripley, B.D., 2004. Selecting amongst large classes of models. In: Adams, N., Crowder, M., Hand, D.J., Stephens, D. (Eds.), Methods and models in statistics. Imperial College Press, London, pp. 155–170.

Rizzo, M.L., 2008. Statistical computing with R. Chapman & Hall, Boca Raton.

Robert, C.P., Casella, G., 2009. Introducing Monte Carlo methods with R. Springer, London.

Roth, T., Amrhein, V., 2010. Estimating individual survival using territory occupancy data on unmarked animals. Journal of Applied Ecology 47, 386–392.

Roth, T., Sprau, P., Naguib, M., Amrhein, V., 2012. Sexually selected signaling in birds: a case for Bayesian change-point analysis of behavioral routines. The Auk 129, 660–669.

Royle, J.A., 2004. N-mixture models for estimating population size from spatially replicated counts. Biometrics 60, 108–115.

Royle, J.A., 2008. Modeling individual effects in the Cormack–Jolly–Seber Model: a state–space formulation. Biometrics 64, 364–370.

Royle, J.A., 2009. Analysis of capture-recapture models with individual covariates using data augmentation. Biometrics 65, 267–274.

Royle, J.A., Chandler, R.B., Sollmann, R., Gardner, B., 2014. Spatial capture-recapture. Academic Press, Boston.

Royle, J.A., Dorazio, R.M., 2008. Hierarchical modelling and inference in ecology. Academic Press, London.

Royle, J.A., Link, W.A., 2002. Random effects and shrinkage estimation in capture-recapture models. Journal of Applied Statistics 29, 329–351.

Sarkar, D., 2008. Lattice: multivariate data visualization with R. Springer, New York.

Schaub, M., Gimenez, O., Sierro, A., Arlettaz, R., 2007. Use of integrated modeling to enhance estimates of population dynamics obtained from limited data. Conservation Biology 21, 945–955.

Schaub, M., Lebreton, J.-D., 2004. Testing the additive versus the compensatory hypothesis of mortality from ring recovery data using a random effects model. Animal Biodiversity and Conservation 27, 73–85.

Schielzeth, H., 2013. Simple means to improve the interpretability of regression coefficients. Methods in Ecology and Evolution 1, 103–113.

Schielzeth, H., Forstmeier, W., 2009. Conclusions beyond support: overconfident estimates in mixed models. Behavioural Ecology 20, 416–420.

Schofield, M.R., Barker, R.J., Taylor, P., 2013. Modeling individual specific fish length from capture-recapture data using the von Bertalanffy growth curve. Biometrics 69, 1012–1021.

Schwarz, G., 1978. Estimating the dimension of a model. Annals of Statistics 6, 461–464.

Seber, G.A.F., 1965. A note on the multiple-recapture census. Biometrika 52, 249–259.

Shmueli, G., 2010. To explain or to predict? Statistical Science 25, 289–310.

Smith, A.C., Koper, N., Francis, C.M., Fahrig, L., 2009. Confronting collinearity: Comparing methods for distentangling the effects of habitat loss and fragmentation. Landscape Ecology 24, 1271–1285.

Smith, A.F.M., Skene, A.M., Shaw, J.E.H., Naylor, J.C., Dransfield, M., 1985. Progress with numerical and graphical methods for Bayesian statistics. Communications in Statistics—Theory and Methods 14, 1079–1102.

Sokal, R.R., Rohlf, F.J., 2003. Biometry, 3rd Ed. W.H. Freeman and Company, New York.

Solymos, P., 2010. dclone: Data cloning in R. The R Journal 2, 29–37.

Spiegelhalter, D.J., Best, N.G., Carlin, B.P., van der Linde, A., 2002. Bayesian measures of model complexity and fit. Journal of the Royal Statistical Society, Series B 64, 1–34.

Spiegelhalter, D., Thomas, A., Best, N., 2003. WinBUGS user manual, Version 1.2. MCR Biostatistics Unit, Cambridge.

Spiegelhalter, D., Thomas, A., Best, N., Lunn, D., 2007. OpenBUGS user manual, Version 3.0.2, September 2007. www.mrc-bsu.cam.ac.uk/bugs.

Stahel, W.A., 2002. Statistische Datenanalyse: Eine Einführung für Naturwissenschaftler [Statistical data analysis: An introduction for scientists], 4th ed. Vieweg, Wiesbaden.

Stan Development Team, 2014. Stan modeling language users guide and reference manual. Version 2.2. http://mc-stan.org.

Stauffer, H.B., 2008. Contemporary Bayesian and frequentist statistical research methods for natural resource scientists. Wiley, Hoboken.

Stone, M., 1977. An asymptotic equivalence of choice of model by cross-validation and Akaike's criterion. Journal of the Royal Statistical Society, Series B 39, 44–47.

Sturtz, S., Ligges, U., Gelman, A., 2005. R2WinBUGS: A package for running WinBUGS from R. Journal of Statistical Software 12, 1–16.

Su, Y.-S., Yajima, M., 2012. R2jags: A Package for running JAGS from R. http://cran.r-project.org/web/packages/R2jags/R2jags.pdf.

Swanson, A.K., Dobrowski, S.Z., Finley, A.O., Thorne, J.H., Schwartz, M.K., 2013. Spatial regression methods capture prediction uncertainty in species distribution model projections through time. Global Ecology and Biogeography 22, 242–251.

Tanner, M., Wong, W., 1987. The calculation of the posterior distributions by data augmentation. Journal of the American Statistical Association 82, 528–549.

Thompson, S.K., 2002. Sampling, 2nd Ed. John Wiley & Sons, New York.

Tibshirani, R., 1996. Regression shrinkage and selection via LASSO. Journal of the Royal Statistical Society B 58, 267–288.

Tobin, J., 1958. Estimation of relationships for limited dependent variables. Econometrica 26, 24–36.

Tobler, W., 1970. A computer movie simulating urban growth in the Detroit region. Economic Geography 46, 234–240.

Vaida, F., Blanchard, S., 2005. Conditional Akaike information for mixed-effects models. Biometrika 92, 351–370.

Valliant, R., Dormann, A.H., Royall, R.M., 2000. Finite population sampling and inference: A prediction approach. John Wiley & Sons, New York.

van de Pol, M., Wright, J., 2009. A simple method for distinguishing within-versus between-subject effects using mixed models. Animal Behaviour 77, 753–758.

Venables, W.N., Ripley, B.D., 2002. Modern applied statistics with S. Springer, New York.

Venables, W.N., Smith, D.M., the R Core Team, 2014. An introduction to R. Notes on R: a programming environment for data analysis and graphics. Network Theory Limited.

Venzon, D.J., Moolgavkar, S.H., 1988. A method for computing profile-likelihood-based confidence intervals. Applied Statistics 37, 87–94.

Verbeke, G., Molenberghs, G., 2000. Linear mixed models for longitudinal data. Springer, Berlin.

Walters, G., 2012. Customary fire regimes and vegetation structure in Gabon's Bateke Plateaux. Human Ecology 40, 943–955.

Wan, A.T.K., 2002. On generalized ridge regression estimators under collinearity and balanced loss. Applied Mathematics and Computation 129, 455–467.

Wasserman, L., 2000. Bayesian model selection and model averaging. Journal of Mathematical Psychology 44, 92–107.

Watanabe, S., 2010. Applicable information criterion in singular learning theory. Journal of Machine Learning Research 11, 3571–3594.

White, F.C., Burnham, K.P., 1999. Program MARK: survival estimation from populations of marked animals. Bird Study 46, S120–189.

Whittingham, M.J., Stephens, P.A., Bradbury, R.B., Freckleton, R.P., 2006. Why do we still use stepwise modelling in ecology and behaviour? Journal of Animal Ecology 75, 1182–1189.

Wickham, H., 2009. ggplot2: Elegant graphics for data analysis. Springer, New York.

Williams, B.K., Nichols, J.D., Conroy, J.C., 2002. Analysis and management of animal populations: Modeling, estimation, and decision making. Academic Press, San Diego.

Wolfson, L.J., Kadane, J.B., Small, M.J., 1996. Bayesian environmental policy decisions: two case studies. Ecological Applications 6, 1056–1066.

Wood, S.N., 2006. Generalized additive models, an introduction with R. Chapman & Hall/CRC, London.

Wright, S.S., 1921. Correlation and causation. J. Agric. Res. 20, 557–585.

Yandell, B.S., 1997. Practical data analysis for designed experiments. Chapman & Hall, London.

Yokomizo, H., Coutts, S.R., Possingham, H.P., 2014. Decision sciences for effective management of populations subject to stochasticity and imperfect knowledge. Population Ecology 56, 41–53.

Zeileis, A., Kleiber, C., Jackman, S., 2008. Regression models for count data in R. Journal of Statistical Software 27, 1–25.

Zeugner, S., 2011. Bayesian Model Averaging with BMS. Tutorial to the R-package BMS 1–30.

Zipkin, E.F., Thorson, J.T., See, K., Lynch, H.J., Grant, E.H.C., Kanno, Y., et al., 2014. Modeling structured population dynamics using data from unmarked individuals. Ecology 95, 22–29.

Zollinger, J.-L., Birrer, S., Zbinden, N., Korner-Nievergelt, F., 2013. The optimal age of sown field margins for breeding farmland birds. Ibis 155, 779–791.

Zuur, A.F., Ieno, E.N., Elphick, C.S., 2010. A protocol for data exploration to avoid common statistical problems. Methods in Ecology and Evolution 1, 3–14.

Zuur, A., Ieno, E.N., Meesters, E., 2009. A beginner's guide to R. Springer, New York.

Zuur, A.F., Ieno, E.N., Walker, N.J., Saveliev, A.A., Smith, G.M., 2009. Mixed effects models and extensions in ecology with R. Springer, New York.

Zuur, A.F., Saveliev, A.A., Ieno, E.N., 2012. Zero-inflated models and generalized linear mixed models with R. Highland Statistics Ltd, Newburgh.

Index

Note: Page numbers followed by "b", "f" and "t" indicate boxes, figures and tables respectively.

M

Q

R

S

Printed in the United States
By Bookmasters